Springer Series in Statistics

Springer Series in Statistics

Andrews/Herzberg: Data: A Collection of Problems from Many Fields for the Student and Research Worker.
Anscombe: Computing in Statistical Science through APL.
Berger: Statistical Decision Theory and Bayesian Analysis, 2nd edition.
Brémaud: Point Processes and Queues: Martingale Dynamics.
Brockwell/Davis: Time Series: Theory and Methods, 2nd edition.
Daley/Vere-Jones: An Introduction to the Theory of Point Processes.
Dzhaparidze: Parameter Estimation and Hypothesis Testing in Spectral Analysis of Stationary Time Series.
Farrell: Multivariate Calculation.
Fienberg/Hoaglin/Kruskal/Tanur (Eds.): A Statistical Model: Frederick Mosteller's Contributions to Statistics, Science,and Public Policy.
Goodman/Kruskal: Measures of Association for Cross Classifications.
Grandell: Aspects of Risk Theory.
Härdle: Smoothing Techniques: With Implementation in S.
Hartigan: Bayes Theory.
Heyer: Theory of Statistical Experiments.
Jolliffe: Principal Component Analysis.
Kres: Statistical Tables for Multivariate Analysis.
Leadbetter/Lindgren/Rootzén: Extremes and Related Properties of Random Sequences and Processes.
Le Cam: Asymptotic Methods in Statistical Decision Theory.
Le Cam/Yang: Asymptotics in Statistics: Some Basic Concepts.
Manoukian: Modern Concepts and Theorems of Mathematical Statistics.
Miller, Jr.: Simultaneous Statistical Inference, 2nd edition.
Mosteller/Wallace: Applied Bayesian and Classical Inference: The Case of *The Federalist Papers*.
Pollard: Convergence of Stochastic Processes.
Pratt/Gibbons: Concepts of Nonparametric Theory.
Read/Cressie: Goodness-of-Fit Statistics for Discrete Multivariate Data.
Reiss: Approximate Distributions of Order Statistics: With Applications to Nonparametric Statistics.
Ross: Nonlinear Estimation.
Sachs: Applied Statistics: A Handbook of Techniques, 2nd edition.
Seneta: Non-Negative Matrices and Markov Chains.
Siegmund: Sequential Analysis: Tests and Confidence Intervals.
Tong: The Multivariate Normal Distribution.
Vapnik: Estimation of Dependences Based on Empirical Data.
West/Harrison: Bayesian Forecasting and Dynamic Models.
Wolter: Introduction to Variance Estimation.
Yaglom: Correlation Theory of Stationary and Related Random Functions I: Basic Results.
Yaglom: Correlation Theory of Stationary and Related Random Functions II: Supplementary Notes and References.

Rupert G. Miller, Jr.

Simultaneous Statistical Inference

Second Edition

With 25 Figures

Springer-Verlag
New York Berlin Heidelberg London
Paris Tokyo Hong Kong Barcelona

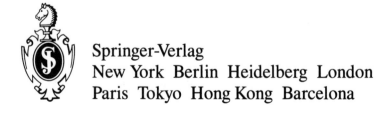

Rupert G. Miller, Jr. (1933-1986)
Department of Statistics
Stanford University
Stanford, CA 94305
USA

Mathematical Subject Classification: 62F99, 62G99, 62J15

Library of Congress Cataloging in Publication Data

Miller, Rupert G
Simultaneous statistical inference.

(Springer series in statistics)
Bibliography: p.
Includes indexes.
1. Mathematical statistics. I. Title. II. Series.
QA276.M474 1980 519.5'4 80-26170

Printed and bound by Edwards Brothers, Ann Arbor, Michigan.
Printed in the United States of America.

9 8 7 6 5 4 3

ISBN 0-387-90548-0 Springer-Verlag New York
ISBN 3-540-90548-0 Springer-Verlag Berlin Heidelberg

DEDICATED TO THE MEMORY OF DR. HARVEY A. SMITH

Preface to the Second Edition

Simultaneous Statistical Inference, which was published originally in 1966 by McGraw-Hill Book Company, went out of print in 1973. Since then, it has been available from University Microfilms International in xerox form. With this new edition Springer-Verlag has republished the original edition along with my review article on multiple comparisons from the December 1977 issue of the *Journal of the American Statistical Association*. This review article covered developments in the field from 1966 through 1976. A few minor typographical errors in the original edition have been corrected in this new edition.

A new table of critical points for the studentized maximum modulus is included in this second edition as an addendum. The original edition included the table by K. C. S. Pillai and K. V. Ramachandran, which was meager but the best available at the time. This edition contains the table published in *Biometrika* in 1971 by G. J. Hahn and R. W. Hendrickson, which is far more comprehensive and therefore more useful.

The typing was ably handled by Wanda Edminster for the review article and Karola Decleve for the changes for the second edition. My wife, Barbara, again cheerfully assisted in the proofreading.

Fred Leone kindly granted permission from the American Statistical Association to reproduce my review article. Also, Gerald Hahn, Richard Hendrickson, and, for *Biometrika*, David Cox graciously granted permission to reproduce the new table of the studentized maximum modulus.

The work in preparing the review article was partially supported by NIH Grant R01 GM21215.

Rupert G. Miller, Jr.

Preface to the First Edition

Slightly over a decade and a half have passed since the great spurt of interest and research in multiple comparisons and simultaneous confidence intervals began. Between the late forties and mid-fifties the principal ideas of simultaneous inference were formulated. From then until the present these ideas have been expanded and applied in different contexts to produce a large battery of simultaneous statistical procedures.

The descriptions of these procedures, and the explorations into their properties, are scattered over a multitude of technical journals. There is no single book or article which describes, studies, and unifies all of this material. Yet there is a need for such a document both as a reference source and as a textbook.

The three statisticians primarily responsible for the fundamental ideas of multiple comparisons and simultaneous confidence intervals are David Duncan, Henry Scheffé, and John Tukey. It would be natural for one of them to produce an opus on this subject, but this has not come to pass. In 1953 Tukey wrote the fundamental treatise "The Problem of Multiple Comparisons," but this was never published and has since been somewhat outdated. Scheffé discusses several principal techniques in "The Analysis of Variance," but he does not begin to cover the field. Other textbooks have brief descriptions of some of the important techniques, but there has been no major work on the subject.

I have taken it upon myself to remedy (successfully or unsuccessfully) this situation. If to no one else, at least to me this will be an aid. To students I can now cite one source where they can begin to study this field. To my clients I can now give a reference on the procedure I applied to their beloved data. This reference may or may not be intelligible to them, but at least it is not in some obscure (to them) statistical journal. They can cite this reference in their paper far more easily than they can the statistical journal.

My hope is that this monograph can be used both as a reference book and as a textbook. The opening section on each simultaneous procedure contains a cookbook-type description of the procedure. This should make the book usable as a reference to a wider audience than just mathematical statisticians. However, some knowledge of mathematics and statistics is required to understand the cookbook descriptions. As a textbook this book could serve as a basic text in a one-quarter course on simultaneous inference, or as a supplementary text in a course on the analysis of variance. It would supplement other texts such as F. A. Graybill, "An Introduction to Linear Statistical Models," vol. I, and H. Scheffé, "The Analysis of Variance." To fully understand all the material presented, the reader must have had a basic course in matrix algebra and a basic course in probability and statistics from a text such as A. M. Mood and F. A. Graybill, "Introduction to the Theory of Statistics." Of course, the more background the reader has, the more he can extract from the material.

The numbering system used in this book starts afresh in each chapter. The numbers for Sections, Theorems, Lemmas, Figures, Tables, and displayed expressions do not have a numeral indicating the chapter attached to them. When reference is made to a Section, Theorem, etc., within the same chapter, no chapter numeral is attached, either. However, when the reference is to a Theorem, Lemma, Figure, Table, or displayed expression in a *different* chapter, a chapter numeral is prefixed to the number. When a Section in a different chapter is referred to, the chapter containing the Section is explicitly mentioned, and no prefix numeral is attached.

ACKNOWLEDGMENTS

I am grateful to the National Science Foundation for its financial support during my 1963–1964 visit at The Johns Hopkins University where I started work on this book. To the Office of Naval Research and the National Institutes of Health I am grateful as well for financial support while completing the book at Stanford University. I appreciate the support and encouragement given to me by the directors of those programs: Allyn Kimball, Geoffrey Watson, Herman Chernoff, Herbert

Solomon, and Lincoln Moses. I also thank both universities, Johns Hopkins and Stanford, for supporting my endeavors.

Four individuals read all or part of the manuscript in detail and contributed innumerable improvements. Any blunders, inaccuracies, or awkward passages that remain cannot be laid at their doorsteps. David Blackwell made several helpful suggestions. John Tukey indicated a number of valuable changes. Maxwell Layard thoroughly checked the manuscript for errors. He found many; hopefully they have all been corrected. Barbara Miller (my wife) also corrected mistakes and contributed greatly to improving the style and consistency of the book. I deeply appreciate the time, effort, and interest these individuals devoted to my book.

Special thanks go to Barbara for voluntarily helping with the entire proofreading. Virginia Currey and Babette Doyle also helped with the proofreading.

I had worthwhile discussions on various aspects of the book with David Duncan, Geoffrey Watson, Bradley Efron, and Myles Hollander.

Lillian Cohen typed Chapter 1 and Chapter 2, Sec. 1, while I was at The Johns Hopkins University. Lill's ceaseless efforts on my behalf did much to get the book under way. Virginia Currey assumed the typing duties upon my return to Stanford. Ginnie tirelessly and lovingly typed, illustrated, compiled the bibliography, indexed, retyped, etc. Her contribution borders on a coauthorship. Ginnie and Lill, I thank you.

Susan Boyle of Stanford University performed the special computations required in Chapter 1 and programmed the computation of the new tables in Appendix B. Finally, I thank the authors and editors who kindly allowed me to reproduce their tables in Appendix B.

Rupert G. Miller, Jr.

Table of Contents

CHAPTER 1

Introduction

To whom the first thoughts on simultaneous statistical inference or multiple comparisons should be attributed is an obscure historical point which is not of prime importance to this monograph. Likely as not, it was a nonstatistician. What one can be certain of is that simultaneous inference did not burst into existence one fine morning in full, completed form, but rather evolved slowly from the treatment of special cases until the general applicability and merit of the underlying principles were recognized and molded into a general theory and philosophy.

Under a broad interpretation the work of Irwin (1925), on what is now termed the *rejection of outliers*, could be considered a rudimentary beginning. This was inspired by some earlier work of Galton (1902) and K. Pearson (1902) on the distribution of the distance between the largest and the next to largest sample observations. Using the results of E. S. Pearson (1926) and Tippett (1925) on the normal range, Student (1927) proposed the range as a criterion for rejecting and repeating observations in routine analyses. Without question this is an unequivocal instance of simultaneous inference. Shortly thereafter, an unrelated innovation came from Working and Hotelling (1929), who derived the confidence band for a simple straight line regression $\alpha + \beta x$, which is equivalent to simultaneously bracketing the expected values $E\{Y_x\} = \alpha + \beta x$ for all values of x. Fisher (1935) proposed the use of individual t tests following an analysis-of-variance F test for picking out those effects creating the statistical significance. This idea later acquired the label of an LSD (least significant difference) test. In 1939 Newman, using some additional ideas of Student for the use of the range, formulated and illustrated the first multiple range test for an analysis-of-variance problem.

This comprised the early work which would serve as a foundation for the later developments. None of the techniques or ideas appeared in as polished a form as they can be found today, nor was there a complete awareness of what the authors had in hand. (It did not appear in print at any rate.)

The general principles of multiple comparisons were forged into their current structure between 1947 and 1955 by three principal investigators: Duncan, Scheffé, and Tukey. There was not complete agreement then about what was best, nor is there now. To this author the principal disagreement seems to revolve around whether the consumer needs a test of significance or a confidence interval, a debate which is not confined solely to problems of simultaneous inference. Duncan (1947,1951,1952, 1955) studied and developed the use of multiple stage tests. These are primarily tests of significance and cannot be adapted to the construction of confidence intervals without a loss of structure which damages their appeal. Tukey (1949) began his advocacy of multiple comparisons procedures with his gap-straggler-variance test, and is perhaps the man most primarily responsible for the incorporation of multiple comparisons techniques into statistical practice. This was followed by his oft-referred-to Blacksburg, Virginia, I.M.S. address (1952a) and his classic (but, unfortunately, unpublished) treatise (1953b) on multiple comparisons. Scheffé (1953) provided the much needed link between multiple comparisons and the general linear hypothesis by giving a simultaneous confidence interval interpretation to the customary analysis-of-variance tests.

Other names perhaps should be included in this briefest of lists, as there were those who before, during, and after the aforementioned dates contributed to our knowledge of simultaneous inference. Since this is not an historical hall of fame, they shall have to be content to find themselves mentioned either in the body of the text or in the (hopefully) complete bibliography at the end. For those missed entirely, a meager reward—the author's apologies.

1 CASE OF TWO MEANS

To illustrate essentially all the general principles and techniques involved in simultaneous inference, it is pedagogically convenient to take the simplest possible case. Namely, let

$$Y_1 \sim N(\mu_1,1) \qquad Y_2 \sim N(\mu_2,1) \\ Y_1 \text{ and } Y_2 \text{ be independent.}^1 \tag{1}$$

[1] $Y \sim N(\mu,\sigma^2)$ means Y is a random variable with a normal distribution, mean μ, and variance σ^2.

The null hypothesis to be considered is

$$H_0: \quad \mu_1 = 0 \qquad \mu_2 = 0. \tag{2}$$

Why, one could ask, should the two means be considered together (simultaneously), instead of attacking separately the two null hypotheses

$$H_0^1: \quad \mu_1 = 0 \qquad H_0^2: \quad \mu_2 = 0? \tag{3}$$

This question, which lies at the tap root of multiple comparisons theology, will be discussed in detail in Sec. 5 after some preliminary concepts on error rates have been introduced. Suffice it to say here that one could attack H_0 by separately testing H_0^1 and H_0^2, but this is only one of a number of possible procedures that could be applied to H_0. Whether one wants to use the separate procedures can be determined only by an objective and subjective evaluation of their performance as compared with the other possibilities.

If the null hypothesis is to be discarded on the basis of the data, the classical Neyman-Pearson theory of testing hypotheses gives the nebulous alternative *not* H_0, that is, either μ_1 or μ_2 or both differ from zero. A prime example of this is the prima donna of the classical theory—the χ^2 test (when σ^2 is unknown, the F test). The credentials of the χ^2 test comprise an impressive portfolio:

1. A likelihood ratio test
2. A uniformly most powerful invariant test
3. A most stringent test
4. A minimax test
5. A test which maximizes average power over spheres (among similar tests for σ^2 unknown)
6. A uniformly most powerful test among tests whose power is constant over spheres

and less gloriously, but no less importantly,

7. An easily computable test
8. A well-tabulated null distribution
9. A fairly well-tabulated power function
10. A robust test

However, the multiple comparisonist, although he is fond of these medals of honor, needs a decision not merely between H_0 and *not* H_0, but between

$$\begin{aligned} H_0&: \quad \mu_1 = 0 \qquad \mu_2 = 0 \\ H_1&: \quad \mu_1 \neq 0 \qquad \mu_2 = 0 \\ H_2&: \quad \mu_1 = 0 \qquad \mu_2 \neq 0 \\ H_3&: \quad \mu_1 \neq 0 \qquad \mu_2 \neq 0. \end{aligned} \tag{4}$$

In fact, this list could be expanded to a sequence of nine hypotheses if, as is likely, a decision is also demanded on the sign, positive or negative, of the mean or means which differ from zero.

For simultaneous statistical inference it is not sufficient to stop with a decision against the null hypothesis should the data indicate statistical significance. A conclusion should be drawn about which means are creating the significance. Once an experimenter is aware that not everything agrees with the null hypothesis it is reasonable, logical, and likely that he will

1. Want to say (publish) what it is (which mean or means) that differs from nullity
2. Possibly pursue the effects of those components (means) which are showing nonnull activity

Hence, the experimenter and statistician find themselves, willing or not, being pushed in the direction of multiple inference.

In addition to a choice between the nine hypotheses the experimenter may also require some statement as to the size of the nonnull effect. This is the junction at which tests of significance and confidence intervals part company. For some problems a nonnull effect is worthy of consideration only if the magnitude of the effect is sufficient to produce a scientific, technologic, philosophic, or social change. Mere statistical significance is not enough to warrant its notice; the effect must also be biologically, physically, socially, etc., significant as well. To answer the question of the apparent magnitude of the effect, the experimenter needs point *and* interval estimates; tests of significance will not suffice.

In other situations it may be that any difference from nullity, no matter how small or large, is of importance. If the existence of any nonnull effect will demolish an existing theory or give a clue to understanding some previously unexplained phenomenon, then a test of significance is what is required. Confidence intervals here would be a luxury and afterthought, and nothing, in particular power, should be sacrificed to gain them.

There are those who might argue against the existence of either one of these two situations, but this author is inclined to concede that both can arise.

To provide a multiple comparisons answer to the choice between H_0 and its alternatives, the professional (or armchair) statistician must make two statements, S_1 and S_2, about μ_1 and μ_2, respectively. S_1 and S_2 may be statements concerning hypotheses (for example, S_1 might be $\mu_1 = 0$, or $\mu_1 < 0$, or $\mu_1 > 0$), or they might be confidence interval statements (for example, $a \leq \mu_1 \leq b$). In fact, whether one is a complete, partial, or non-multiple comparisonist, if you are forced to examine two means in related or unrelated problems within your lifetime, you will have to make two statements. One principle on which all statisticians are likely to agree is that correct statements are good and desirable, and incorrect statements are bad and to be avoided. If oracular statements which are correct under any eventuality are ruled out, the statistician has a chance of being incorrect on each statement, S_1 and S_2. A reasonable method of determining how the statistician should act when confronted with the possibility of making zero, one, or two misstatements is to examine the frequency with which his mistakes occur.

Some statisticians would like to go one step farther and append the thought that slight mistakes are not as bad as big ones. The inevitable end of this garden path is the loss function. Despite its seductive appeal the author does not care to venture down this path. In practice (for which this monograph is intended) this author never knows his loss function and would be unwilling to employ any procedure which was not fairly good for a wide class of loss functions. Studies of this sort of robustness are few, and for a natural loss function the winner (when the mathematical smoke clears) is frequently an old friend of the author, whose merit could have been established on intuitive, distributional, or simplicity grounds.

2 ERROR RATES

Consider a family of statements $\mathfrak{F} = \{S_f\}$ where $N(\mathfrak{F})$ is the number of statements in the family. In the previous section $\mathfrak{F} = \{S_1,S_2\}$ and $N(\mathfrak{F}) = 2$. The statements $S_f \in \mathfrak{F}$ may be independent, or dependent upon one another. Just what statements should be grouped to form a family is the topic of Sec. 5.

Let N_w $(\mathfrak{F})$ be the number of incorrect statements in the family. The error rate for the family is

$$Er\{\mathfrak{F}\} = \frac{N_w(\mathfrak{F})}{N(\mathfrak{F})}. \tag{5}$$

Assume for the moment $N(\mathfrak{F})$ is finite so that $Er\{\mathfrak{F}\}$ is clearly well defined. The error rate is a random variable whose distribution depends upon the procedure utilized in making the family of statements and the underlying probability structure. To use it to assess the overall merit of any procedure, some global, nonrandom parameter of its distribution must be selected. Two criteria naturally present themselves: the *probability of a nonzero family error rate* and the *expected family error rate*. The term *naturally* is used since both have an intuitive feel to them and in most situations are susceptible to mathematical treatment.

The standard *modus operandi* in tests of significance is to control one or both of these error rate criteria under the null hypothesis, and then see what can be done about error rates and power under the alternatives. For confidence intervals the error rate criteria are controlled for all the underlying distributions, and then the lengths of the intervals are examined for shortness.

In the ensuing discussion, reference to error rates in hypothesis testing will mean error rates *under the null hypothesis* unless specified to the contrary. Error rates under alternative hypotheses will be specifically labeled. There is, of course, no distinction for confidence intervals.

Since these error rate criteria are the backbone of the analysis to come, their properties and relationship will now be studied in detail.

2.1 Probability of a nonzero family error rate

This is the quantity most frequently and easily controlled by multiple comparisons techniques. It will be denoted by

$$P\{\mathfrak{F}\} = P\{N_w(\mathfrak{F})/N(\mathfrak{F}) > 0\} = P\{N_w(\mathfrak{F}) > 0\} \tag{6}$$

and will be referred to as the *probability error rate.* Frequently, when a confidence or significance probability is equal to $1 - \alpha$ or α, respectively, $P\{\mathfrak{F}\}$ is in fact α, and the two symbols will be used interchangeably.

This criterion makes no distinction between families with just one incorrect statement and those with more than one, or between families with a small number of statements and those with a large number. It is oblivious to the number of errors other than whether the number is zero or positive. Thus, the two families $\mathfrak{F}_1$ and $\mathfrak{F}_2$ with $N(\mathfrak{F}_1) = 5$, $N_w(\mathfrak{F}_1) = 1$, and $N(\mathfrak{F}_2) = 100$, $N_w(\mathfrak{F}_2) = 2$, are guilty of the same amount of error. The larger the number of statements $N(\mathfrak{F})$ the easier it will customarily be for one of them to be incorrect, so the statements will have to be weaker (smaller critical regions, larger confidence intervals) in order to achieve the same $P\{\mathfrak{F}\}$.

In practice, if $P\{\mathfrak{F}\}$ is small (say, .01 to .05), $\mathfrak{F}$ is not too large (say, $N(\mathfrak{F}) = 5$ to 10), and the dependence between the individual statements S_f is not too weird, then the expected number of incorrect statements is roughly equal to $P\{\mathfrak{F}\}$. Take, for example, $N(\mathfrak{F}) = 7$ and $P\{\mathfrak{F}\} = .03$. The bounds on the expected number of incorrect statements are

$$.03 \le E\{N_w(\mathfrak{F})\} \le .21. \tag{7}$$

The lower and upper bounds are achieved, respectively, when the dependence is such that one incorrect statement implies that the other six are all correct, or all incorrect. In the event the seven statements are independent and identically distributed, then

$$E\{N_w(\mathfrak{F})\} = 7(1 - \sqrt[7]{.97}) = .0304. \tag{8}$$

In most practical situations the statements S_f involve statistics whose numerators are independent, or nearly so, but whose denominators contain a common random variable. Some quick, crude approximations or bounds will frequently give a good idea of $E\{N_w(\mathfrak{F})\}$ since the expectation operator is additive regardless of the dependence; that is,

$$E\{N_w(\mathfrak{F})\} = E\{I(S_1)\} + \cdots + E\{I(S_{N(\mathfrak{F})})\} \tag{9}$$

where

$$I(S_f) = \begin{cases} 1 & \text{if } S_f \text{ is incorrect} \\ 0 & \text{if } S_f \text{ is correct.} \end{cases} \tag{10}$$

The probability error rate criterion has an advantage in that when written in the form $P\{\mathfrak{F}\} = P\{N_w(\mathfrak{F}) > 0\}$ it is well-defined for infinite families as well as for finite families. This would include both families with cardinality $N(\mathfrak{F})$ equal to c (the continuum) and $\aleph_0$ (the integers). Infinite families become a theoretical consideration in the Scheffé- and Tukey-type procedures. In these situations the error rate concept may be fraught with definitional difficulties and is conveniently replaced by the dichotomy between zero and nonzero numbers of errors.

The frequency interpretation of the probability $P\{\mathfrak{F}\}$ is quite simple. If in his consulting lifetime a statistician faces an infinite sequence of independent families $\mathfrak{F}_1, \mathfrak{F}_2, \ldots$, and on each he uses the probability error rate $P\{\mathfrak{F}\}$, then in $100(1 - P\{\mathfrak{F}\})$ percent of the families he will make no misstatements, and in $100P\{\mathfrak{F}\}$ percent he will make one or more mistakes with probability 1. In other words, the strong law of large numbers applies to the sequence of families. Each family, not statement, is a Bernoulli trial, and is counted as good or bad on the great tally sheet in the sky. Since the probability error rate creates an all-or-nothing situation *for families*, some care and thought should be exercised on just what constitutes a family. An attempt at this will be made in Sec. 5.

How the overall probability $P\{\mathfrak{F}\}$ is related to the frequencies with which the individual statements S_f are incorrect cannot, in general, be determined without a knowledge of the joint probabilities of incorrectness, which are generally mathematically unobtainable in statistical problems. When couched in terms of families of statements, the standard law for the expansion of the probability of a union in terms of the probabilities of the intersections becomes

$$\begin{aligned} P\{\mathfrak{F}\} &= P\{\bigcup_f [I(S_f) = 1]\} \\ &= \sum_f P\{I(S_f) = 1\} - \sum_{f_1<f_2} P\{I(S_{f_1})I(S_{f_2}) = 1\} + \cdots \\ &\qquad + (-1)^{N(\mathfrak{F})-1} P\{I(S_1) \cdots I(S_{N(\mathfrak{F})}) = 1\} \end{aligned} \tag{11}$$

and the joint probabilities from $P\{I(S_{f_1})I(S_{f_2}) = 1\}$ on are usually beyond the computable. In one special case, *independence*, there is an exact, simple relation:

$$1 - P\{\mathfrak{F}\} = \prod_{1}^{N(\mathfrak{F})} (1 - P\{I(S_f) = 1\}). \tag{12}$$

Independence is rarely found in pure form, but when it is, (12) should be remembered.

A crude bound relates $P\{\mathfrak{F}\}$ to the individual probabilities and is called a *Bonferroni inequality*. Let $\alpha_f = P\{I(S_f) = 1\}$, $f = 1, \ldots, N(\mathfrak{F})$. Then,

$$1 - P\{\mathfrak{F}\} \geq 1 - \alpha_1 - \cdots - \alpha_{N(\mathfrak{F})}$$

that is,

$$P\{\bigcap_f [I(S_f) = 0]\} \geq 1 - \sum_f P\{I(S_f) = 1\}. \tag{13}$$

This follows readily from a Venn set diagram, or Boole's inequality $(P\{A \cup B\} \leq P\{A\} + P\{B\})$, and was known and utilized well before the modern theory of multiple comparisons emerged. It is still an extremely useful practical tool, especially when more sophisticated ideas lead to distributional difficulties. Furthermore, it is not as crude as one might think, provided $N(\mathfrak{F})$ is not too large (for example, 5) and the α_f are all small (for example, .01).

If (as is unlikely) more knowledge of the joint probabilities is available, there are other Bonferroni inequalities for assessing how many of the statements may be incorrect [see Feller (1957, p. 100)].

2.2 Expected family error rate

The second criterion, whose numerator has been discussed partially in the previous subsection, will be referred to as the *expected error rate* and will be denoted by

$$E\{\mathfrak{F}\} = E\{N_w(\mathfrak{F})/N(\mathfrak{F})\}. \tag{14}$$

Immediately this global error rate faces existence difficulties in the case of infinite $N(\mathfrak{F})$, so its applicability will only be considered in the finite case.

The expected error rate has the nice property of being directly related to the marginal performances of each individual statement S_f in the family $\mathfrak{F}$. Let $\alpha_f = P\{I(S_f) = 1\} = E\{I(S_f)\}$, $f = 1, 2, \ldots, N(\mathfrak{F})$. Then,

$$E\{\mathfrak{F}\} = \frac{\alpha_1 + \cdots + \alpha_{N(\mathfrak{F})}}{N(\mathfrak{F})}. \tag{15}$$

Statements whose dependence is difficult to assess can be grouped together in a family, and the family's expected error rate will be known exactly from the separate behavior of the individual components. Thus, if an

$E\{\mathfrak{F}\} = p$ is desired, constructing the procedure so that each S_f has a probability $1 - p$ of being correct will produce the desired effect. In fact, there is no need for all the error probabilities to be alike; any combination for which $\alpha_1 + \cdots + \alpha_{N(\mathfrak{F})} = pN(\mathfrak{F})$ will yield the appropriate expected error rate. This is an important point which will be discussed further in the next section.

In the special case where the family $\mathfrak{F}$ consists of a single statement, the probability error rate and expected error rate are identical. For in this case the error rate $Er\{\mathfrak{F}\}$ becomes a Bernoulli random variable (that is, $Er\{\mathfrak{F}\} = 0$ or 1), and the expected value of a Bernoulli random variable is equal to the probability of the value 1. The proper terminology for this error rate is *statement error rate* (or *error rate per statement*) rather than the term *family error rate*. This is the error rate which is of prime importance to the nonmultiple comparisonist, for it measures the performance of an individual statement rather than a group of them.

A frequency interpretation can be given to the expected error rate as well as to the probability error rate. Consider an infinite sequence of independent families $\mathfrak{F}_1, \mathfrak{F}_2, \ldots,$ each of which has an expected error rate of $E\{\mathfrak{F}\} = p$. Let the actual error rates in the families be $P_i = Er\{\mathfrak{F}_i\} = N_w(\mathfrak{F}_i)/N(\mathfrak{F}_i)$, $i = 1, 2, \ldots.$ By assumption the P_i are independent random variables. $E\{P_i\} = p$, but otherwise there need be no distributional similarity whatsoever between them. Then, with probability 1, $\bar{P} = \sum_1^n P_i/n$ converges to p as $n \to \infty$, and, under the additional assumption that there exists an N, maybe quite large, for which $N(\mathfrak{F}_i) \leq N$ for all i, $\bar{\bar{P}} = \sum_1^n N(\mathfrak{F}_i)P_i/\sum_1^n N(\mathfrak{F}_i)$ also converges to p as $n \to \infty$ (see Appendix A for proof). $\bar{P}$ is the unweighted mean of the P_i, $\bar{\bar{P}}$ is the weighted mean, and the strong law of large numbers applies to both. The unweighted mean $\bar{P}$ represents the average of the actual proportions of incorrect statements in the families as the statistician progresses through his sequence of families. The weighted average $\bar{\bar{P}}$ is the ratio of incorrect statements to total statements when family ties are ignored and all statements are lumped together. This should be of some comfort to the nonmultiple comparisonist who wants to make each statement separately and is leery of families.

If there is workable knowledge of the structure of the dependence between the statements in a family, it may be possible to achieve better bounds, but in general the only relation between $P\{\mathfrak{F}\}$ and $E\{\mathfrak{F}\}$ is

$$E\{\mathfrak{F}\} \leq P\{\mathfrak{F}\} \leq N(\mathfrak{F}) \cdot E\{\mathfrak{F}\}. \tag{16}$$

The first bound is usually quite good since both quantities are kept small, and for small numbers of statements the second provides adequate information. In addition, the first inequality in (16) produces another frequency interpretation for sequences of independent families with the same $P\{\mathfrak{F}\}$ (but not necessarily the same $E\{\mathfrak{F}\}$) besides the one already given on the convergence of the proportion of incorrect families to $P\{\mathfrak{F}\}$. With probability 1 the limit superiors of $\bar{P}$, the average over families of the proportions of incorrect statements, and of $\bar{\bar{P}}$, the proportion of incorrect statements, are less than or equal to $P\{\mathfrak{F}\}$. The limits of $\bar{P}$ and $\bar{\bar{P}}$ may not exist, but in the limit their fluctuations will remain below $P\{\mathfrak{F}\}$. [The argument is a slight modification of Appendix A using (16).]

Despite the fact that the expected error rate has a frequency interpretation and is conveniently related to the component error rates, this author continually finds himself drawn back to the probability error rate as his security blanket for feelings he cannot entirely analyze. The probability error rate is applicable for both finite and infinite numbers of statements in families. In a sense it does not put trust in a mean value without precise knowledge as to the dispersion. The proportion of incorrect families has a binomial distribution with mean $P\{\mathfrak{F}\}$, variance $P\{\mathfrak{F}\}(1 - P\{\mathfrak{F}\})/n$, where n is the number of families, and the rate of convergence to $P\{\mathfrak{F}\}$ is better understood than the convergence of $\bar{P}$ and $\bar{\bar{P}}$ to $p = E\{\mathfrak{F}\}$. For splicing together dependent statements into one family, the expected error rate gives exact results through (15), but thanks to (13) there is also an adequate bound for $P\{\mathfrak{F}\}$ for small numbers of statements or subfamilies. Control of the probability error rate gives a known degree of protection for the entire family and an upper bound on the expected proportion of mistakes. And finally, to the author's clients the thought that all of their statements are correct with high probability seems to afford them a greater serenity and tranquility of mind than a discourse on their expected number of mistakes.

Fortunately, leaning more heavily on one rather than the other does not mean the other has been banished into exile. Their close connection makes it essentially a matter of taste. The reader is invited to feast upon whichever satisfies him best.

2.3 Allocation of error

Within a family there may be natural groupings of statements which create subfamilies. As an illustration, consider a two-way classification design, fixed effects model, with equal, multiple numbers of observations per cell (a lovely, if rarely found, situation). The experimenter and statistician might like to consider all statements concerning this two-way experiment as a family and have an error rate $P\{\mathfrak{F}\}$ less than or equal to .05. If one way in the classification represents treatments and the other blocks, the statements readily divide themselves into treatment statements, block statements, and interaction statements. By the orthogonality of the design the numerators of the statement statistics will be independent between groups, but they will have the same residual variance s^2 in common in their denominators, thereby creating dependence between groups. Within any subfamily (treatments, blocks, interactions) the numerators of the statistics will usually be dependent as well.

Most likely the experimenter's interest is centered on treatments with secondary interest in blocks and interactions, provided they are not large. Differences between blocks and existence of interactions affect the universal applicability of the treatments and are of importance if they are sufficiently large to seriously influence the net effect of the treatments. Otherwise, the experimenter is probably content to notice their existence or nonexistence with passing interest.

Block effects could be conveniently investigated by Tukey's studentized range (Sec. 1 of Chap. 2) to pick up any possible differences. Existence of interactions could be studied by Scheffé's F projections (Sec. 2 of Chap. 2) or by Bonferroni t statistics (Sec. 3 of Chap. 2), depending upon which would give shorter intervals. If straight differences between pairs of treatment means were the only comparisons of interest, then Tukey's studentized range would probably give the shortest confidence intervals. On the other hand, if differences between weighted averages of separate groups of means as well as straight differences were important, then Scheffé's F projections procedure would most likely be best.

Whatever technique is applied to each subfamily, it will produce a probability error rate for the subfamily: α_T for treatments, α_B for blocks, and α_I for interactions. If $\alpha_T = \alpha_B = \alpha_I = .016$, then the probability error rate for the whole family will be less than or equal to .048 by an application of (13) to subfamilies instead of statements, and the expected error rate for the proportion of incorrect subfamilies will be .016. However, there is no law that insists on α_T, α_B, and α_I being equal. Two other possible choices are $\alpha_T = .03$, $\alpha_B = .01$, $\alpha_I = .01$ and $\alpha_T = .01$, $\alpha_B = .02$, $\alpha_I = .02$, both of which give $P\{\mathfrak{F}\} \leq .05$. And indeed, it is likely one of these would be a more logical choice in this experiment than equal α's. If the experimenter was very anxious to detect any conceivable differences between treatment means, he would have better power with $\alpha_T = .03$ than $\alpha_T = .016$, but he would still have his overall $P\{\mathfrak{F}\} \leq .05$. On the other hand, if the situation required a good deal of caution, then $\alpha_T = .01$ instead of .016 would give him better protection against erroneous statements about treatments.

In general, it seems unlikely that one would want to weigh treatments equally with blocks and interactions. If the two-way classification had been treatments vs. treatments (i.e., both classes constituting different types of treatments), there would be a plausible argument for treating each α with equal care.

Before leave is taken of this section on error rates, it should be noted that when the statements S_f are tests of hypotheses, another concept, other than $P\{\mathfrak{F}\}$ or $E\{\mathfrak{F}\}$, will play a role. This is Duncan's idea of *p-mean significance levels.* However, until some of the techniques have been discussed in the next section, the motivation would be lacking, and an abstract discussion would only tend to confuse. This topic will be taken up at the appropriate time in Sec. 4.

3 BASIC TECHNIQUES

The basic techniques of multiple comparisons divide themselves into two groups: those which can provide confidence intervals or corresponding tests of hypotheses, and those which are essentially only tests of hypotheses because of their multistage structure. For distinction the former are labeled *confidence regions,* and the latter *significance tests.*

CONFIDENCE REGIONS

3.1 Repeated normal statistics

Suppose $\alpha = .05$ is the probability error rate *per statement* that a nonsimultaneous statistician is willing to tolerate. For the null hypothesis (2) the acceptance region based on separately testing the hypotheses in (3) is

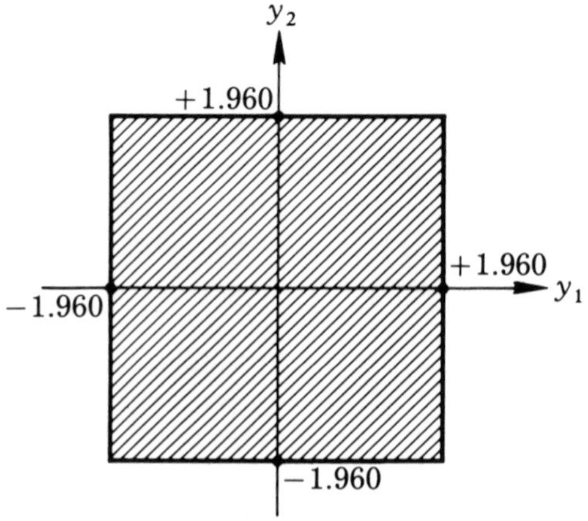

Figure 1

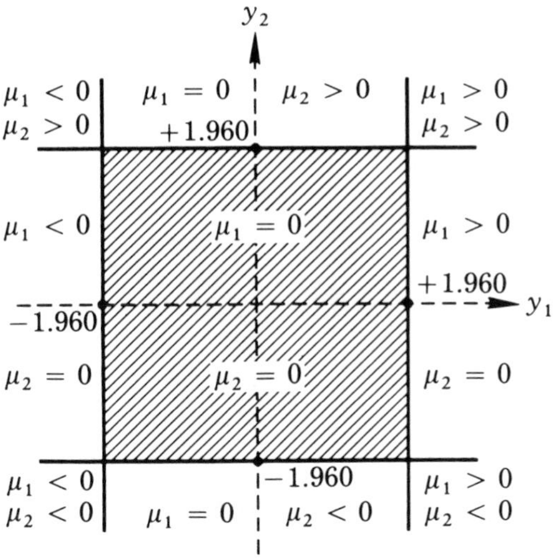

Figure 2

the square-shaped region centered at (0,0) in Fig. 1. The critical points are set at ± 1.960, the two-tailed 5 percent points of a normal distribution. For a confidence region the square is centered at (Y_1, Y_2) in the (μ_1,μ_2)-plane.

If the sample point fell outside the acceptance region in a test of significance, the statistician would be led to one of eight possible conclusions depending upon which of the alternative regions in Fig. 2 contained the sample point.

For each statement on a mean value, of which there are two, the probability error rate and expected error rate are exactly .05 under the joint (2) or separate (3) null hypotheses. Viewed as a family consisting of a pair, the family probability error rate under the joint null hypothesis (2) is $1 - (.95)^2 = .0975$, and the family expected error rate is .05.†

The pivotal statistics (Y_1, Y_2) involved in the construction of the region are normally distributed by assumption. If σ was unknown, then (Y_1, Y_2) would have a common sample standard deviation s inserted in their denominators, and the intervals would become t intervals. For example, with 10 d.f.‡ the critical points in the (y_1,y_2)-plane [or (μ_1,μ_2)-plane] are $\pm 2.228s$.

For the t region the probability and expected error rates for each statement are still .05. The pair expected error rate remains at .05, but the

† For significance probabilities, four-decimal accuracy, and for critical points, three-decimal accuracy will be maintained for comparative purposes. It is not implied that this degree of accuracy is required, nor necessarily desirable, in applications to data.

‡ d.f. = degrees of freedom.

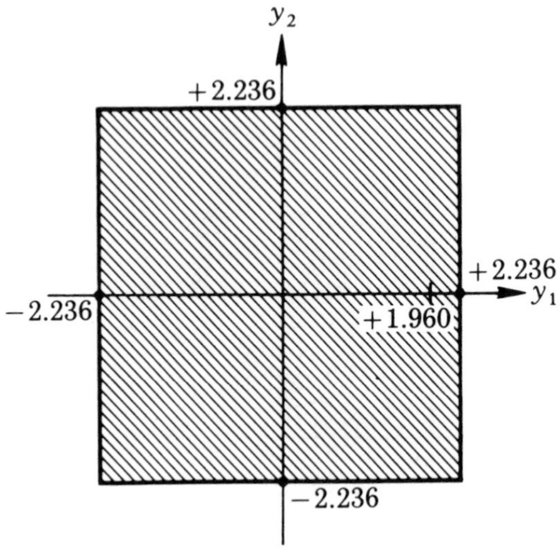

Figure 3

pair probability error rate is .0945.¶ The inequality (13) gives an upper bound of .10 for the pair probability error rate.

3.2 Maximum modulus (Tukey)

This is the prototype of the Tukey-type procedures to be studied in detail in Secs. 1, 4, and 5 of Chap. 2.

The probability .0975, approximately $\frac{1}{10}$, of at least one statement being incorrect under the preceding procedure might worry some statisticians as being too large for certain situations. This can be reduced to a pair probability error rate of .05 under the null hypothesis (2) by increasing the probability coverage of the square region from .9025 to .95. This is equivalent to finding the constant c such that

$$P\{\max\{|Y_1|, |Y_2|\} \leq c\} = .95. \tag{17}$$

By virtue of the independence of Y_1 and Y_2, c is given by the two-tailed 2.53 percent ($= 100(1 - \sqrt{.95})$ percent) point of the normal distribution, that is, 2.236 [see (12)].[1] The acceptance region appears in Fig. 3. The corresponding confidence region is centered at (Y_1, Y_2) in the (μ_1, μ_2)-plane.

¶ Some significance probabilities (such as .0945) and critical points in this chapter have been specially computed. The reader is not expected to be able to derive them in his head, nor obtain them directly from a table.

[1] Tukey coined the word *allowance* for the constant (such as 2.236) which gives one-half the width of the confidence interval.

For a test of significance, rejection of the null hypothesis would lead to acceptance of one of the eight alternatives, the choice being determined by the region in which the sample point falls. The diagram of the one null and eight alternative regions is identical with that in Fig. 2 except that the inner square is enlarged to intersect the axes at ± 2.236 instead of at ± 1.960.

With this procedure the probability error rate for the pair is the desired .05. The expected error rate for the pair, and the expected and probability error rates for each statement, under the separate null hypotheses (3), have now been reduced to .0253.

For σ^2 unknown, the appropriate t region expands to $\pm 2.609s$ for 10 d.f. The probability error rate for the pair under this procedure is .05, and the expected (pair and statement) error rates and statement probability error rates are .0261.

3.3 Bonferroni normal statistics

This is the analog of the approximate, rough-and-ready procedure to be described in Sec. 3 of Chap. 2.

The independence of Y_1 and Y_2 permitted the easy calculation of the critical point for the simultaneous confidence region in the preceding section. However, in numerous practical situations, the critical point would not be determinable because of messy dependence between Y_1 and Y_2. For such problems the Bonferroni inequality (13) will provide a region possessing probability greater than that demanded. For two statements

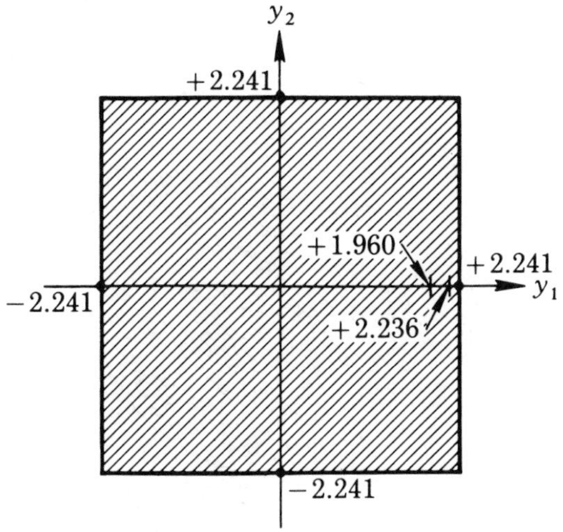

Figure 4

the desired significance level .05 should be halved (.025) to obtain the individual critical points. The two-tailed 2.5 percent normal critical points are ± 2.241, which give the region in Fig. 4. Note that the increase of .005 between this technique and the preceding exact one is not apprecia-ble. This will customarily be the case when α is small and the number of statements is not large.

For the region in Fig. 4 the probability error rate for the pair is exactly $1 - (.975)^2 = .0494$, which is quite close to the bound .05. The pair expected error rate, and the probability and expected error rates for individual statements drop from .0253 for the maximum modulus tech-nique to exactly .025.

With 10 d.f. the appropriate t intervals are $\pm 2.634s$, which again is not a dramatic increase over $2.609s$ for the exact maximum modulus technique. The exact probability error rate for the pair is .0479, while the other error rates remain at .025.

3.4 χ^2 projections (Scheffé)

This is the prototype of the Scheffé procedures to be discussed in Sec. 2 of Chap. 2.

If a classicist were to attack the joint hypothesis (2), he would use the χ^2 statistic $Y_1^2 + Y_2^2$ with 2 d.f. to construct a confidence region or sig-nificance test. The acceptance region for the null hypothesis (2) is circular as depicted in Fig. 5. For a level of significance $\alpha = .05$, the

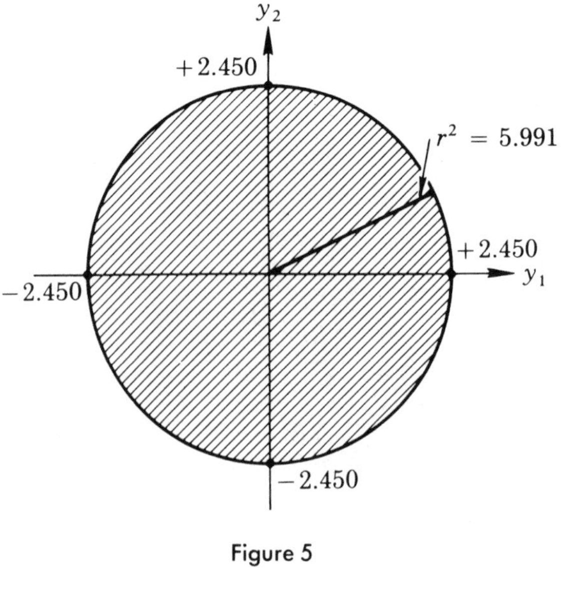

Figure 5

acceptance region is defined by $Y_1^2 + Y_2^2 \leq 5.991$, which intersects the axes at ± 2.450.

In an attempt to adapt his old friend to his purposes, a multiple comparisonist might circumscribe a square on the spherical acceptance region and construct his one null and eight alternative regions as in Fig. 6. The acceptance region has now been enlarged from the circle to the circumscribed square which accordingly lowers the level of significance.

Circumscription of the square on the circle is equivalent to projecting the circular χ^2 region onto the Y_1 and Y_2 axes to obtain individual acceptance intervals and then pasting these together to form a joint region. Each of the preceding three techniques had a common motif: construction (by their respective principles) of individual acceptance or confidence intervals on Y_1 and Y_2 and formation of their product. This technique is no different. Now the individual intervals are obtained by projections of the bivariate χ^2 region.

A square was circumscribed on the circle rather than inscribed because this was the conservative procedure. Without any calculation it follows that the probability error rate for the pair is less than .05. Had the square been inscribed then the rate would be raised to an amount exceeding .05.

In most practical problems further evaluation of the probability error rate would be impossible or exceedingly difficult. However, in this simple example the probability error rate is readily calculated to be .0284.

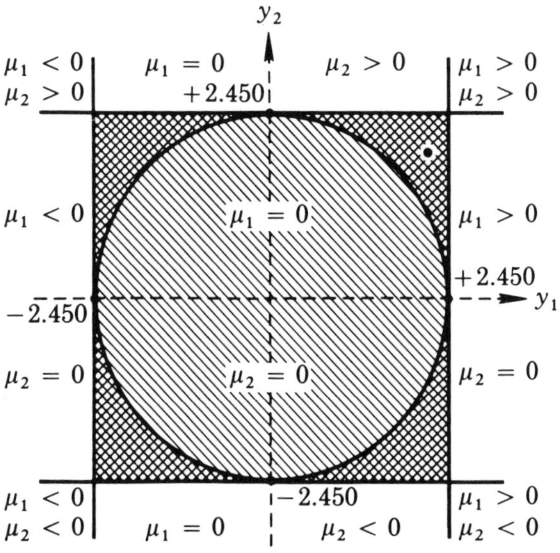

Figure 6

The expected pair error rate, and the expected and probability individual error rates are .0143.

Both this technique and that of the preceding section, the Bonferroni statistics, are conservative procedures in the sense that $P\{\mathfrak{F}\} \leq .05$. Comparison of the two shows that in this example the acceptance region based on the Bonferroni normal statistics is the more sensible choice since it is smaller (2.241 vs. 2.450). With the same conservative bound of .05 the Bonferroni region achieves better power. In applied problems the Bonferroni region will frequently be smaller than the projected χ^2 region when the number of statements is small. When the number of statements or comparisons is large, the reverse will hold true. In any problem it is ethical as well as theoretically sound for the statistician to compute the critical points for both regions and select whichever region is smaller. The choice is independent of the observed data and is therefore perfectly valid. Of course, if an exact maximum modulus technique of the type discussed in Sec. 3.2 is available, it will yield smaller regions than either of these conservative procedures and would hence be preferable.

Without further interpretation the four crosshatched regions in Fig. 6 would be somewhat disturbing to a simultaneous inference devotee. They constitute sample points which would reject the null hypothesis for the χ^2 test, but would not declare either μ_1 or μ_2 different from zero

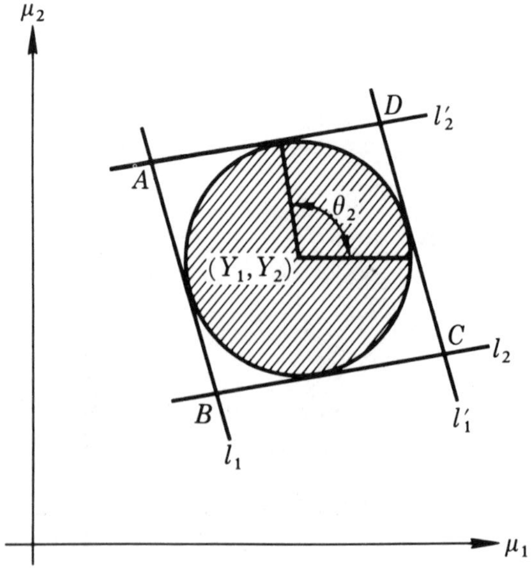

Figure 7

for the derived simultaneous procedure. If the umbilical cord to the χ^2 test is still strong, some explanation of how (μ_1,μ_2) differs from (0,0) for these points might be sought. There is such an explanation which resolves this dilemma.

Consider the confidence region for (μ_1,μ_2) centered at (Y_1,Y_2) in Fig. 7. Let l_1 and l_1' be a pair of parallel lines tangent to the confidence circle. The mean pairs (μ_1,μ_2) which lie between the lines l_1 and l_1' comprise a slanted infinite strip. For another pair of parallel lines, l_2 and l_2', tangent to the circle, another infinite strip of points (μ_1,μ_2) lying between l_2 and l_2' is created. The intersection of these two strips is the parallelogram $ABCD$. It consists of all mean points (μ_1,μ_2) which simultaneously fall between the pairs l_1, l_1' and l_2, l_2'.

Now consider all possible pairs of parallel lines tangent to the circle. These can be generated by letting the point of tangency rotate around the circle. Visual contemplation should convince one that the intersection of all the infinite strips corresponding to the pairs of lines is exactly the circle itself.

This correspondence between the circle and the intersection of all the strips suggests the following interpretation of the points in the crosshatched regions of Fig. 6. The equations of the lines l_1 and l_1' are

$$\begin{aligned} l_1&: a_1\mu_1 + b_1\mu_2 = c_1 \\ l_1'&: a_1\mu_1 + b_1\mu_2 = c_1' \end{aligned} \tag{18}$$

where $c_1 < c_1'$.† All points in the infinite strip between l_1 and l_1' satisfy

$$c_1 \leq a_1\mu_1 + b_1\mu_2 \leq c_1'. \tag{19}$$

Let the family of all pairs of parallel, tangent lines be indexed by θ, $0 \leq \theta < \pi$, where, relative to the circle with (Y_1,Y_2) as origin, θ is the angle between the point of tangency of the upper line and the μ_1 axis. In terms of the equations of the lines, the family consists of

$$\mathcal{L} = \{(a_\theta,b_\theta,c_\theta,c_\theta'),\ 0 \leq \theta < \pi\}.$$

Since the circle is the intersection of all the strips between the tangent, parallel lines

$$\begin{aligned} 1 - \alpha &= P\{(\mu_1^*,\mu_2^*) \in \text{circle}\} \\ &= P\{c_\theta \leq a_\theta\mu_1^* + b_\theta\mu_2^* \leq c_\theta', \text{ for all } 0 \leq \theta < \pi\} \end{aligned} \tag{20}$$

under the hypothesis (μ_1^*,μ_2^*) is the mean of (Y_1,Y_2).

For the point in the upper, right crosshatched area in Fig. 6, the con-

† c_1, c_1' are functions of Y_1, Y_2, a_1, b_1, and r^2 (the upper 100α percent point of the χ^2 distribution with 2 d.f.). Their dependence on these quantities is suppressed for notational simplicity.

fidence circle centered at it appears in Fig. 8. The null hypothesis (2) is that the mean of (Y_1, Y_2) is (0,0). Although (0,0) falls in the strips for $\theta = 0$ and $\theta = \pi/2$, there exists at least one strip $\theta = \theta_0$ (actually many strips) which does not contain (0,0). For the strips $\theta = 0$ and $\theta = \pi/2$,

$$\begin{aligned} c_0 < a_0 \cdot 0 + b_0 \cdot 0 = 0 < c_0' \\ c_{\pi/2} < a_{\pi/2} \cdot 0 + b_{\pi/2} \cdot 0 = 0 < c_{\pi/2}' \end{aligned} \tag{21}$$

but for $\theta = \theta_0$

$$a_{\theta_0} \cdot 0 + b_{\theta_0} \cdot 0 = 0 < c_{\theta_0} < c_{\theta_0}'. \tag{22}$$

The conclusion to be drawn is that although the data is not strong enough to decide that $\mu_1 \neq 0$ by itself or $\mu_2 \neq 0$ by itself, it is sufficient to declare that $a_{\theta_0}\mu_1 + b_{\theta_0}\mu_2 > 0$.

The null hypothesis (2) is equivalent to the null hypothesis

$$H_0^{\mathfrak{L}}: \quad a\mu_1 + b\mu_2 = 0 \qquad \text{for all } a, b. \tag{23}$$

In view of (20) the χ^2 test is in fact a simultaneous test of an *infinite* number of linear combinations of the means. For each combination (a,b) there is a corresponding confidence interval (c,c') for the parametric combination $a\mu_1 + b\mu_2$. The χ^2 test rejects the null hypothesis if and only if for some combination (a_θ,b_θ) the value $a_\theta \cdot 0 + b_\theta \cdot 0 = 0$ does not belong to the confidence interval (c_θ,c_θ'). Hence, the χ^2 test gives, and is given by, the infinite family of statements about the values of $a\mu_1 + b\mu_2$.

When the family of statements under consideration is taken to be the infinite family corresponding to $\mathfrak{L}$ rather than the specific pair μ_1 and μ_2, the family probability error rate is .05, the significance level of the χ^2 test. If a finite number of linear combinations other than μ_1 and μ_2 were of inter-

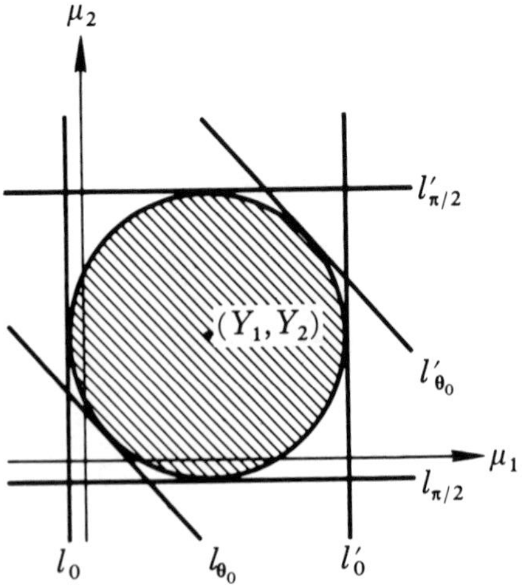

Figure 8

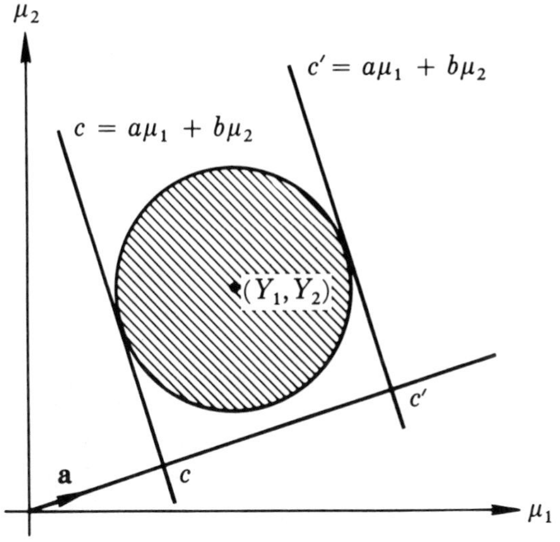

Figure 9

est, but not the entire family of all linear combinations, the probability error rate would be something between .0284 and .05. The exact probability would be the density in the (Y_1, Y_2)-plane outside the polyhedron cut out by the finite family of tangent, parallel lines. The more lines there are, the closer the probability will approach .05.

Depending upon the practical problem, this interpretation may have some merit, or it may be so much garbage. In certain problems linear combinations of the means have a physical basis. They could represent drug mixtures, metal alloy compounds, etc. On the other hand, they may be just mathematical artifacts and have no correspondence with reality. For the former type of experiment the simultaneous interpretation of the χ^2 test may be useful. In the latter type where only the two comparisons $\mu_1 = 0$? and $\mu_2 = 0$? have any physical import, it is likely that one of the preceding techniques would give shorter intervals for μ_1 and μ_2. The statistician is paying for confidence intervals for all linear combinations but only using two.

Before leave is taken of this interpretation of the χ^2 test, it is worthwhile to note that χ^2 *projections* is not merely a suggestive title for this technique. It is mathematically precise. Suppose the linear combination $\mathbf{a} = (a,b)$ is normalized so that its length is one, that is, $a^2 + b^2 = 1$. Then, the confidence interval for $a\mu_1 + b\mu_2$ is exactly the interval on the line spanned by **a** obtained by projecting the χ^2 confidence circle onto this line, as in Fig. 9.[1]

[1] *Proof:* From $a(ca) + b(cb) = c(a^2 + b^2) = c$, it follows that $c\mathbf{a}$ lies on the lower tangent line to the circle. Similarly, $c'\mathbf{a}$ lies on the upper tangent line. ||

If σ^2 is unknown, but s^2 is an independent estimate of σ^2 based on a χ^2 distribution with ν d.f., all the preceding discussion carries over with the square of the radius replaced by $s^2 \times 2 \times$ (upper 100α percent point of an F distribution on 2 and ν d.f.).[1] With 10 d.f. for s^2 this replaces 5.991 by $8.206s^2$ or the radius 2.450 by $2.865s$. The acceptance interval for Y_1 or Y_2 is $\pm 2.865s$, which should be compared with $\pm 2.634s$—the largest interval of the preceding techniques.

The entire preceding discussion on interpretation, applicability, etc., can be repeated for the F test. The only difference is that now the radius of the confidence circle is random (a multiple of s) whereas before it was fixed.

For the pair of statements on μ_1 and μ_2 the probability error rate is .0324, while the other three error rates are .0267. Considered as an infinite family of statements on all possible linear combinations, the probability error rate is exactly .05, and the individual statement error rates remain at .0267.

3.5 Allocation

For the repeated normal, maximum modulus, and Bonferroni normal techniques the confidence or acceptance region was always taken to be square. There is no compelling need for this, however. If neither mean was more important or crucial than the other, then this is the most sensible procedure. But there may be occasions when it is desirable to have greater protection for one mean under the null hypothesis, or greater power under the alternative. In these instances rectangular regions are the answer.

For normal statistics (σ^2 known) the appropriate rectangular regions are readily constructed. If two separate normal intervals are constructed for Y_1 and Y_2 with significance levels α_1 and α_2, respectively, then the rectangle has a base with endpoints at the two-tailed $100\alpha_1$ percent points of the normal distribution and an altitude with endpoints at the $100\alpha_2$ percent point. For such a rectangle the pair probability error rate is $1 - (1 - \alpha_1)(1 - \alpha_2)$, and the expected error rate is $(\alpha_1 + \alpha_2)/2$. The individual probability and expected error rates vary from α_1 to α_2 for Y_1 and Y_2, respectively. The case $\alpha_1 = .01$, $\alpha_2 = .04$ is illustrated in Fig. 10.

For the repeated normal (Sec. 3.1) and Bonferroni normal (Sec. 3.3), once α_1 and α_2 have been selected, the region can be easily obtained from normal tables and the error rates computed directly. For the maximum

[1] $\dfrac{Y_1^2 + Y_2^2}{s^2} \sim 2F_{2,\nu}$.

modulus (Sec. 3.2) with a predetermined probability error rate α for the pair, the statistician can select either of two schemes. If he wants the individual error rates to have an assigned ratio (that is, $\alpha_1/\alpha_2 = \rho$), he can solve the equation $\alpha = 1 - (1 - \rho\alpha_2)(1 - \alpha_2)$ for α_2 (and α_1) and then obtain his intervals from normal tables. If he wants the lengths of the intervals to have a prescribed ratio, he will be forced into a trial-and-error search procedure in the normal tables. Any fancier ideas on the relative selection of intervals for Y_1 and Y_2 subject to a fixed $P\{\mathfrak{F}\} = \alpha$ will probably also lead to a search procedure.

For t statistics when σ^2 is unknown it is theoretically possible to do the same thing. For repeated t statistics (Sec. 3.1) and Bonferroni t statistics (Sec. 3.3) the selection of α_1 and α_2 immediately gives the region. However, the exact evaluation of $P\{\mathfrak{F}\}$ for the pair is beyond the scope of any existing tables. It would require the tabulation of the probabilities

$$P\left\{\frac{|Y_1|}{s} < c_1, \frac{|Y_2|}{s} < c_2\right\} \qquad \text{for } c_1 \neq c_2 \tag{24}$$

and this has not been done. Bounds on $P\{\mathfrak{F}\}$ are of course available. The lack of these tables also prevents one from starting with a predetermined α for the maximum modulus technique (Sec. 3.2) and obtaining the region.

Presumably relative allocation could also be accomplished for the projected χ^2 technique (Sec. 3.4). This would be based on an elliptical region determined from the quadratic form $Y_1^2 + \rho^2 Y_2^2$. Percentage points of quadratic forms of this type are available: Grad and Solomon (1955) and Solomon (1960). However, since the other three techniques are

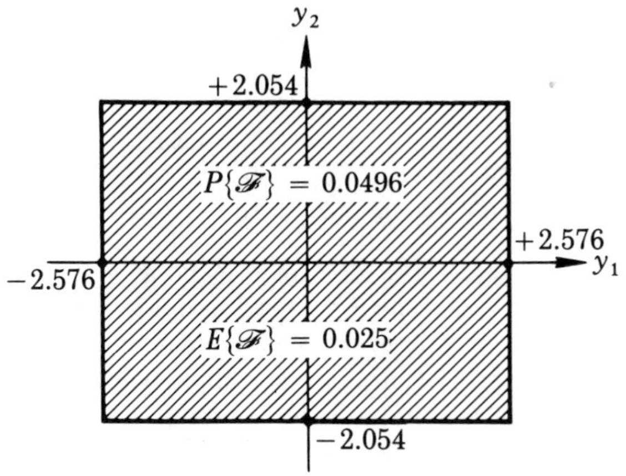

Figure 10

better for just the two statements on μ_1 and μ_2, the practical motivation for the use of these quadratic forms in this context appears to be scanty.

SIGNIFICANCE TESTS

3.6 Multiple modulus tests (Duncan)

For a 5 percent level maximum modulus test (Sec. 3.2) it is necessary for both Y_1 and Y_2 to exceed 2.236 in absolute value in order for both μ_1 and μ_2 to be inferred as different from zero. Had each variable been treated separately at the 5 percent level the requirement would have been lowered to 1.960. Suppose, for illustration, the first variable was quite large, say, $Y_1 = 4.37$, leaving little doubt that $\mu_1 > 0$. In order for μ_2 to be judged different from zero it must still exceed 2.236, whereas if it were now considered by itself the critical point would drop to 1.960. Despite the performance of Y_1 in vastly exceeding the limits for the maximum of two independent, unit normal random variables, Y_2 must still exceed the limits for the *maximum of two* random variables rather than the limits for an *individual* mean. It would seem that this should give sufficient grounds for picketing the maximum modulus test as unfair to the lesser mean.

A remedy for this is to perform the testing in successive stages, i.e., a multiple modulus test, or multiple range test when applied to differences. The prototype of a multiple modulus test is conducted in the following manner.

Stage 1 Compare max $\{|Y_1|,|Y_2|\}$ with 2.236.

(*a*) If max $\{|Y_1|,|Y_2|\} \leq 2.236$, decide in favor of H_0: $\mu_1 = 0$, $\mu_2 = 0$.

(*b*) If max $\{|Y_1|,|Y_2|\} = |Y|_{(2)} > 2.236$, decide $|\mu|_{(2)} \neq 0$, and proceed to stage 2.†

Stage 2 Compare min $\{|Y_1|,|Y_2|\} = |Y|_{(1)}$ with 1.960.

(*a*) If $|Y|_{(1)} \leq 1.960$, decide $|\mu|_{(1)} = 0$.

(*b*) If $|Y|_{(1)} > 1.960$, decide $|\mu|_{(1)} \neq 0$.

In the event of significance, signs can also be attached to the nonzero population means from the corresponding sample values.

This two-stage testing procedure is equivalent to determining which of the nine regions in Fig. 11 contains the sample point (Y_1, Y_2).

The difference between the multiple modulus test and the maximum modulus test is the enlargement of the four decision regions $\{\mu_1 > 0,$

† $|Y|_{(1)} \leq |Y|_{(2)}$ are the order statistics corresponding to $|Y_1|$, $|Y_2|$, that is, the ordered absolute values. $|\mu|_{(i)}$ is the true mean corresponding to $|Y|_{(i)}$, $i = 1,2$.

$\mu_2 > 0\}$, $\{\mu_1 > 0, \mu_2 < 0\}$, $\{\mu_1 < 0, \mu_2 > 0\}$, and $\{\mu_1 < 0, \mu_2 < 0\}$ at the expense of the four regions $\{\mu_1 > 0, \mu_2 = 0\}$, $\{\mu_1 < 0, \mu_2 = 0\}$, $\{\mu_1 = 0, \mu_2 > 0\}$, and $\{\mu_1 = 0, \mu_2 < 0\}$. Since the null hypothesis acceptance regions are identical for the two tests, the multiple modulus test has the same probability error rate under the null hypothesis as the maximum modulus. The expected null error rate will be somewhat higher for the multiple modulus test since the chance of deciding that both means differ from zero is increased. The behavior of the two tests under the alternatives will also be different because of the changed alternative decision regions. A partial analysis of their performance under the alternative hypotheses is undertaken in the discussion of the *p-mean significance levels* in the next section.

For σ^2 unknown the corresponding t intervals replace the normal intervals, but the structure of the multistage test remains the same. The t constant $2.609s$ on 10 d.f. replaces 2.236, and $2.228s$ should be substituted for 1.960.

When the reader comes to study multiple range tests in detail in Sec. 6 of Chap. 2, he will learn that although Duncan would agree with the general shape of the region in Fig. 11 (for normal and t tests), he would not agree with its exact dimensions. The size is in agreement rather with the principles of Newman and Keuls.

Duncan would argue as follows. For two unrelated, independent statistical tests, statisticians are often willing to use separate .05 level tests on each. For the pair the probability error rate is $.0975 = 1 - (.95)^2$.

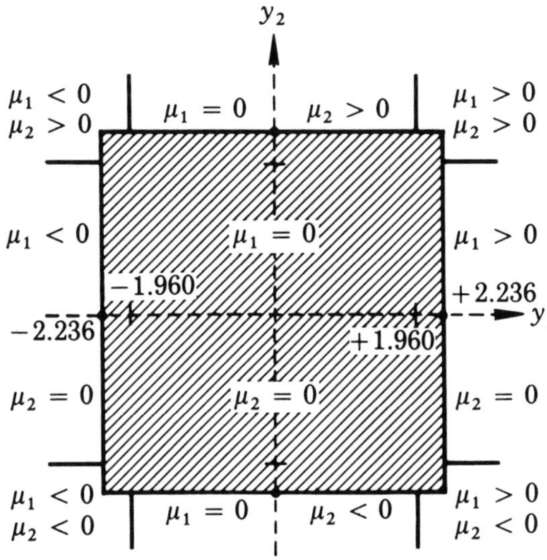

Figure 11

Why, then, for any pair of tests, dependent or independent, related or unrelated, should the statistician have a probability error rate different from .0975? Thus, for the normal tests Duncan would use 1.960 in both stages 1 and 2, and for the t tests on 10 d.f. he would use $2.209s$ (giving $P\{\mathfrak{F}\} = .0975$) in stage 1, and $2.228s$ in stage 2.

With this philosophy the family probability error rate for dependent or independent statements is allowed to increase with family size in agreement with the rate for an allied number of independent statements (see Sec. 6 of Chap. 2 for elaboration). Suffice it to say here, this author prefers to maintain a preassigned low probability error rate for his families because they are formed of related statements requiring increased overall protection under the null hypothesis. Grouping of statements into families should depend on the need for protection and not on statistical dependence.

In closing it should be noted that the technique of this section is a testing technique and cannot be adapted to the construction of confidence intervals. The multiple stages have no analog in the confidence domain. They depend upon successive comparisons of the sample values with the hypothesized ones. For confidence regions there are no standard values for comparison purposes. What is sought is a region, depending on the sample point, which has a specified probability of containing the parameters, no matter what their values.

3.7 Least significant difference test (Fisher)[1]

Historically this test precedes the multiple modulus test. It was proposed by Fisher (1935), and was a predecessor of the later multiple stage tests, such as the multiple range tests and multiple F tests.

The least significant difference (LSD) test proceeds in stages.

Stage 1 Compare $Y_1^2 + Y_2^2$ with the critical value 5.991, the upper 5 percent point of a χ^2 distribution with 2 d.f.

(*a*) If $Y_1^2 + Y_2^2 \leq 5.991$, accept the null hypothesis H_0: $\mu_1 = 0$, $\mu_2 = 0$.

(*b*) If $Y_1^2 + Y_2^2 > 5.991$, proceed to stage 2.

[1] This title, which is historically applied to this test, is not a terribly fortunate choice, but it seems to be permanently affixed. The term *least significant difference* refers to the normal or t critical value, such as 1.960 or $2.228s$ (10 d.f.), respectively, which an individual observation must exceed in order to be judged significant. When the observation is considered to be a member of a family, the critical value for significance will increase, or at least not decrease, depending on the technique. Thus, to be significant, the observation has to *at least* exceed 1.960 or $2.228s$. Tukey has applied the term *wholly significant difference* to the critical value of the maximum modulus statistic with the obvious connotations.

Stage 2

(*a*) If $|Y_1| > 1.960$, decide $\mu_1 \neq 0$. Otherwise, decide $\mu_1 = 0$.
(*b*) If $|Y_2| > 1.960$, decide $\mu_2 \neq 0$. Otherwise, decide $\mu_2 = 0$.

In other words, a preliminary χ^2 test precedes repeated normal tests. If the χ^2 test is nonsignificant, nothing is judged significant. If the χ^2 test is significant, repeated individual normal tests are applied to evaluate each component.

The LSD test is equivalent to partitioning the sample space into the regions depicted in Fig. 12. Sample points in the four crosshatched regions are in the unhappy position of being indicative of significance under stage 1 but failing to declare either μ_1 or μ_2 different from zero at stage 2. The obvious recourse for these points is an interpretation of the type introduced in Sec. 3.4—although neither μ_1 nor μ_2 can be declared nonzero, it is likely some linear combination $a\mu_1 + b\mu_2$ differs from zero.

Comparison of this technique with the projected χ^2 (Sec. 3.4) shows that the center square (see Fig. 6) has been reduced in size and no longer circumscribes the circle. The four annoying crosshatched regions have been reduced in size, and the regions for declaring the μ_i's different from zero have increased.

When σ^2 is unknown, the χ^2 and normal tests are correspondingly replaced by F and t tests.

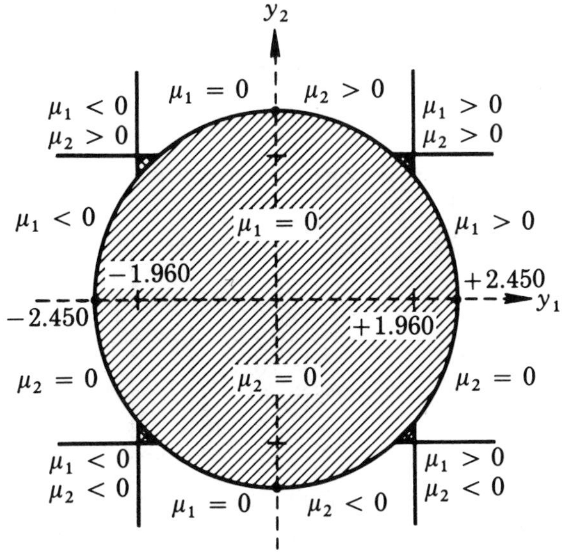

Figure 12

4 p-MEAN SIGNIFICANCE LEVELS

The multiple modulus test (Sec. 3.6) has in effect lifted the lid on Pandora's box. A whole flock of multistage tests bursts forth, each of which has the same probability error rate under the null hypothesis, but greatly differing null expected error rates and performance under the alternative hypotheses. Consequently, some additional criteria are needed to sort out the reasonable tests from the unreasonable ones.

At the two extremes of the gamut of Pandora's tests are the *all-or-nothing* and *only one* tests. These tests have the same first stage as the multiple modulus test, but at the second stage they are diametrically divergent.

Stage 2 (*all-or-nothing*) Decide both μ_1 and μ_2 differ from zero.

Stage 2 (*only one*) Decide only $|\mu|_{(2)}$ differs from zero (i.e., the mean corresponding to the observation with the larger absolute value).

Figures 13 and 14 give the partitionings of the sample space for the all-or-nothing and only one tests, respectively.

Since all three tests have the same first stage, namely, to decide in favor of the null hypothesis if and only if $\max\{|Y_1|,|Y_2|\} \leq 2.236$, the probability error rate under the null hypothesis must be identical for each.

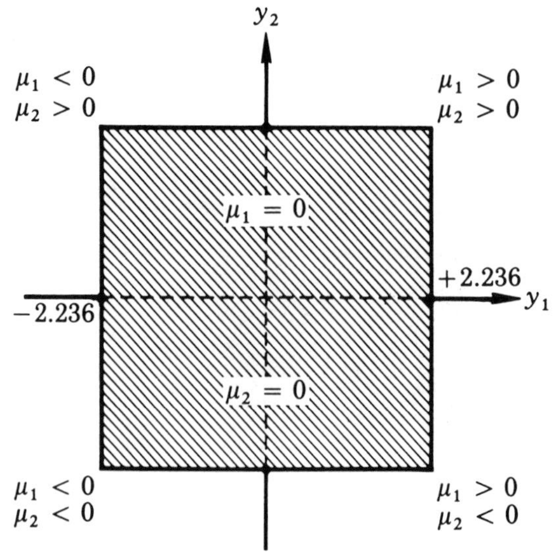

Figure 13

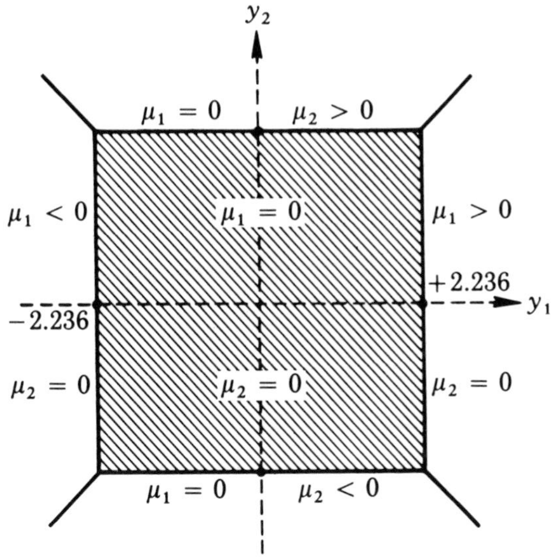

Figure 14

Obviously, however, the expected error rates under the null hypothesis differ, with the all-or-nothing test having the maximum error rate, the only one having the minimum rate, and the multiple modulus falling somewhere in between. Also, the performances of the three tests are considerably different under the alternative hypotheses.[1]

In an attempt to numerically evaluate the different behavior of these tests Duncan (1955) introduced the concept of a p-mean significance level. Let $D(\mu_1 \neq 0)$ stand for the decision that μ_1 differs from zero, and consider the probability of incorrectly making this decision when, in fact, $\mu_1 = 0$, and μ_2 is anything, that is,

$$P\{D(\mu_1 \neq 0)|\mu_1 = 0, \mu_2\}. \tag{25}$$

Define
$$\alpha(\mu_1) = \sup_{\mu_2} P\{D(\mu_1 \neq 0)|\mu_1 = 0, \mu_2\}. \tag{26}$$

The probability $\alpha(\mu_1)$ is a *one-mean significance level.*[2] There is a corresponding one-mean significance level for μ_2, namely, $\alpha(\mu_2)$.

Let $D(\mu_1 \neq 0 \cup \mu_2 \neq 0)$ denote the decision that either μ_1 or μ_2 or both differ from zero. For just two means the *two-mean significance level*

$$\alpha(\mu_1,\mu_2) = P\{D(\mu_1 \neq 0 \cup \mu_2 \neq 0)|\mu_1 = 0, \mu_2 = 0\} \tag{27}$$

[1] The multiple modulus, all-or-nothing, and only one tests are biased in the sense of Lehmann (1957). A test such as the maximum modulus could be unbiased (depending on the losses).

[2] The notation is designed to suggest α for significance, and μ_1 for μ_1 being equal to zero. μ_2 is unspecified since the probability has been maximized over μ_2.

reduces to the ordinary significance level α, since there are no additional variables to be maximized out. Had there been a third mean, with observation Y_3, the probability on the right in (27) would have been maximized over μ_3 to obtain the two-mean significance level.

For a group of m means the p-mean significance levels for $p = 1, 2, \ldots, m$ are defined analogously. The one m-mean significance level always reduces to the ordinary probability error rate under the null hypothesis. For a fixed p there are a total of $\binom{m}{p}$ different p-mean significance levels depending upon which p means are fixed equal to zero. Because of symmetry in the problem and in the test procedure, frequently all these p-mean significance levels for fixed p have the same value.

For the three tests used here as illustrations the two-mean significance levels are all equal to .05. The one-mean significance levels are, correspondingly,

only one:	$\alpha(\mu_1) = \alpha(\mu_2) = .025$
multiple modulus:	$\alpha(\mu_1) = \alpha(\mu_2) = .05$
all-or-nothing:	$\alpha(\mu_1) = \alpha(\mu_2) = 1.$

For the only one test the supremum in (26) is achieved when $\mu_2 = 0$, since for any value of μ_2, $|Y_1|$ has the same chance of exceeding 2.236, but for $\mu_2 = 0$ it has the greatest chance $(\frac{1}{2})$ of being the larger observation in absolute value. For the multiple modulus and all-or-nothing tests the supremum in (26) is achieved when $\mu_2 = \pm\infty$. This guarantees with absolute certainty that Y_2 will exceed 2.236, so for the multiple modulus test Y_1 only has to exceed 1.960 to be judged significant, and for the all-or-nothing test it is already automatically significant. By symmetry the value of $\alpha(\mu_2)$ equals that of $\alpha(\mu_1)$ for all three.

The p-mean significance levels represent an attempt to numerically evaluate the marginal performances of simultaneous tests of hypotheses. When a subset of the means has values equal to zero, what is the chance that at least one of these means will be declared different from zero when the whole group containing this subset is processed by the test? The p-mean significance levels evaluate this chance of a wrong decision at the worst alternative.[1]

Unfortunately, these additional significance levels do not seem to be an entirely satisfactory solution to the problem. The main drawback is the supremum operator. To produce a number, some operator must be applied to $P\{D(\mu_1 \neq 0)|\mu_1 = 0, \mu_2\}$, which is a function of μ_2. The supremum operator selects the largest value, or least upper bound, which

[1] There are other probabilities of wrong decisions that can be of interest and could lead to different systems of significance levels. See, for example, Duncan (1955, p. 31).

customarily occurs at $\mu_2 = \pm\infty$. But $\mu_2 = \pm\infty$, or any relatively large value, is completely unrealistic from the practical point of view. Any mean whose value is extremely large is not likely to be so unknown as to be included in a simultaneous test of a group of means. Such a spectacular effect could go unnoticed before its inclusion in an experiment with other variables, but this type of occurrence in simultaneous inference is rare. Consequently, the p-mean significance levels are determined by mean values totally unrelated to the practical problem. A more satisfactory criterion would somehow utilize the values of $P\{D(\mu_1 \neq 0)|\mu_1 = 0, \mu_2\}$ for reasonable candidates of μ_2 in the vicinity of zero.

For the three examples the numbers 1, .05, and .025 give some indication of the behavior of the three tests, but give by no means a complete picture. The one-mean significance level $\alpha(\mu_1) = 1$ for the all-or-nothing test is a warning beacon of something strange about this test, but the difference between .05 and .025 gives little sign of the different character of the tests involved. These last two levels are evaluated at two very different values of μ_2, namely, $\pm\infty$ and 0, respectively.

In handling simultaneous statistical tests Duncan advocates allowing the p-mean significance levels to increase in size above 5 percent as p increases. This is what occurs for a set of unrelated, independent tests. Also, from a Bayesian point of view, he argues that it is unlikely for all the m means to be simultaneously zero, and therefore one should not be too concerned about the p-mean significance levels for p large. Newman and Keuls keep all the p-mean significance levels equal to 5 percent as in the multiple modulus test. Tukey's procedure, the maximum modulus test, is the most conservative procedure with $\alpha(\mu_1,\mu_2) = .05$ and

$$\alpha(\mu_1) = \alpha(\mu_2) = .0253.$$

5 FAMILIES

Time has now run out. There is nowhere left for the author to go but to discuss just what constitutes a family. This is the hardest part of the entire book because it is where statistics takes leave of mathematics and must be guided by subjective judgment.

Two extremes of behavior are open to anyone involved in statistical inference. A nonmultiple comparisonist regards each separate statistical statement as a family, and does not give increased protection to any group of statements through group error rates. At the other extreme is the ultraconservative statistician who has just a single family consisting of every statistical statement he might make during his lifetime. If all statisticians operated in this latter fashion at the 5 percent level, then 95 percent of the world's statisticians would never falsely reject a null hypothesis, and 5 percent would be guilty of some sin against nullity. There are a few statisticians who would adhere to the first principle, but the author has never met one of the latter variety. As will be seen, there are good reasons for this. Most statisticians fall somewhere in between these two extremes, but have no well-formulated principles on family size or constitution.

The basic premise of simultaneous statistical inference is to give increased protection to a null hypothesis involving a group of parameters. But this protection applies only to the null hypothesis, and bears no heed to the number of errors that may occur under the alternative. Or, if confidence regions are being used, the requirement of simultaneous inclusion of all the parameters disregards the size of the region necessary to accomplish this. Yet it is not always the null hypothesis which in fact is true, and attention must also be paid to the error rates under the alternative hypotheses. This entails consideration of the power function of the test, or the size of the confidence region.

The ultraconservative statistician, who wants a high chance of having all his null statements correct, soon finds that his confidence intervals are fantastically wide, and his tests have extremely large critical points. Provided he meets only null hypotheses in his lifetime, or never requires tight confidence intervals, he is doing fine. Without implying any semblance of Bayesian leanings, it would be safe to say that the chance of always encountering null hypotheses is rather slim. (To be any kind of Bayesian, one should be able to give a less vague probability structure to the occurrence of the null and alternative hypotheses.) Although tests perform extraordinarily well for null hypotheses under this conservative doctrine, the error rates for alternative hypotheses are discouragingly high when the alternatives occur. The ultraconservative statistician is like a man who looks only in one direction—straight ahead at the null hypothesis—and never bothers to glance over his shoulder to see what is happening behind him. He gazes out on Elysian fields, oblivious of the chaos, destruction, and poverty behind him.

The antithesis of the ultraconservative statistician is the nonmultiple comparisonist, who only controls the statement error rate. His contention is that if a statistician is willing to tolerate an error rate of α for separate, independent statements, there is no reason to change one's philosophy for a sequence of dependent (or independent) statements just because some of them happen to be related. Some mathematical substantiation can be given to this position if the total loss for a sequence of statements is the sum of the component losses, and a Bayesian approach is taken [see Duncan (1955,1961,1965) and Lehmann (1957)]. This result is not surprising because of the loss structure. For different losses (viz., under the null hypothesis: zero for no mistakes, a constant for one or more mistakes) procedures of the multiple comparisons type are likely to emerge as optimal.

Provided the nonsimultaneous statistician and his client are well aware of their error rates for groups of statements, and feel the group rates are either satisfactory or unimportant, the author can find no quarrel with them. Every man should get to pick his own error rates. Simultaneous techniques certainly do not apply, or should not be applied, to every problem. However, there are numerous problems in which the context of the problem calls for greater protection of the joint null hypothesis, and in which the component losses are not additive.

The statistician is a man inescapably strapped to a teeter-totter (or seesaw). As the error rates are forced down in one direction they must increase in the other. This is congenital with problems of statistical inference. The only reprieve is through sequential experimentation or preselection of the sample size, when the statistician is fortunate enough to be able to do this. Here the statistician can simultaneously bend down both ends of the teeter-totter. But, once again, his attention cannot be focused exclusively on one item, the amount of bend, for the more he bends the greater the expected sample size will be.

Even for the classic Neyman-Pearson model of simple hypothesis vs. simple alternative the problem of the relative choice of α and β (the probabilities of errors of the first and second kind, respectively) is unsolved. There have been various attempts to remedy this. Some have been of non-Bayesian variety [Lehmann (1958)] while others have been along Bayesian lines. (The author does not consider Bayes strategies to be a practical answer since he finds it almost universally impossible to specify a priori probabilities.) But none seem to have been adopted into practice with any degree of regularity. Still predominantly guiding our statistical fortunes is an old rule of thumb: if $\alpha = .01$ gives reasonable power against reasonable alternatives, use it; otherwise drop back to $\alpha = .05$, or even .10 if necessary.[1]

The introduction of families complicates this already unsolved issue. As family size is increased, confidence intervals are widened and critical test points increased, provided the same family probability error rate is maintained. This correspondingly reduces the power of the test. To increase power either the family size must be reduced, the error rate increased, or the sample size increased. In this instance, increased generality (created by the introduction of families) does not resolve the issue.

For fixed sample size problems, a procedure which in a sense avoids the

[1] Levels below .01 are generally unsatisfactory because of sensitivity to the assumed form of the tails of the distribution.

issue of α vs. β has found frequent acceptance in practice. This is to report the P value, that is, the level at which the experimental data are just significant. To a lesser extent this can also be used in simultaneous problems. If the P value is used as a summary device for reporting the degree of significance of the data, it is a very acceptable procedure. However, it should not be used as a dodge for shifting the burden of drawing conclusions from the writer to the reader. The experimenter is far more familiar with the data, its virtues and vagaries, than the reader, so it is his prime responsibility to draw the main conclusions.

In the succeeding discussion of what types of families have been useful to the author, it is important to distinguish between two types of experiments. The first is the preliminary, search-type experiment concerned with uncovering leads that can be pursued further to determine their relevance to the problem. The second is the final, more definitive experiment from which conclusions will be drawn and reported. Most experiments will involve a little of both, but it is conceptually convenient to think of them as being basically distinct. The statistician does not have to be as conservative for the first type as the second, but simultaneous techniques are still quite useful for the first in keeping the number of leads that must be traced within reasonable bounds. In the latter type multiple comparisons techniques are very helpful in avoiding public pronouncements of red herrings simply because the investigation was quite large. The subsequent remarks will be principally aimed at experiments of the latter variety, although they do apply to a degree to preliminary experiments as well.

The *natural family* for the author *in the majority of instances* is the *individual experiment* of a *single researcher*. This would include, for example, a two-way classification analysis of variance, comparison of a half-dozen mean values, a regression analysis from which a fixed number of predictions will be made, a standard curve regression analysis from which an indeterminate number of discriminations will be made, etc.

By *natural* it is meant that this produces a conceptually clear and convenient method of division (e.g., the law of large numbers is easily interpretable), and it gives peace of mind to both the statistician and his client. Included is the requirement of reasonable power against reasonable alternatives with reasonable protection for the available sample size. For without this there can be no peace of mind.

An *individual experiment* means an obviously related group of observations which were collected through an autonomous experiment (i.e., the same experimental operation except for varying conditions such as intensity, heat, soil, etc.), and whose statistical analysis will fall into a single mathematical framework (e.g., analysis of variance, regression analysis, etc.). The *single researcher* refers simply to not grouping together different experimenters unless they are collaborating on the same experiment. It is doubtful that any experimenter would care to sacrifice sensitivity of his experiment so that when considered with someone else (e.g., a departmental colleague), the pair would have a better joint error rate. If the experimenter is at all active, it is likely he will perform numerous single experiments in his lifetime. He may even have a number of single experiments comprising one global investigation into a particular question. For the sake of power and sensitivity, it will usually be better to treat these smaller units separately.

The loophole is of course the clause *in the majority of instances*. Whether or not this rule of thumb applies will depend upon the size of the individual experiment. Large single experiments cannot be treated as a whole without an unjustifiable loss in sensitivity. For illustration, consider a two-way classification (fixed effects) analysis-of-variance problem with multiple observations per cell. If the number of rows and columns is not excessive, and the degrees of freedom are adequate, the author might like to treat all statements on row effects, column effects, and interaction effects as one family. For a large two-way design, however, this may not be practical. A less conservative grouping would be the row effects statements as one family, columns effects as another, and interactions another. Or rows and columns might be combined and grouped separately from interactions. It is unlikely that the author would subdivide further and treat each row statement separately in a nonsimultaneous fashion unless there were just a few specific comparisons spotlighted before the start of the experiment. *There are no hard-and-fast rules for where the family lines should be drawn, and the statistician must rely on his own judgment for the problem at hand.*

CHAPTER 2

Normal Univariate Techniques

This chapter contains a number of multisample and regression techniques whose distribution theory assumes an underlying normal distribution. It includes, in particular, the fundamental work of Duncan, Scheffé, and Tukey. Those techniques which are peculiar primarily to regression (e.g., prediction and discrimination) are discussed in Chapter 3, even though they are directly related, and in some instances special cases of the methods in this chapter. The nonparametric analogs of the techniques in this chapter are covered in Chapter 4.

In this and later chapters the format for each section on a principal technique will be: (1) *Method* (general cookbook description); (2) *Applications*; (3) *Comparison* (with other, alternative techniques); (4) *Derivation* (mathematical verification of the statements given in Sec. 1); and (5) *Distributions and Tables* (distributions needed for the computation of critical points; description and source of available tables).

The distinction is still maintained between those methods which can be used for tests of hypotheses and confidence regions (*Confidence Regions*), and those which are just testing techniques (*Significance Tests*). Sections 1 to 5 are confidence regions, and Secs. 6 and 7 are significance tests.

1 STUDENTIZED RANGE (TUKEY)

This technique was proposed by Tukey (1952a,1953b).

1.1 Method

Let $\{Y_{ij};\, i = 1, \ldots, r,\, j = 1, \ldots, n\}$ be r independent samples of n independently, normally distributed random variables with common variance σ^2 and expectations $E\{Y_{ij}\} = \mu_i, i = 1, \ldots, r, j = 1, \ldots, n$. This is a standard one-way classification design with equal numbers of observations in each class.

For this design the differences $\mu_i - \mu_{i'}$, $i \neq i'$, between class means are frequently the parametric quantities of interest. This problem might be formulated as a test of the hypothesis

$$H_0\colon \quad \mu_1 = \mu_2 = \cdots = \mu_r \tag{1}$$

or equivalently,

$$H_0\colon \quad \mu_i - \mu_{i'} = 0 \qquad i \neq i'. \tag{2}$$

Or the problem might be to bracket the differences $\mu_i - \mu_{i'}$ in confidence intervals. In either case, the studentized range technique is embodied in the following probability statement:

$$P\left\{|(\bar{Y}_{i\cdot} - \bar{Y}_{i'\cdot}) - (\mu_i - \mu_{i'})| \leq q^{\alpha}_{r,r(n-1)}\frac{s}{\sqrt{n}}, i, i' = 1, \ldots, r\right\}$$
$$= 1 - \alpha \tag{3}$$

where $$\bar{Y}_{i\cdot} = \frac{1}{n}\sum_j Y_{ij} \qquad s^2 = \frac{1}{r(n-1)}\sum_{i,j}(Y_{ij} - \bar{Y}_{i\cdot})^2 \tag{4}$$

and $q^{\alpha}_{r,r(n-1)}$ is the upper 100α percent point of the studentized range distribution with r, $r(n-1)$ for parameters [i.e., the distribution of the range of r independent, unit normal, random variables divided by the square root of an independent χ^2_ν/ν variable with $\nu = r(n-1)$ d.f.]. The family probability error rate is α for the following $\binom{r}{2}$ confidence statements:

$$\mu_i - \mu_{i'} \in \bar{Y}_{i\cdot} - \bar{Y}_{i'\cdot} \pm q^{\alpha}_{r,r(n-1)}\frac{s}{\sqrt{n}}$$
$$i, i' = 1, \ldots, r \qquad i \neq i'.\dagger \tag{5}$$

The corresponding significance test consists of examining the confidence intervals for the inclusion of zero in each. For any interval not containing zero the paired mean values are inferred to be different, with the direction of the difference given by the sample values.

The only change between this technique and its prototype discussed in Sec. 3.2 in Chap. 1 is that differences between mean values are considered rather than the mean values themselves. The pivotal statistic becomes a studentized range rather than a studentized maximum modulus.

A geometric setting can be given to the intervals in (5). If the points

† This notation is to signify

$$\bar{Y}_{i\cdot} - \bar{Y}_{i'\cdot} - q^{\alpha}_{r,r(n-1)}\frac{s}{\sqrt{n}} \leq \mu_i - \mu_{i'} \leq \bar{Y}_{i\cdot} - \bar{Y}_{i'\cdot} + q^{\alpha}_{r,r(n-1)}\frac{s}{\sqrt{n}}$$

and will be adopted throughout the book.

are plotted in the $\binom{r}{2}$-dimensional space of the paired differences $(\mu_1 - \mu_2, \ldots, \mu_{r-1} - \mu_r)$, then those parametric differences contained in all the intervals in (5) form a square region with center at $(\bar{Y}_{1.} - \bar{Y}_{2.}, \ldots, \bar{Y}_{r-1,.} - \bar{Y}_{r.})$ and sides at a distance $q^{\alpha}_{r,r(n-1)}s/\sqrt{n}$ from the center. If, instead, the points are plotted in the original r-dimensional space of $(\mu_1, \ldots, \mu_r)$, then the region is transformed to a slanted infinite cylinder with polyhedral base and $\binom{r}{2}$ pairs of parallel sides formed from the planes given by the equations $\mu_i - \mu_{i'} = \bar{Y}_{i.} - \bar{Y}_{i'.} \pm q^{\alpha}_{r,r(n-1)}s/\sqrt{n}$, $i \neq i'$. If the points are plotted in an $(r-1)$-dimensional space of a basis for the contrasts, then the region is a convex polyhedron.

A generalization of the pairwise comparisons of means is the study of *contrasts.* A contrast is a linear combination of the means, $\sum_1^r c_i\bar{Y}_{i.}$ for the sample and $\sum_1^r c_i\mu_i$ for the parameters, for which $\sum_1^r c_i = 0$. This includes the previous pairwise comparisons and also the differences of weighted averages such as $\mu_1 - [(\mu_2 + \mu_3)/2]$. A contrast $(c_1, \ldots, c_r)$ will be denoted by the vector $\mathbf{c}$, and the linear space of the totality of all contrasts by $\mathcal{L}_c$.[1,2]

The probability statement governing the family of contrasts is

$$P\left\{\sum_1^r c_i\mu_i \in \sum_1^r c_i\bar{Y}_{i.} \pm q^{\alpha}_{r,r(n-1)} \frac{s}{\sqrt{n}} \sum_1^r \frac{|c_i|}{2}, \forall \mathbf{c} \in \mathcal{L}_c\right\} = 1 - \alpha.\dagger \quad (6)$$

This is primarily intended for confidence interval use on those $\mathbf{c} \in \mathcal{L}_c$ which are of interest. This form is not practical for a test of significance of (1) or (2) since *all* contrasts would have to be checked for inclusion of zero. However, it is algebraically equivalent to the test derived from (5) (see Sec. 1.4 for proof), and thus can be considered to be a test of significance. Note that for a pairwise mean comparison $\sum_1^r |c_i|/2 = 1$, so the intervals remain the same as in (5) for pairwise comparisons.

[1] All vectors will be taken as column vectors, except when written coordinatewise in a sentence where they will appear as row vectors.

[2] Henceforth, throughout this book, the following notation will be adhered to:

$\mathcal{L}$ = any arbitrary linear space or subspace

$\mathcal{L}_c$ = linear space of contrasts $\{(c_1, \ldots, c_r): \sum_1^r c_i = 0\}$

$\mathcal{L}_{lc}$ = linear space of all linear combinations $\{(l_1, \ldots, l_r): l_1, \ldots, l_r \text{ arbitrary}\}$

$\mathcal{L}_a$ = any affine subspace such as $\{(a_1, \ldots, a_r): a_1 \text{ fixed}, a_2, \ldots, a_r \text{ arbitrary}\}$.

† $\forall$ means *for all.*

Contrasts only permit comparisons *between* means, and not comparisons of means with theoretical values. To include the latter it is necessary to enlarge the space of **c**'s to $\mathfrak{L}_{lc}$, the space of all possible linear combinations, so as to include the μ_i themselves in the family.[1] This requires more drastic modification of the probability statement:

$$P\left\{\sum_1^r l_i\mu_i \in \sum_1^r l_i\bar{Y}_{i\cdot} \pm q'^{\alpha}_{r,r(n-1)} \frac{s}{\sqrt{n}} L_l,\ \forall l \in \mathfrak{L}_{lc}\right\} = 1 - \alpha \tag{7}$$

where

$$\begin{aligned} L_l &= \max\left\{\sum_1^r l_i^+, -\sum_1^r l_i^-\right\} \\ l_i^+ &= \max\{0,l_i\} \qquad l_i^- = \min\{0,l_i\} \end{aligned} \tag{8}$$

and $q'^{\alpha}_{r,r(n-1)}$ is the upper 100α percent point of the studentized augmented range distribution with parameters r, $r(n-1)$. The studentized augmented range with parameters r,ν is the random variable

$$Q'_{r,\nu} = \frac{\max\{|M|_r, R_r\}}{\sqrt{\chi^2_\nu/\nu}} \tag{9}$$

where $|M|_r = \max\limits_i \{|Y_i|\}$, $R_r = \max\limits_{i,i'}\{|Y_i - Y_{i'}|\}$; $Y_1, \ldots, Y_r$ are independent, unit normal, random variables, and χ^2_ν is an independent χ^2 variable on ν d.f. An alternative, equivalent method of defining the studentized augmented range is

$$Q'_{r,\nu} = \frac{\max\limits_{i,i'=0,1,\ldots,r}\{|Y_i - Y_{i'}|\}}{\sqrt{\chi^2_\nu/\nu}} \tag{10}$$

where $Y_1, \ldots, Y_r$, χ^2_ν are as defined previously, but Y_0 is an independent random variable with $Y_0 = 0$. For individual mean values, pairwise mean comparisons, and comparison of weighted means, the constant L_l reduces to one.

The expression (7) is primarily intended for confidence interval use. However, it also yields a test of significance for the hypothesis

$$H_0\colon\ \mu_1 = \cdots = \mu_r = 0. \tag{11}$$

Simultaneous inclusion of zero in the intervals $\sum_1^r l_i\bar{Y}_{i\cdot} \pm q'^{\alpha}_{r,r(n-1)}(s/\sqrt{n})L_l$ for all $l \in \mathfrak{L}_{lc}$ is algebraically equivalent (see Sec. 1.4) to the inclusion of

[1] Tukey (1953b) distinguishes between groups of statements concerned only with differences between means and those concerned only with comparisons of individual means with theoretical values. The former he calls *families*, and the latter, *batches*.

zero in the $\binom{r+1}{2}$ intervals

$$\begin{aligned} &\bar{Y}_{i\cdot} - \bar{Y}_{i'\cdot} \pm q'^{\alpha}_{r,r(n-1)} \frac{s}{\sqrt{n}} \qquad i, i' = 1, \ldots, r \qquad i \neq i' \\ &\qquad \bar{Y}_{i\cdot} \pm q'^{\alpha}_{r,r(n-1)} \frac{s}{\sqrt{n}} \qquad i = 1, \ldots, r. \end{aligned} \tag{12}$$

The same trick used in the definition (10) of the studentized augmented range can be applied to the test (12). Define $\bar{Y}_0 \equiv 0$. Then, inclusion of zero in all the intervals (12) is identical to the intervals

$$\bar{Y}_{i\cdot} - \bar{Y}_{i'\cdot} \pm q'^{\alpha}_{r,r(n-1)} \frac{s}{\sqrt{n}} \qquad i, i' = 0, 1, \ldots, r \tag{13}$$

all containing zero.

In the preceding discussion the probability statements and intervals were structured with reference to the one-way classification. This is the principal area of application of studentized range procedures, but the intervals and probability statements really involve only r independent, normal random variables and an independent estimate of their variance, so the results could be stated more generally. In fact, a certain type of correlation between the variables can even be permitted.

Let $(Y_1, \ldots, Y_r)$ have a multivariate normal distribution with mean $\boldsymbol{\mu} = (\mu_1, \ldots, \mu_r)$ and covariance matrix $\boldsymbol{\Sigma} = \sigma^2 \boldsymbol{\Sigma}$ where $\sigma^2 > 0$ and

$$\boldsymbol{\Sigma} = \begin{pmatrix} 1 & \rho & \cdots & & \rho \\ \rho & \cdot & \cdot & \cdot & \cdot \\ \cdot & \cdot & \cdot & \cdot & \cdot \\ \cdot & \cdot & \cdot & \cdot & \rho \\ \rho & & \cdots & \rho & 1 \end{pmatrix}. \tag{14}$$

The condition $-1/(r-1) < \rho < +1$ is a necessary and sufficient condition for the matrix (14) to be positive-definite, and, hence, a nonsingular covariance matrix. Let s^2 be a random variable independent of $(Y_1, \ldots, Y_r)$ for which $\nu s^2/\sigma^2$ has a χ^2 distribution with ν d.f. The parameter ρ is assumed to be known; $\boldsymbol{\mu}, \sigma^2$ are unknown.

The previous one-way classification is included within this framework with $\nu = r(n-1)$, $\rho = 0$, and the current σ^2 being the previous σ^2/n.

For pairwise comparisons of the means in this more general setup, the appropriate probability statement is

$$P\{\mu_i - \mu_{i'} \in Y_i - Y_{i'} \pm q^{\alpha}_{r,\nu} s \sqrt{1-\rho},\, i, i' = 1, \ldots, r\} = 1 - \alpha. \tag{15}$$

The only change in the structure of the intervals from the one-way classification is the replacement of the d.f. $r(n-1)$ by a general ν and the insertion of the multiplier $\sqrt{1-\rho}$ to correct for the correlation. Likewise, adjustments can be made for contrasts and linear combinations. They are, respectively,

$$P\left\{\sum_1^r c_i\mu_i \in \sum_1^r c_i\bar{Y}_i \pm q^{\alpha}_{r,\nu}s\sqrt{1-\rho}\sum_1^r \frac{|c_i|}{2},\ \forall\, c \in \mathcal{L}_c\right\} = 1-\alpha \tag{16}$$

and $$P\left\{\sum_1^r l_i\mu_i \in \sum_1^r l_i\bar{Y}_i \pm q'^{\alpha}_{r,\nu}s\sqrt{1-\rho}\,L_l,\ \forall\, l \in \mathcal{L}_{lc}\right\} = 1-\alpha \tag{17}$$

where L_l is given by (8). The probability statements (15) to (17) can be used for simultaneous confidence intervals, or for testing the hypotheses (1) or (11) by examination of the appropriate intervals for inclusion or exclusion of zero.

1.2 Applications

The principal application of the Tukey studentized range is pairwise comparison of r means $\bar{Y}_{1.}, \ldots, \bar{Y}_{r.}$ to determine if any of the population means $\mu_1, \ldots, \mu_r$ differ. Frequently, these means will arise from a simple one-way classification, so expressions (3), (6), and (7) were stated in the form correct for this design.

Applications to other designs are also possible. Consider, for example, a two-way classification with r rows, c columns, and n observations per cell. A fixed effects model is assumed [that is,

$$E\{Y_{ijk}\} = \mu_{ij} = \mu + \alpha_i + \beta_j + (\alpha\beta)_{ij}$$
$$\sum_i \alpha_i = \sum_j \beta_j = \sum_i (\alpha\beta)_{ij} = \sum_j (\alpha\beta)_{ij} = 0].$$

For pairwise comparisons of the row means it should be clear from the general form (15) that the appropriate intervals are

$$(\mu+\alpha_i) - (\mu+\alpha_{i'}) = \alpha_i - \alpha_{i'} \in \bar{Y}_{i..} - \bar{Y}_{i'..} \pm q^{\alpha}_{r,rc(n-1)}\frac{s}{\sqrt{cn}} \qquad i, i' = 1, \ldots, r \tag{18}$$

where $\bar{Y}_{i..} = \sum_{j,k} Y_{ijk}/cn$ and s^2 is the mean square residual error with $rc(n-1)$ d.f. With just a single observation per cell ($n = 1$), the mean square for interactions s_I^2 is an estimate of the error when no interactions are present, and (18) becomes

$$\alpha_i - \alpha_{i'} \in \bar{Y}_{i.} - \bar{Y}_{i'.} \pm q^{\alpha}_{r,(r-1)(c-1)}\frac{s_I}{\sqrt{c}} \qquad i, i' = 1, \ldots, r. \tag{19}$$

If a single degree of freedom for Tukey's test of additivity is separated from s_I^2 and found to be nonsignificant, then the denominator degrees of freedom in the studentized range should be decreased by 1 to correspond to the new error variance.

The author cannot recall any instance of a practical application of the studentized range procedure where $\rho \neq 0$. [In a sense the two-way classification furnishes an example, since

$$\hat{\alpha}_i - \hat{\alpha}_{i'} = (\bar{Y}_{i..} - \bar{Y}_{...}) - (\bar{Y}_{i'..} - \bar{Y}_{...}),$$

but here the correlation is trivially removed by reconsidering the difference as $(\hat{\mu} + \hat{\alpha}_i) - (\hat{\mu} + \hat{\alpha}_{i'}) = \bar{Y}_{i..} - \bar{Y}_{i'..}$.] There may at some time be an application to a multivariate analysis problem in which the statistician has n independent observation vectors,

$$\mathbf{Y}_j = (Y_{1j}, \ldots, Y_{rj}),$$

$j = 1, \ldots, n$, each of which has a multivariate normal distribution specified by $\boldsymbol{\mu}$ and $\boldsymbol{\Sigma} = \sigma^2\boldsymbol{\Sigma}$ with $\boldsymbol{\Sigma}$ given by (14). The mean $\boldsymbol{\mu}$ is estimated by the vector average $\bar{\mathbf{Y}}_. = (\bar{Y}_{1.}, \ldots, \bar{Y}_{r.})$, and σ^2 by

$$s^2 = \frac{1}{r(n-1)} \sum_j (\mathbf{Y}_j - \bar{\mathbf{Y}}_.)^T \boldsymbol{\Sigma}^{-1} (\mathbf{Y}_j - \bar{\mathbf{Y}}_.). \tag{20}$$

The variable $r(n-1)s^2/\sigma^2$ has a χ^2 distribution with $r(n-1)$ d.f., and is independent of $\bar{\mathbf{Y}}_.$.

A disadvantage of the studentized range is the requirement that all the sample means be based on the same number of observations so that they have the same variance σ^2/n. This requirement cannot be relaxed because of the nonexistence of appropriate tables, but due to people dropping test tubes, animals cantankerously dying, and, in general, lack of control over the number of observations, the statistician often finds himself faced with imbalance. In practice, if the departure from balance is slight, it is sensible and safe to use (5) with an average or median value of n inserted under the square root sign. This would certainly be preferable to chucking out observations to bring all samples down to the same common size. For gross departures the statistician has no alternative but to turn to either Scheffé's procedure or Bonferroni t statistics, each of which applies to both the balanced and unbalanced cases.

1.3 Comparison

The utility of the studentized range seems to be essentially limited to the construction of simultaneous confidence intervals on pairwise mean differences $\mu_i - \mu_{i'}$. The extensions to contrasts and linear combinations, although theoretically exact, are not as useful because in the majority of instances their competitors, the Scheffé, Bonferroni, and maximum modulus procedures, give shorter intervals. A more thorough discussion of this point will be given when these techniques are studied.

As a simultaneous test of either hypothesis (1) or (11), the single studentized range studied here is eclipsed by the multistage studentized range tests of Newman-Keuls and Duncan (Sec. 6). The Tukey test is a little unnecessarily conservative, and can be liberalized by using multiple ranges with an accompanying increase in performance under the alternative. A more detailed comparison will be presented in Sec. 6.

1.4 Derivation

The proofs for the results of the method section will be given in terms of the general random vector $(Y_1, \ldots, Y_r)$ and s^2, which were introduced in the paragraph containing (14). This includes the one-way classification statistics as a special case.

Consider first the case $\rho = 0$, that is, $Y_1, \ldots, Y_r$ are independent. A necessary and sufficient condition that the inequalities

$$\frac{|(Y_i - Y_{i'}) - (\mu_i - \mu_{i'})|}{s} \leq c \tag{21}$$

be satisfied for all $i \neq i'$ is for

$$\frac{\max_{i,i'} \{|(Y_i - \mu_i) - (Y_{i'} - \mu_{i'})|\}}{s} \leq c \tag{22}$$

to hold. The numerator $R_r = \max_{i,i'} \{|(Y_i - \mu_i) - (Y_{i'} - \mu_{i'})|\}$ is the range of r independent, $N(0,\sigma^2)$ random variables, so this immediately establishes the assertion (3).

The system of inequalities

$$\left|\sum_1^r c_i(Y_i - \mu_i)\right| \leq cs \sum_1^r \frac{|c_i|}{2} \qquad \forall \mathbf{c} \in \mathcal{L}_c \tag{23}$$

implies that (21) holds for all $i \neq i'$, by the appropriate choice of $\mathbf{c}$'s. To establish (6) it remains to show the reverse, namely, that (21) for all $i \neq i'$ implies (23). The proof is given in the form of a lemma where $Y_i - \mu_i$, $i = 1, \ldots, r$ should be identified with the constants y_i, $i = 1, \ldots, r$, and cs with the constant c.

Lemma 1 If $|y_i - y_{i'}| \leq c$ for all $i, i' = 1, \ldots, r$, then

$$\left|\sum_1^r c_i y_i\right| \leq c \sum_1^r \frac{|c_i|}{2} \tag{24}$$

for all $\mathbf{c} \in \mathcal{L}_c$.

Proof If $c_i \equiv 0$, $i = 1, \ldots, r$, then (24) is trivially true.

Suppose $c_i \not\equiv 0$, $i = 1, \ldots, r$. Let $P = \{i: c_i > 0\}$ and $N = \{i: c_i < 0\}$. Denote $\sum_1^r |c_i|/2$ by g. Then

$$g = \tfrac{1}{2} \sum_{i \epsilon P} c_i + \tfrac{1}{2} \sum_{i \epsilon N} (-c_i)$$

and since

$$0 = \sum_i c_i = \sum_{i \epsilon P} c_i + \sum_{i \epsilon N} c_i$$

it follows that

$$\sum_{i \epsilon P} c_i = \sum_{i \epsilon N} (-c_i) = g.$$

Multiplication of numerator and denominator of $\sum_1^r c_i y_i$ by g gives

$$\begin{aligned}
\sum_1^r c_i y_i &= \frac{\sum_{i' \epsilon N} (-c_{i'}) \sum_{i \epsilon P} c_i y_i + \sum_{i \epsilon P} c_i \sum_{i' \epsilon N} c_{i'} y_{i'}}{g} \\
&= \frac{\sum_{i \epsilon P} \sum_{i' \epsilon N} c_i(-c_{i'}) y_i + \sum_{i \epsilon P} \sum_{i' \epsilon N} c_i c_{i'} y_{i'}}{g} \\
&= \frac{\sum_{i \epsilon P} \sum_{i' \epsilon N} c_i(-c_{i'})(y_i - y_{i'})}{g}.
\end{aligned}$$

But for $i \epsilon P$, $i' \epsilon N$,

$$|c_i(-c_{i'})(y_i - y_{i'})| = c_i(-c_{i'})|y_i - y_{i'}| \le c_i(-c_{i'}) \cdot c$$

so

$$\left| \sum_1^r c_i y_i \right| \le \frac{\sum_{i \epsilon P} \sum_{i' \epsilon N} c_i(-c_{i'}) c}{g} = \frac{g^2 c}{g} = cg$$

which is just (24). ||

The above argument not only establishes the validity of (6), but also shows that the events (22) and (23) are identical. Thus, although (23) has a form which is impractical for a test of significance, it can be interpreted as one because of its equivalence to (22). If there exists a $\mathbf{c} \epsilon \mathcal{L}_c$ for which $\left| \sum_1^r c_i Y_i \right| > cs \sum_1^r |c_i|/2$, there exists a pair i, i' for which $|Y_i - Y_{i'}| > cs$, and conversely.

The proof of (7) proceeds as follows: Take $Y_0 \equiv 0$, to be independent of $(Y_1, \ldots, Y_r)$ and take $\mu_0 = 0$. The linear combination $\sum_1^r l_i Y_i$ can then be written as a contrast $\sum_0^r c_i Y_i$ where $c_i = l_i$, $i = 1, \ldots, r$,

$c_0 = -\sum_1^r l_i$. By Lemma 1 and the converse implication just before it, the two events

$$\max_{i,i'=0,1,\ldots,r} \{|(Y_i - Y_{i'}) - (\mu_i - \mu_{i'})|\} \leq cs \tag{25}$$

and

$$\left|\sum_0^r c_i(Y_i - \mu_i)\right| \leq cs \sum_0^r \frac{|c_i|}{2} \tag{26}$$

are identical. The distribution of $\max_{i,i'=0,1,\ldots,r} \{|(Y_i - Y_{i'}) - (\mu_i - \mu_{i'})|\}/s$ is the studentized augmented range defined by (10) or, equivalently, (9). But from the definition of the c_i above

$$\begin{aligned}\sum_0^r |c_i| &= |c_0| + \sum_1^r |c_i| = \left|\sum_1^r l_i\right| + \sum_1^r |l_i| \\ &= \left|\sum_{i\epsilon P} l_i + \sum_{i\epsilon N} l_i\right| + \sum_{i\epsilon P} l_i + \sum_{i\epsilon N} (-l_i) \\ &= 2L_l\end{aligned} \tag{27}$$

where $P = \{i: l_i > 0\}$, $N = \{i: l_i < 0\}$, and L_l is defined in (8). The last equality is most easily seen by considering the two cases $\sum_{i\epsilon P} l_i \gtrless \sum_{i\epsilon N} (-l_i)$. Hence, (7) follows from the equivalence of (25) and (26).

Consider now the dependent case $\rho \neq 0$. This case can be reduced to the one of independence by a trick. Define $Z_i = Y_i - \gamma\bar{Y}_.$, $i = 1, \ldots, r$, where $\bar{Y}_.$ is the mean of the Y_i's. Clearly, $(Z_1, \ldots, Z_r)$ has a multivariate normal distribution with means $\eta_i = \mu_i - \gamma\bar{\mu}$, $i = 1, \ldots, r$ $\left(\text{where } \bar{\mu} = \sum_1^r \mu_i/r\right)$, and variances and covariances

$$\begin{aligned}\operatorname{Var}(Z_i) &= \sigma^2 - 2\gamma\frac{\sigma^2}{r}[1 + (r-1)\rho] + \gamma^2\frac{\sigma^2}{r}[1 + (r-1)\rho] \\ &\qquad\qquad i = 1, \ldots, r \\ \operatorname{Cov}(Z_i,Z_{i'}) &= \rho\sigma^2 - 2\gamma\frac{\sigma^2}{r}[1 + (r-1)\rho] + \gamma^2\frac{\sigma^2}{r}[1 + (r-1)\rho] \\ &\qquad\qquad i \neq i'.\end{aligned} \tag{28}$$

By selecting γ to be a root (real) of the quadratic equation in γ obtained by putting $\operatorname{Cov}(Z_i,Z_{i'}) = 0$, that is,

$$\gamma = 1 \pm \sqrt{\frac{1-\rho}{1 + (r-1)\rho}} \tag{29}$$

the Z_i become independent with variance $\sigma^2(1 - \rho)$. But any pairwise comparison between the Y_i's is the identical comparison between the Z_i's since

$$\begin{aligned}Z_i - Z_{i'} &= (Y_i - \gamma\bar{Y}_.) - (Y_{i'} - \gamma\bar{Y}_.) = Y_i - Y_{i'} \\ \eta_i - \eta_{i'} &= (\mu_i - \gamma\bar{\mu}) - (\mu_{i'} - \gamma\bar{\mu}) = \mu_i - \mu_{i'}.\end{aligned} \tag{30}$$

Consequently,

$$\frac{\max_{i,i'} \{|(Y_i - Y_{i'}) - (\mu_i - \mu_{i'})|\}}{s\sqrt{1-\rho}} = \frac{\max_{i,i'} \{|(Z_i - Z_{i'}) - (\eta_i - \eta_{i'})|\}}{s\sqrt{1-\rho}} \sim Q_{r,\nu} \tag{31}$$

so (15) is valid. By an argument from (31) identical to the independence case, expressions (16) and (17) can be established as well.

1.5 Distributions and tables

The cumulative probability integral of the studentized range $Q_{r,\nu}$ can be written down directly:

$$\begin{aligned} P\left\{Q_{r,\nu} = \frac{R_r}{\sqrt{\chi^2_\nu/\nu}} \le q\right\} &= \int_0^\infty P\left\{R_r \le qx \,\middle|\, \sqrt{\frac{\chi^2_\nu}{\nu}} = x\right\} dP\left\{\sqrt{\frac{\chi^2_\nu}{\nu}} \le x\right\} \\ &= \int_0^\infty \left\{r \int_{-\infty}^{+\infty} \varphi(y)[\Phi(y) - \Phi(y - qx)]^{r-1}\, dy\right\} \\ &\qquad \cdot \frac{\nu^{\nu/2}}{\Gamma\left(\frac{\nu}{2}\right) 2^{(\nu/2)-1}} x^{\nu-1} e^{-\nu x^2/2}\, dx \end{aligned} \tag{32}$$

where

$$\begin{aligned} \varphi(x) &= \frac{1}{\sqrt{2\pi}} e^{-x^2/2} \\ \Phi(x) &= \frac{1}{\sqrt{2\pi}} \int_{-\infty}^{x} e^{-u^2/2}\, du. \end{aligned} \tag{33}$$

No simplification of (32) is possible which will facilitate the calculation of the critical points of $Q_{r,\nu}$. The computations have been performed by obtaining, for fixed r and ν, the value of the integral in (32) for selected values of q, and then inversely interpolating in the table of r, ν, and q. The evaluation of the integral is either by numerical quadrature, or series expansion in powers of $1/\nu$ about $\nu = +\infty$. The latter method runs into difficulty for small values of ν and large values of q.

In book form tables of the critical points of the studentized range are available in Pearson and Hartley (1962), "Biometrika Tables for Statisticians," vol. I (upper $\alpha = .05, .01$; $r = 2(1)20$; $\nu = 1(1)20, 24, 30, 40, 60, 120, +\infty$; lower $\alpha = .01, .05$; $r = 2(1)20$; $\nu = 10(1)20, 24, 30, 40, 60, 120, +\infty$); Owen (1962), "Handbook of Statistical Tables" ($\alpha = .005, .01, .025, .05, .10, .90, .95, .975, .99, .995$; $r = 2(1)20, 24, 30, 40, 60, 100$; $\nu = 1, 3, 5, 10, 15, 20, 60, +\infty$); and Scheffé (1959), "The Analysis of Variance" (upper $\alpha = .10, .05, .01$; $r = 2(1)20$; $\nu = 1(1)20, 24, 30, 40, 60, 120, +\infty$). The first and third tables report two decimals or three significant figures; the second three decimals or four significant figures.

In journal form a very complete set of tables is given by Harter (1960a) (upper $\alpha = .10, .05, .025, .01, .005, .001$; $r = 2(1)20(2)40(10)100$;

$\nu = 1(1)\ 20, 24, 30, 40, 60, 120, +\infty$) and these are the tables appearing in Owen (1962). The accuracy of these tables is reported to be within a unit in the last place of the entries, which have either four significant figures or three decimal places. The tables for $\alpha = .05, .01$ are reproduced in the back of this book in Table I of Appendix B. Even more extensive tables are available in a less widely circulated source: Harter and Clemm (1959), and Harter, Clemm, and Guthrie (1959).

Other tables have appeared in journals but have been superseded in time or incorporated in one of the aforementioned tables [viz., Newman (1939), Pearson and Hartley (1943), May (1952), Pillai (1952), Hartley (1953), and Pachares (1959)].

An expression similar to (32) could be written down for the probability integral of the studentized augmented range, but it is not worthwhile. The integral is no simpler than (32), no tables have been computed for it, and in most instances techniques other than those based on the augmented range are preferable. Tukey (1953b) has shown that for $r > 2$, $\alpha \leq .05$, $q^{\alpha}_{r,\nu}$ is a good approximation to $q'^{\alpha}_{r,\nu}$.

2 F PROJECTIONS (SCHEFFÉ)

Scheffé (1953) proposed this method for handling contrasts in an analysis of variance. The extension to arbitrary linear (or affine) spaces is immediate and requires little change in the original proof. A general discussion can be found in Roy and Bose (1953).

2.1 Method

Let $\mathbf{Y} = (Y_1, \ldots, Y_n)$ be a vector of n independently, normally distributed random variables with common variance σ^2. Let the mean vector $\boldsymbol{\mu}$ be given by

$$\boldsymbol{\mu} = \mathbf{X}\boldsymbol{\beta} \tag{34}$$

where

$$\mathbf{X} = \begin{pmatrix} x_{11}, & \ldots, & x_{1p} \\ \cdots & \cdots & \cdots \\ x_{n1}, & \ldots, & x_{np} \end{pmatrix}$$

$$\text{Rank } \mathbf{X} = p \qquad (p < n) \tag{35}$$

$$\boldsymbol{\beta} = (\beta_1, \ldots, \beta_p).\dagger$$

† The number of columns p is assumed to be strictly less than the number of rows n so that there will be some degrees of freedom for estimating the error σ^2. The *nonsingular* case where $\mathbf{X}$ has full rank (column rank $= p$) is assumed for simplicity. Since the *singular* case (column rank $< p$) can always be reduced to the nonsingular by redefinition of the regression parameters $\boldsymbol{\beta}$ or imposition of linear constraints, it is assumed this reduction has already been performed.

The matrix **X** of independent variables is known, but the regression parameters $\boldsymbol{\beta}$ and error variance σ^2 are unknown. The observation variables **Y** are sometimes referred to as the dependent variables.

The least squares and maximum likelihood estimator of $\boldsymbol{\beta}$ is

$$\hat{\boldsymbol{\beta}} = (\mathbf{X}^T\mathbf{X})^{-1}\mathbf{X}^T\mathbf{Y}.$$

The unbiased χ^2 estimator of σ^2 is $s^2 = \mathbf{Y}^T(\mathbf{I} - \mathbf{X}(\mathbf{X}^T\mathbf{X})^{-1}\mathbf{X}^T)\mathbf{Y}/(n - p)$. These estimators are distributed as follows:

$$\hat{\boldsymbol{\beta}} \sim N(\boldsymbol{\beta},\sigma^2(\mathbf{X}^T\mathbf{X})^{-1})$$
$$\frac{(n-p)s^2}{\sigma^2} \sim \chi^2 \qquad \text{with } n - p \text{ d.f.} \tag{36}$$
$$\hat{\boldsymbol{\beta}} \text{ and } s^2 \text{ independent.}$$

Let $\mathfrak{L} = \{\boldsymbol{l} = (l_1, \ldots, l_p)\}$ be any fixed d-dimensional linear subspace of p-dimensional space. The Scheffé technique will give confidence intervals on the linear combinations $\boldsymbol{l}^T\boldsymbol{\beta} = \sum_1^p l_i\beta_i$ for all $\boldsymbol{l} \in \mathfrak{L}$. The confidence intervals will be simultaneous in the sense that with probability $1 - \alpha$ the parametric linear combinations $\boldsymbol{l}^T\boldsymbol{\beta}$ will be contained in their corresponding confidence intervals for all $\boldsymbol{l} \in \mathfrak{L}$; i.e., the family of statements $\mathfrak{F}$ is equivalent to $\mathfrak{L}$ and $P\{\mathfrak{F}\} = \alpha$.

The space $\mathfrak{L}$ can be almost anything, including contrasts, arbitrary linear combinations, individual coordinates, etc. The wide applicability of the Scheffé technique is due to the very general regression framework for the distribution of **Y**, and the generality of the linear subspace $\mathfrak{L}$. The many possible choices for **X** and $\mathfrak{L}$ will be illustrated and discussed in the next section on applications.

The method is summarized in the following probability statement:

$$P\{\boldsymbol{l}^T\boldsymbol{\beta} \in \boldsymbol{l}^T\hat{\boldsymbol{\beta}} \pm (dF^{\alpha}_{d,n-p})^{\frac{1}{2}}s(\boldsymbol{l}^T(\mathbf{X}^T\mathbf{X})^{-1}\boldsymbol{l})^{\frac{1}{2}},\ \forall \boldsymbol{l} \in \mathfrak{L}\} = 1 - \alpha \tag{37}$$

where $F^{\alpha}_{d,n-p}$ is the upper 100α percent point of the F distribution with d d.f. in the numerator, $n - p$ in the denominator.

The variance of the estimator $\boldsymbol{l}^T\hat{\boldsymbol{\beta}}$ is $\sigma^2\boldsymbol{l}^T(\mathbf{X}^T\mathbf{X})^{-1}\boldsymbol{l}$, so the term $s^2\boldsymbol{l}^T(\mathbf{X}^T\mathbf{X})^{-1}\boldsymbol{l}$ appearing in (37) is just the estimate of the variance of $\boldsymbol{l}^T\hat{\boldsymbol{\beta}}$ and could be written as $\widehat{\text{Var}}(\boldsymbol{l}^T\hat{\boldsymbol{\beta}})$. Frequently, for problems involving qualitative variables, $\boldsymbol{l}^T(\mathbf{X}^T\mathbf{X})^{-1}\boldsymbol{l}$ is already known from classical analysis-of-variance results or can be calculated directly from the specific form of $\boldsymbol{l}^T\hat{\boldsymbol{\beta}}$ in terms of **Y**—thereby avoiding the more cumbersome matrix inversion and multiplication (see Sec. 2.2 for examples).

If just a single confidence interval on one linear combination $\boldsymbol{l}^T\boldsymbol{\beta}$ had been desired, classical regression theory would have prescribed the interval

$$\boldsymbol{l}^T\boldsymbol{\beta} \in \boldsymbol{l}^T\hat{\boldsymbol{\beta}} \pm t^{\alpha/2}_{n-p}s(\boldsymbol{l}^T(\mathbf{X}^T\mathbf{X})^{-1}\boldsymbol{l})^{\frac{1}{2}} \tag{38}$$

where $t_{n-p}^{\alpha/2}$ is the upper $100\alpha/2$ percent point of the t distribution on $n - p$ d.f. Thus, the only change from the interval in (38) to the intervals in (37) is to increase the critical constant from $t_{n-p}^{\alpha/2}$ to $(dF_{d,n-p}^{\alpha})^{\frac{1}{2}}$. In the special case that $\mathfrak{L}$ has dimension one (that is, $\mathfrak{L}$ is spanned by a single linear combination), (37) reduces to (38) since $(t_{n-p})^2 \sim F_{1,n-p}$.

The proof of (37) is little more than the Cauchy-Schwarz inequality, and will be given in the derivation section. The argument hinges entirely on the fact that $\boldsymbol{\beta}$ is contained in the region

$$E_{\boldsymbol{\beta}} = \{\boldsymbol{\beta}\colon (\mathbf{L}\boldsymbol{\beta} - \mathbf{L}\hat{\boldsymbol{\beta}})^T(\mathbf{L}(\mathbf{X}^T\mathbf{X})^{-1}\mathbf{L}^T)^{-1}(\mathbf{L}\boldsymbol{\beta} - \mathbf{L}\hat{\boldsymbol{\beta}}) \leq dF_{d,n-p}^{\alpha}s^2\} \tag{39}$$

if and only if $\boldsymbol{\beta}$ satisfies

$$\boldsymbol{l}^T\boldsymbol{\beta} \in \boldsymbol{l}^T\hat{\boldsymbol{\beta}} \pm (dF_{d,n-p}^{\alpha})^{\frac{1}{2}}s(\boldsymbol{l}^T(\mathbf{X}^T\mathbf{X})^{-1}\boldsymbol{l})^{\frac{1}{2}} \qquad \forall\, \boldsymbol{l} \in \mathfrak{L} \tag{40}$$

where $\mathbf{L}$ is any $d \times p$ matrix whose linearly independent rows form a basis for $\mathfrak{L}$ [cf. (41)].

As it is presented in (37), the Scheffé procedure is strictly a confidence procedure. No hypothesis has been formulated so none is tested. However, for linear hypotheses on $\boldsymbol{\beta}$, the corresponding simultaneous intervals (37) furnish a test which is equivalent to the likelihood ratio test (F test) of the linear hypothesis. This relationship is now investigated in detail.

Consider $d(d \leq p)$ linearly independent linear combinations

$$\boldsymbol{l}_i = (l_{i1}, \ldots, l_{ip}), \qquad i = 1, \ldots, d,$$

and the matrix $\mathbf{L}$ with rows composed of these combinations:

$$\mathbf{L} = \begin{pmatrix} l_{11}, & \ldots, & l_{1p} \\ \cdot & \cdots & \cdot \\ l_{d1}, & \ldots, & l_{dp} \end{pmatrix}. \tag{41}$$

By the assumption of linear independence $\mathbf{L}$ has rank d. A linear hypothesis on the regression parameters $\boldsymbol{\beta}$ has the form

$$H_0\colon\ \mathbf{L}\boldsymbol{\beta} = \boldsymbol{\gamma}^0 \tag{42}$$

where $\boldsymbol{\gamma}^0 = (\gamma_1^0, \ldots, \gamma_d^0)$ is a specified set of constants. Frequently, $\boldsymbol{\gamma}^0 = \mathbf{0} = (0, \ldots, 0)$, as, for instance, when $\mathbf{X}$ is the design matrix for a one-way classification, and $\mathbf{L}$ is the matrix giving mean differences $\mu_i - \mu_{i'}$ (see Sec. 2.2 for details). The vector estimator $\mathbf{L}\hat{\boldsymbol{\beta}}$ of $\mathbf{L}\boldsymbol{\beta}$ has the distribution

$$\mathbf{L}\hat{\boldsymbol{\beta}} \sim N(\mathbf{L}\boldsymbol{\beta}, \sigma^2\mathbf{L}(\mathbf{X}^T\mathbf{X})^{-1}\mathbf{L}^T) \tag{43}$$

and the likelihood ratio test of (42) is identical to the F test based on (43):

$$\frac{1}{d}\,\frac{(\mathbf{L}\hat{\boldsymbol{\beta}} - \boldsymbol{\gamma}^0)^T(\mathbf{L}(\mathbf{X}^T\mathbf{X})^{-1}\mathbf{L}^T)^{-1}(\mathbf{L}\hat{\boldsymbol{\beta}} - \boldsymbol{\gamma}^0)}{s^2} \begin{cases} > F_{d,n-p}^{\alpha} & \text{reject } H_0 \\ \leq F_{d,n-p}^{\alpha} & \text{accept } H_0. \end{cases} \tag{44}$$

Graphically, this procedure is equivalent to checking whether the point

$\boldsymbol{\gamma}^0$ is contained in the $100(1-\alpha)$ percent confidence ellipsoid for $\boldsymbol{\gamma}$ centered at $\hat{\boldsymbol{\gamma}} = \mathbf{L}\hat{\boldsymbol{\beta}}$, that is, the ellipsoid

$$E_{\gamma} = \{\boldsymbol{\gamma}: (\boldsymbol{\gamma} - \hat{\boldsymbol{\gamma}})^T(\mathbf{L}(\mathbf{X}^T\mathbf{X})^{-1}\mathbf{L}^T)^{-1}(\boldsymbol{\gamma} - \hat{\boldsymbol{\gamma}}) \leq dF^{\alpha}_{d,n-p}s^2\}. \tag{45}$$

Let $\mathcal{L}$ be the linear subspace spanned by the row vectors of $\mathbf{L}$, and consider the family of simultaneous confidence intervals (37) for this $\mathcal{L}$:

$$\boldsymbol{l}^T\boldsymbol{\beta} \in \boldsymbol{l}^T\hat{\boldsymbol{\beta}} \pm (dF^{\alpha}_{d,n-p})^{\frac{1}{2}}s(\boldsymbol{l}^T(\mathbf{X}^T\mathbf{X})^{-1}\boldsymbol{l})^{\frac{1}{2}} \qquad \forall\, \boldsymbol{l} \in \mathcal{L}. \tag{46}$$

Any vector $\boldsymbol{l} \in \mathcal{L}$ can be written as a linear combination of the rows of $\mathbf{L}$, that is, $\boldsymbol{l} = \sum_1^d \lambda_i \boldsymbol{l}_i$ where $\boldsymbol{l}_i$ is the ith row of $\mathbf{L}$. If the hypothesized value for $\boldsymbol{l}_i^T\boldsymbol{\beta}$ is γ_i^0, then the hypothesized value for $\boldsymbol{l}^T\boldsymbol{\beta}$, $\boldsymbol{l} \in \mathcal{L}$, would be $\gamma_l^0 = \sum_1^d \lambda_i\gamma_i^0$. By virtue of the equivalence of (39) and (40), the test (44) is equivalent to checking whether

$$\gamma_l^0 \in \boldsymbol{l}^T\hat{\boldsymbol{\beta}} \pm (dF^{\alpha}_{d,n-p})^{\frac{1}{2}}s(\boldsymbol{l}^T(\mathbf{X}^T\mathbf{X})^{-1}\boldsymbol{l})^{\frac{1}{2}} \qquad \forall\, \boldsymbol{l} \in \mathcal{L}. \tag{47}$$

Thus, the Scheffé simultaneous confidence intervals provide a test which is the likelihood ratio F test. The test would, of course, be applied in the form (44), and not (47), because of the impossibility of checking an infinity of confidence intervals.

The same difficulty may be encountered here as appeared in connection with Fig. 1.6. It can happen that for $i = 1, \ldots, d$,

$$\gamma_i^0 \in \boldsymbol{l}_i^T\hat{\boldsymbol{\beta}} \pm (dF^{\alpha}_{d,n-p})^{\frac{1}{2}}s(\boldsymbol{l}_i^T(\mathbf{X}^T\mathbf{X})^{-1}\boldsymbol{l}_i)^{\frac{1}{2}} \tag{48}$$

but there exists an $\boldsymbol{l}^* \in \mathcal{L}$ for which

$$\gamma_{l^*}^0 \notin \boldsymbol{l}^{*T}\hat{\boldsymbol{\beta}} \pm (dF^{\alpha}_{d,n-p})^{\frac{1}{2}}s(\boldsymbol{l}^{*T}(\mathbf{X}^T\mathbf{X})^{-1}\boldsymbol{l}^*)^{\frac{1}{2}}. \tag{49}$$

This occurs when the confidence ellipsoid E_γ does not contain $\boldsymbol{\gamma}^0$, but the rectangle circumscribed on E_γ with sides parallel to the axes does contain $\boldsymbol{\gamma}^0$. The individual projected confidence intervals each include their respective γ_i^0, but the overall F test rejects the hypothesis H_0: $\boldsymbol{l}_i^T\boldsymbol{\beta} = \gamma_i^0$, $i = 1, \ldots, d$, since some mixture $\sum_1^d \lambda_i\boldsymbol{l}_i$ has a value not included in its interval. Whether this mixture has any realistic meaning will depend upon the physical problem at hand.

The label F *Projections* at the heading of this section is not spurious. It is meant to indicate that the simultaneous confidence interval for any $\boldsymbol{l} \in \mathcal{L}$ is the projection of the confidence ellipsoid onto the one-dimensional subspace generated by $\boldsymbol{l}$. For a d-dimensional subspace $\mathcal{L}$, let $\boldsymbol{l}_i$, $i = 1, \ldots, d$, be a basis for $\mathcal{L}$, and $\mathbf{L}$ be the matrix with the $\boldsymbol{l}_i$ as row vectors. (Here $\mathcal{L}$ precedes $\mathbf{L}$; in the preceding discussion the direction was reversed: $\mathbf{L}$ to $\mathcal{L}$.) Any vector $\boldsymbol{l} \in \mathcal{L}$ can be expressed in terms of its basis as

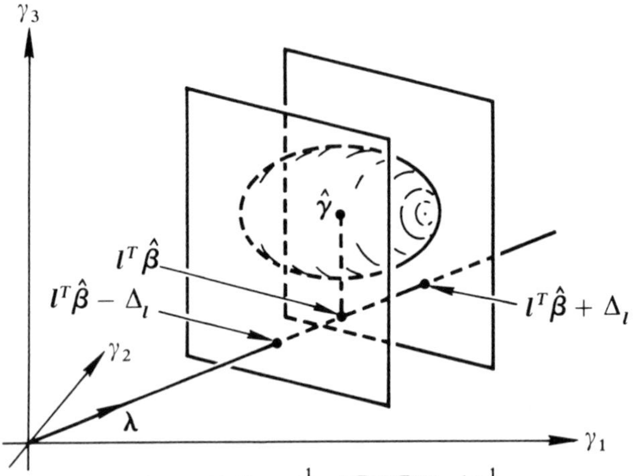

Figure 1

$l = \sum_1^d \lambda_i l_i$, and is uniquely represented by the vector $\boldsymbol{\lambda} = (\lambda_1, \ldots, \lambda_d)$. Assume for simplicity that l has been normalized so that $\boldsymbol{\lambda}$ has unit length. Then, the confidence interval for $l^T\boldsymbol{\beta}$ is the interval on the line generated by $\boldsymbol{\lambda}$ cut out by the two planes (or hyperplanes) which are perpendicular to $\boldsymbol{\lambda}$ and tangent (on opposite sides) to the ellipsoid (45).

Figure 1 illustrates this projection in three dimensions for $|\boldsymbol{\lambda}| = 1$. If l does not have $\boldsymbol{\lambda}$ of unit length, the same idea is correct, but the values of the points in the projected interval must be multiplied by the length of $\boldsymbol{\lambda}$.

Before turning to the applications of the Scheffé F projections it is worthwhile mentioning two extensions or amplifications of the preceding material.

Dwass (1955) pointed out that, free of charge, the intervals in (37) give a confidence interval on the numerator of the noncentrality parameter (of the noncentral F distribution). For the metric

$$\rho(\boldsymbol{\beta}_1,\boldsymbol{\beta}_2) = \sqrt{(\boldsymbol{\beta}_1 - \boldsymbol{\beta}_2)^T[\mathbf{L}^T(\mathbf{L}(\mathbf{X}^T\mathbf{X})^{-1}\mathbf{L}^T)^{-1}\mathbf{L}](\boldsymbol{\beta}_1 - \boldsymbol{\beta}_2)} \tag{50}$$

the inclusion of $\boldsymbol{\beta}$ in the region

$$E_{\boldsymbol{\beta}} = \{\boldsymbol{\beta}\colon \rho(\boldsymbol{\beta},\hat{\boldsymbol{\beta}}) \le (dF^{\alpha}_{d,n-p}s^2)^{\frac{1}{2}}\} \tag{51}$$

[cf. (39)] implies by the triangle inequality that for any $\boldsymbol{\beta}^0$

$$\rho(\boldsymbol{\beta}^0,\boldsymbol{\beta}) \in \rho(\boldsymbol{\beta}^0,\hat{\boldsymbol{\beta}}) \pm (dF^{\alpha}_{d,n-p}s^2)^{\frac{1}{2}} \tag{52}$$

or
$$\sqrt{(\boldsymbol{\beta} - \boldsymbol{\beta}^0)^T[\mathbf{L}^T(\mathbf{L}(\mathbf{X}^T\mathbf{X})^{-1}\mathbf{L}^T)^{-1}\mathbf{L}](\boldsymbol{\beta} - \boldsymbol{\beta}^0)}$$
$$\in \sqrt{(\hat{\boldsymbol{\beta}} - \boldsymbol{\beta}^0)^T[\mathbf{L}^T(\mathbf{L}(\mathbf{X}^T\mathbf{X})^{-1}\mathbf{L}^T)^{-1}\mathbf{L}](\hat{\boldsymbol{\beta}} - \boldsymbol{\beta}^0)} \pm \sqrt{dF^{\alpha}_{d,n-p}s^2}. \tag{53}$$

Therefore, still with exact probability $1 - \alpha$, the statements (40) *and* (53) hold simultaneously.

The quantity

$$\Delta^2 = (\boldsymbol{\beta} - \boldsymbol{\beta}^0)^T[\mathbf{L}^T(\mathbf{L}(\mathbf{X}^T\mathbf{X})^{-1}\mathbf{L}^T)^{-1}\mathbf{L}](\boldsymbol{\beta} - \boldsymbol{\beta}^0) \tag{54}$$

is invariant under changes in $\mathbf{L}$ so long as each $\mathbf{L}$ spans the same space $\mathfrak{L}$. In other words, it is $\mathfrak{L}$ which determines Δ^2 and not a particular choice of a basis $\mathbf{L}$. For any $\boldsymbol{\beta}^0$ such that $\boldsymbol{\gamma}^0 = \mathbf{L}\boldsymbol{\beta}^0$, Δ^2 can be written as

$$\Delta^2 = (\boldsymbol{\gamma} - \boldsymbol{\gamma}^0)^T(\mathbf{L}(\mathbf{X}^T\mathbf{X})^{-1}\mathbf{L}^T)^{-1}(\boldsymbol{\gamma} - \boldsymbol{\gamma}^0). \tag{55}$$

Under the null hypothesis H_0: $\boldsymbol{\gamma} = \boldsymbol{\gamma}^0$ the ratio in (44) has a central F distribution with d.f. $d, n - p$, but under an alternative $\boldsymbol{\gamma} \neq \boldsymbol{\gamma}^0$ it has a noncentral F distribution with d.f. $d, n - p$ and noncentrality parameter $\delta^2 = \Delta^2/\sigma^2$.

The crux of Dwass's addition is that the Scheffé method gives a confidence interval on the distance of the correct parameter value $\boldsymbol{\gamma}$ from the hypothesized null value $\boldsymbol{\gamma}^0$. How useful this is in practice is not clear.

The Scheffé method also extends to the case of correlated and heteroscedastic variables. Let the observation vector $\mathbf{Y} = (Y_1, \ldots, Y_n)$ have the distributional structure assumed at the beginning of this section except for the covariance matrix, which is $\sigma^2\boldsymbol{\Sigma}$ instead of merely $\sigma^2\mathbf{I}$. The matrix $\boldsymbol{\Sigma}$ is assumed known and nonsingular, but σ^2 is unknown. This includes, for instance, the case covered by the extension of Tukey's studentized range to equally correlated variables.

The appropriate estimators of $\boldsymbol{\beta}$ and σ^2 are now

$$\begin{aligned} \hat{\boldsymbol{\beta}} &= (\mathbf{X}^T\boldsymbol{\Sigma}^{-1}\mathbf{X})^{-1}\mathbf{X}^T\boldsymbol{\Sigma}^{-1}\mathbf{Y} \\ s^2 &= \frac{\mathbf{Y}^T[\boldsymbol{\Sigma}^{-1} - \boldsymbol{\Sigma}^{-1}\mathbf{X}(\mathbf{X}^T\boldsymbol{\Sigma}^{-1}\mathbf{X})^{-1}\mathbf{X}^T\boldsymbol{\Sigma}^{-1}]\mathbf{Y}}{n - p} \end{aligned} \tag{56}$$

with the distributional properties

$$\begin{gathered} \hat{\boldsymbol{\beta}} \sim N(\boldsymbol{\beta},\sigma^2(\mathbf{X}^T\boldsymbol{\Sigma}^{-1}\mathbf{X})^{-1}) \\ \frac{(n - p)s^2}{\sigma^2} \sim \chi^2 \qquad \text{with } n - p \text{ d.f.} \\ \hat{\boldsymbol{\beta}} \text{ and } s^2 \text{ independent.} \end{gathered} \tag{57}$$

The correct modification for (37) then becomes

$$P\{\boldsymbol{l}^T\boldsymbol{\beta} \in \boldsymbol{l}^T\hat{\boldsymbol{\beta}} \pm (dF^\alpha_{d,n-p})^{\frac{1}{2}}s(\boldsymbol{l}^T(\mathbf{X}^T\boldsymbol{\Sigma}^{-1}\mathbf{X})^{-1}\boldsymbol{l})^{\frac{1}{2}}, \forall\, \boldsymbol{l} \in \mathfrak{L}\} = 1 - \alpha \tag{58}$$

where $\hat{\boldsymbol{\beta}}$ and s^2 are given by (56). The discussions following (37) on tests of hypotheses, projections, and noncentrality parameters can be adapted as well to the correlated case by the insertion of $\boldsymbol{\Sigma}^{-1}$ in the appropriate places.

2.2 Applications

Due to the great generality of the assumptions for $E\{\mathbf{Y}\}$ and $\mathfrak{L}$, the Scheffé F projection method is a very versatile technique. It can be applied to any linear problem and constitutes a competitor to the other four confidence interval techniques discussed in this chapter. It also provides a basis for the regression techniques of the next chapter.

The great versatility of the Scheffé technique is, at the same time, its major drawback. For some special situations, such as pairwise mean comparisons and Secs. 4 and 5, specialized techniques exist. These techniques cannot be used in general, but for the problems for which they were designed, they do better than the general Scheffé method. A specialized procedure takes advantage of the peculiar nature of the problem to produce better results for that problem.

In order to apply the Scheffé method (37) all that must be specified is the $\mathbf{X}$ matrix and $\mathfrak{L}$. Rather than to somehow attempt an exhaustive enumeration of all the possibilities, a few examples will be given to illustrate the use of the method. Hopefully, the reader will grasp the general scope of the idea by analogy and allusion.

In discussing the illustrations it is convenient to distinguish between *qualitative* and *quantitative* variables in the $\mathbf{X}$ matrix. Qualitative variables are indicator variables used to insert or delete a regression parameter in the mean value of an observation. These variables have no physical meaning per se. Their values are either 1, 0, or -1: 1 when the parameter should be inserted, 0 when it should be deleted, and -1 when, in order to reduce the $\mathbf{X}$ matrix to full rank, a parameter is defined by the negative sum of another group of parameters. This is the type of matrix encountered in an analysis-of-variance problem. Quantitative variables are actual measurements on some physical entity and, except in accidental cases, assume values other than 1, 0, and -1. Problems involving quantitative variables are usually referred to as regression problems. The illustrations are split into three groups, depending upon whether the $\mathbf{X}$ matrix contains only qualitative variables, only quantitative, or both.

Qualitative variables (analysis of variance). Included for illustration are the one-way and two-way classifications. Numerous other designs for application can be found in any textbook on the analysis of variance or design of experiments.

One-way classification (fixed effects); unequal numbers per class. Let $\{Y_{ij};\ i = 1, \ldots, r,\ j = 1, \ldots, n_i\}$ be r independent samples of independently, normally distributed random variables with common variance σ^2 and expectations $E\{Y_{ij}\} = \mu_i, i = 1, \ldots, r, j = 1, \ldots, n_i$. For this design the observation vector $\mathbf{Y}$, the qualitative $\mathbf{X}$ matrix, and the regression parameter vector $\boldsymbol{\beta}$ are analogous to the one presented below for the case $r = 3$, $n_1 = 2$, $n_2 = 3$, $n_3 = 1$:

$$\mathbf{Y} = \begin{pmatrix} Y_{11} \\ Y_{12} \\ Y_{21} \\ Y_{22} \\ Y_{23} \\ Y_{31} \end{pmatrix} \qquad \mathbf{X} = \begin{pmatrix} 1 & 0 & 0 \\ 1 & 0 & 0 \\ 0 & 1 & 0 \\ 0 & 1 & 0 \\ 0 & 1 & 0 \\ 0 & 0 & 1 \end{pmatrix} \qquad \boldsymbol{\beta} = \begin{pmatrix} \mu_1 \\ \mu_2 \\ \mu_3 \end{pmatrix}. \tag{59}$$

The vector estimator $\hat{\boldsymbol{\beta}}$ of the vector $\boldsymbol{\beta}$ has the components

$$\hat{\mu}_i = \bar{Y}_{i.} = \frac{1}{n_i} \sum_{j=1}^{n_i} Y_{ij} \qquad i = 1, \ldots, r. \tag{60}$$

The $\hat{\mu}_i$ are independent, and $\text{Var}(\hat{\mu}_i) = \sigma^2/n_i$. The χ^2 estimator of σ^2 with $N - r$ d.f. is $s^2 = \sum_i \sum_j (Y_{ij} - \bar{Y}_{i.})^2/(N - r)$, where $N = \sum_1^r n_i$.

The parametric combinations customarily of interest in a one-way classification are the pairwise mean comparisons or the contrasts. Since the space of contrasts $\mathfrak{L}_c$ is a linear space, and the pairwise comparisons span $\mathfrak{L}_c$, there is no distinction between the two families in the Scheffé method. The $(r - 1) \times r$ matrix,

$$\begin{pmatrix} 1 & -1 & 0 & \cdots & \cdots & \cdots & 0 \\ 0 & 1 & -1 & 0 & \cdots & \cdots & 0 \\ \cdots & \cdots & \cdots & \cdots & \cdots & \cdots & \cdots \\ 0 & \cdots & \cdots & \cdots & \cdots & 1 & -1 \end{pmatrix} \tag{61}$$

which is the $\mathbf{L}$ matrix for the contrasts $\hat{\mu}_i - \hat{\mu}_{i+1}$, $i = 1, \ldots, r - 1$, has $r - 1$ linearly independent rows which span $\mathfrak{L}_c$. There are innumerable other possible choices of $r - 1$ contrasts which would span the $(r - 1)$-dimensional space $\mathfrak{L}_c$. In order to apply the Scheffé method, it is unnecessary to specify any particular $\mathbf{L}$, but only to know that $\mathfrak{L}_c$ has dimension $r - 1$.

The expression (37) for the one-way classification becomes, with probability $1 - \alpha$,

$$\begin{gathered} \sum_1^r c_i\mu_i \in \sum_1^r c_i\bar{Y}_{i.} \pm [(r - 1)F^{\alpha}_{r-1,N-r}]^{\frac{1}{2}} s \left(\sum_1^r \frac{c_i^2}{n_i}\right)^{\frac{1}{2}} \\ \forall\, (c_1, \ldots, c_r) \text{ such that } \sum_1^r c_i = 0. \end{gathered} \tag{62}$$

For those pairwise comparisons and contrasts of interest to him, the statistician can select confidence intervals from (62) which together have a joint confidence coefficient greater than $1 - \alpha$.

Note that the Scheffé method can be applied when the numbers of observations in each class are *unequal*. This is not possible with the Tukey studentized range (Sec. 1), although it was suggested that the Tukey method would be approximate for mild imbalance. For severe imbalance either the Scheffé method or Bonferroni t statistics (Sec. 3) must be used.

Two-way classification (fixed effects); interactions present; equal numbers per cell. The model for this design is

$$\begin{gathered} Y_{ijk} = \mu + \alpha_i + \beta_j + (\alpha\beta)_{ij} + e_{ijk} \\ i = 1, \ldots, r,\ j = 1, \ldots, c,\ k = 1, \ldots, m \\ 0 = \sum_i \alpha_i = \sum_j \beta_j = \sum_i (\alpha\beta)_{ij} = \sum_j (\alpha\beta)_{ij} \\ e_{ijk} \text{ independent } N(0,\sigma^2). \end{gathered} \tag{63}$$

The linear constraints $\left(0 = \sum_i \alpha_i, \text{ etc.}\right)$ can be incorporated into the $\mathbf{X}$ matrix by defining

$$\alpha_r = -\sum_1^{r-1} \alpha_i \qquad \beta_c = -\sum_1^{c-1} \beta_j$$

$$(\alpha\beta)_{ic} = -\sum_{j=1}^{c-1} (\alpha\beta)_{ij} \qquad i = 1, \ldots, r-1$$

$$(\alpha\beta)_{rj} = -\sum_{i=1}^{r-1} (\alpha\beta)_{ij} \qquad j = 1, \ldots, c-1$$

$$(\alpha\beta)_{rc} = \sum_1^{r-1} \sum_1^{c-1} (\alpha\beta)_{ij}$$

and omitting the columns for the parameters α_r, β_c, $(\alpha\beta)_{1c}, \ldots, (\alpha\beta)_{r-1,c}$, $(\alpha\beta)_{r1}, \ldots, (\alpha\beta)_{r,c-1}$, $(\alpha\beta)_{rc}$. For a 2×2 design with a single observation per cell the vectors and matrices are

$$\mathbf{Y} = \begin{pmatrix} Y_{11} \\ Y_{12} \\ Y_{21} \\ Y_{22} \end{pmatrix} \qquad \mathbf{X} = \begin{pmatrix} 1 & 1 & 1 & 1 \\ 1 & 1 & -1 & -1 \\ 1 & -1 & 1 & -1 \\ 1 & -1 & -1 & 1 \end{pmatrix} \qquad \boldsymbol{\beta} = \begin{pmatrix} \mu \\ \alpha_1 \\ \beta_1 \\ (\alpha\beta)_{11} \end{pmatrix}. \tag{64}$$

There are other $\mathbf{X}$ matrices for this model which may have advantages for certain special tasks, but all such parameterizations are theoretically equivalent.

The least squares estimators of the parameters are

$$\begin{aligned}\hat{\mu} &= \bar{Y}_{...} \\ \hat{\alpha}_i &= \bar{Y}_{i..} - \bar{Y}_{...} \qquad i = 1, \ldots, r \\ \hat{\beta}_j &= \bar{Y}_{.j.} - \bar{Y}_{...} \qquad j = 1, \ldots, c \\ (\widehat{\alpha\beta})_{ij} &= \bar{Y}_{ij.} - \bar{Y}_{i..} - \bar{Y}_{.j.} + \bar{Y}_{...} \qquad i = 1, \ldots, r \\ & \qquad\qquad j = 1, \ldots, c\dagger\end{aligned} \tag{65}$$

and the χ^2 estimator of σ^2 with $rc(m-1)$ d.f. is

$$s^2 = \frac{1}{rc(m-1)} \sum_{ijk} (Y_{ijk} - \bar{Y}_{ij.})^2. \tag{66}$$

In actual computation the estimate s^2 is usually obtained by subtraction in the analysis-of-variance table.

Either the row effects, column effects, or interactions might be of interest. For contrasts within each of these groups the dimensions of the linear spaces are $r-1$, $c-1$, and $(r-1)(c-1)$, respectively. Because of the constraint $\sum_1^r \alpha_i = 0$, an arbitrary linear combination of the α_i, $i = 1, \ldots, r$, is expressible as a contrast among the α_i, $i = 1, \ldots, r$. This equivalence of linear combinations and contrasts also holds for column effects and interactions because of the constraints. In the 2×2 design the $\mathbf{L}$ matrix for the column contrasts is $\mathbf{L} = (0\ 0\ 2\ 0)$. This contrast is $\beta_1 - \beta_2$, since the constraint $\beta_1 + \beta_2 = 0$ gives

$$\beta_1 - \beta_2 = \beta_1 - (-\beta_1) = 2\beta_1.$$

Actually, any nonzero entry in the columns coordinate would generate the column contrasts linear space.

For row contrasts by themselves the probability expression (37) gives the following statement, with probability $1 - \alpha$:

$$\sum_1^r c_i\alpha_i \in \sum_1^r c_i\hat{\alpha}_i \pm [(r-1)F^{\alpha}_{r-1,rc(m-1)}]^{\frac{1}{2}} s \left(\frac{1}{cm} \sum_1^r c_i^2\right)^{\frac{1}{2}} \tag{67}$$

for all contrasts $(c_1, \ldots, c_r)$. The variance term $\sum_1^r c_i^2/cm$ was easily computed mentally by remembering that $\sum_1^r c_i\hat{\alpha}_i = \sum_1^r c_i\bar{Y}_{i..}$ since $(c_1, \ldots, c_r)$ is a contrast. Included in (67) are the $\binom{r}{2}$ pairwise row

† The notation is: $\bar{Y}_{...} = \left(\sum_{ijk} Y_{ijk}\right)/rcm$, $\bar{Y}_{i..} = \left(\sum_{jk} Y_{ijk}\right)/cm$, etc. The symbol "." means summation over that subscript, and "–" means division by the appropriate constants to obtain the average.

mean comparisons:

$$(\mu + \alpha_i) - (\mu + \alpha_{i'}) = \alpha_i - \alpha_{i'} \in \bar{Y}_{i..} - \bar{Y}_{i'..} \pm [(r-1)F^{\alpha}_{r-1,rc(m-1)}]^{\frac{1}{2}} s \left(\frac{2}{cm}\right)^{\frac{1}{2}} \qquad (68)$$
$$i, i' = 1, \ldots, r.$$

If only pairwise comparisons are being made, then the Tukey studentized range and the Bonferroni t statistics are competitors and may be better (see Secs. 2.3 and 3.3).

Statements similar to (67) can be made for column effects and interaction effects by themselves.

If any two or all three of these effect groups are lumped together for increased protection, the corresponding linear spaces should be added. This amounts to the addition of the numerator degrees of freedom in the critical F constant. For example, putting row and column effects together gives

$$\sum_1^r c_i\alpha_i + \sum_1^c c_{r+j}\beta_j \in \sum_1^r c_i\hat{\alpha}_i + \sum_1^c c_{r+j}\hat{\beta}_j \pm [(r+c-2)F^{\alpha}_{r+c-2,rc(m-1)}]^{\frac{1}{2}} s \left(\frac{1}{cm}\sum_1^r c_i^2 + \frac{1}{rm}\sum_1^c c_{r+j}^2\right)^{\frac{1}{2}} \qquad (69)$$

for all $(c_1, \ldots, c_r, c_{r+1}, \ldots, c_{r+c})$ with $\sum_1^r c_i = 0$ and $\sum_1^c c_{r+j} = 0$. Row (column) contrasts are obtained when $c_{r+1} = \cdots = c_{r+c} = 0$ $(c_1 = \cdots = c_r = 0)$. The statistician is paying in the width of his confidence intervals for sums of contrasts between row and column effects which are likely to be of no use or interest to him. For this reason it might be better to combine the row and column effects through the Bonferroni inequality (1.13) with level $\alpha/2$ for each set.

Simultaneous Scheffé intervals are also available for an unbalanced two-way classification, but there are no nice expressions for the estimators and their variances in this case. The general multiple linear regression form (37) will provide intervals, but their expressions will not simplify.

Quantitative variables (regression analysis). There is little to be said for an $\mathbf{X}$ matrix of this type other than that the general expression (37) covers whatever is needed. Since quantitative variables rarely have any regularity in their values, the estimator $\hat{\boldsymbol{\beta}}$ and its covariance matrix $\sigma^2(\mathbf{X}^T\mathbf{X})^{-1}$ rarely simplify.

In multiple regression the $\mathbf{X}$ matrix frequently has a first column consisting entirely of 1's, that is, a qualitative variable. This inserts an overall mean constant into the expected values of the observations.

Logically, this type of matrix should belong to the next section, but it more commonly is treated as belonging to this section.

An important special case is the *simple linear regression* model:

$$Y_i = \alpha + \beta x_i + e_i \qquad i = 1, \ldots, n \\ e_i \text{ independent } N(0,\sigma^2). \tag{70}$$

The estimators of α, β, and σ^2 are:

$$\begin{aligned} \hat{\alpha} &= a = \bar{Y} - b\bar{x} \\ \hat{\beta} &= b = \frac{\sum_i (Y_i - \bar{Y})(x_i - \bar{x})}{\sum_i (x_i - \bar{x})^2} \\ s^2 &= \frac{1}{n-2}\left\{\sum_1^n Y_i^2 - n\bar{Y}^2 - b\left[\sum_1^n (Y_i - \bar{Y})(x_i - \bar{x})\right]\right\}. \end{aligned} \tag{71}$$

The variance estimator s^2 is distributed as $\sigma^2\chi^2_{n-2}/(n-2)$ (where χ^2_{n-2} is a generic χ^2 variable with $n - 2$ d.f.) and is independent of (a,b), which has a bivariate normal distribution with mean (α,β) and covariance matrix

$$\sigma^2 \begin{pmatrix} \dfrac{1}{n} + \dfrac{\bar{x}^2}{\sum_1^n (x_i - \bar{x})^2} & \dfrac{-\bar{x}}{\sum_1^n (x_i - \bar{x})^2} \\ \dfrac{-\bar{x}}{\sum_1^n (x_i - \bar{x})^2} & \dfrac{1}{\sum_1^n (x_i - \bar{x})^2} \end{pmatrix}. \tag{72}$$

Separate confidence intervals for α and β can be constructed by the standard single interval theory, and are equivalent to applying the Scheffé technique with $\mathfrak{L}$ of dimension 1 in each case. For linear combinations of α and β (that is, $l_1\alpha + l_2\beta$, l_1, l_2 arbitrary) the F projections utilize the bivariate elliptical confidence region for (α,β) and give the statements

$$l_1\alpha + l_2\beta \in l_1a + l_2b \pm (2F^{\alpha}_{2,n-2})^{\frac{1}{2}} s \left(\frac{l_1^2}{n} + \frac{(l_2 - l_1\bar{x})^2}{\sum_1^n (x_i - \bar{x})^2}\right)^{\frac{1}{2}} \tag{73}$$

for all l_1, l_2 (with probability $1 - \alpha$). The family of statements (73) includes the intervals for α and β ($l_1 = 1$, $l_2 = 0$, and $l_1 = 0$, $l_2 = 1$, respectively), but for the price of simultaneity the critical constant has been raised from $t^{\alpha/2}_{n-2} = (F^{\alpha}_{1,n-2})^{\frac{1}{2}}$ to $(2F^{\alpha}_{2,n-2})^{\frac{1}{2}}$.

In the special case $\bar{x} = 0$ the covariance term in (72) vanishes. The independent variable mean $\bar{x}$ can always be made to equal zero by reparameterizing the model (70) as

$$Y_i = \alpha' + \beta x_i' + e_i \qquad x_i' = x_i - \bar{x} \qquad i = 1, \ldots, n. \tag{74}$$

In such a reparameterization the one thing which must be watched is that the intercept α' still retains some physical significance for the experimenter, since it is now located at $\bar{x}$ rather than zero.

If $\bar{x} = 0$ and the only statements of interest are the two on α and β, then there is a better technique than the Scheffé. The studentized maximum modulus (Sec. 4) will give shorter intervals for α and β just as the prototype technique in Sec. 4.2 in Chap. 1 was better than Sec. 4.4 in Chap. 1. The property $\bar{x} = 0$ makes a and b independent, so an exact box-shaped confidence region can be constructed which is smaller than the box circumscribed on the football confidence region.

If only the two statements on α and β are required but $\bar{x} \neq 0$, Bonferroni t statistics (Sec. 3) could be applied. Since there are just two statements involved, the Bonferroni intervals will be shorter than the Scheffé for most degrees of freedom and α.

Note that provided $l_1 \neq 0$, (73) can be renormalized by dividing out l_1:

$$\alpha + l\beta \in a + lb \pm (2F^{\alpha}_{2,n-2})^{\frac{1}{2}} s \left[\frac{1}{n} + \frac{(l - \bar{x})^2}{\sum_{1}^{n} (x_i - \bar{x})^2} \right]^{\frac{1}{2}}. \tag{75}$$

The ratio $l = l_2/l_1$ can vary from $-\infty$ to $+\infty$, so if l is identified with x, (75) gives the Working-Hotelling simultaneous confidence band on the regression line. This will be discussed in greater detail in Chapter 3.

Durand (1954) applied the Scheffé F projections to obtain simultaneous confidence intervals on all linear combinations of multiple regression coefficients (except possibly the general mean term).

Qualitative and quantitative variables (analysis of covariance). An $\mathbf{X}$ matrix of this type can be written in the form

$$\begin{gathered} \mathbf{X} = (\mathbf{X}_1, \mathbf{X}_2) \\ \mathbf{X}_1 \; n \times p_1 \qquad \mathbf{X}_2 \; n \times p_2 \\ p_1 > 0 \qquad p_2 > 0 \qquad p_1 + p_2 = p \end{gathered} \tag{76}$$

where $\mathbf{X}_1$ contains the columns for the qualitative variables, $\mathbf{X}_2$ the quantitative. The regression vector $\boldsymbol{\beta}$ is correspondingly divided into $(\boldsymbol{\beta}_1, \boldsymbol{\beta}_2)$.

As with the previous cases of only quantitative or only qualitative variables, this model is just a special case of the multiple linear regression model, and estimates, tests, and confidence intervals can be obtained from the general expressions. Historically, specialized techniques were developed and labeled *the analysis of covariance* to circumvent the inversion of the large matrix $\mathbf{X}^T\mathbf{X}$. With paper and pencil or with a desk calculator, this inversion could be quite time-consuming, and was ingeniously replaced by repeated application of the analysis of variance appropriate for the model $\mathbf{Y} = \mathbf{X}_1\boldsymbol{\beta}_1 + \mathbf{e}$ to different combinations of the observations $\mathbf{Y}$ and concomitant variables $\mathbf{X}_2$. With increasingly easier access to ever faster electronic digital computers, these specialized techniques have become archaic except in rare cases.

Despite the obsolescence of the special analysis-of-covariance techniques, the problems of proper application and interpretation of analyses of covariance still remain. The mathematics is not open to question, just the proper use of it. The problems and pitfalls in this area have been discussed at length elsewhere and will not be touched upon here.

To illustrate the use of the F projections method, consider the *unbalanced one-way classification* with a *single concomitant variable:*

$$Y_{ij} = \mu_i + \beta x_{ij} + e_{ij} \qquad i = 1, \ldots, r \qquad j = 1, \ldots, n_i \qquad e_{ij} \text{ independent } N(0,\sigma^2). \tag{77}$$

The appropriate estimators are

$$\begin{aligned}
\hat\beta &= \frac{\sum_i \sum_j (Y_{ij} - \bar Y_{i\cdot})(x_{ij} - \bar x_{i\cdot})}{\sum_i \sum_j (x_{ij} - \bar x_{i\cdot})^2} \\
\hat\mu_i &= \bar Y_{i\cdot} - \hat\beta \bar x_{i\cdot} \\
s^2 &= \frac{1}{N - r - 1} \Big\{ \sum_i \sum_j Y_{ij}^2 \\
&\qquad - \sum_i n_i \bar Y_{i\cdot}^2 - \hat\beta \Big[\sum_i \sum_j (Y_{ij} - \bar Y_{i\cdot})(x_{ij} - \bar x_{i\cdot}) \Big] \Big\}
\end{aligned} \tag{78}$$

where $N = \sum_1^r n_i$. The usual distribution theory pertains, with $(\hat\mu_1, \ldots, \hat\mu_r, \hat\beta)$ having a multivariate normal distribution, and $(N - r - 1)s^2/\sigma^2$ having a χ^2 distribution. Note that $\hat\mu_i$ and $\hat\mu_{i'}$, $i \neq i'$, are dependent unless $\bar x_{i\cdot}\bar x_{i'\cdot} = 0$, and $\hat\mu_i$ and $\hat\beta$ are dependent unless $\bar x_{i\cdot} = 0$.

Customarily, the parametric comparisons of interest are the contrasts among the μ_i, either just the pairwise comparisons or the more general contrasts. Since $\sum_1^r c_i\hat\mu_i = \sum_1^r c_i\bar Y_{i\cdot} - \hat\beta \sum_1^r c_i\bar x_{i\cdot}$, and $\bar Y_{1\cdot}, \ldots, \bar Y_{r\cdot}, \hat\beta$ are all independent, the variance of $\sum_1^r c_i\hat\mu_i$ is easily evaluated. The general expression (37) becomes, with probability $1 - \alpha$,

$$\sum_1^r c_i\mu_i \in \sum_1^r c_i\hat\mu_i \pm ((r-1)F^{\alpha}_{r-1,N-r-1})^{\frac12} s \left[\sum_1^r \frac{c_i^2}{n_i} + \frac{\left(\sum_1^r c_i\bar x_{i\cdot}\right)^2}{\sum_i \sum_j (x_{ij} - \bar x_{i\cdot})^2} \right]^{\frac12} \tag{79}$$

for all contrasts $(c_1, \ldots, c_r)$.

When $n_i \equiv n$, and $\bar{x}_{i\cdot} \equiv 0$ (or $\bar{x}_{i\cdot} \equiv \bar{x}_{\cdot\cdot}$), the Tukey studentized range procedure becomes a competitor, which does better for pairwise comparisons but worse for more general contrasts.

Halperin and Greenhouse (1958) published this application of the Scheffé method to provide an exact procedure for handling multiple comparisons in the analysis of covariance. Earlier, Kramer (1957) had proposed an inexact method based on the studentized range. The behavior of Kramer's procedure, which is a multiple range test, has never been evaluated because it involves the range of heteroscedastic, unequally correlated, normal variables; nor has the inexactness been shown to be on the conservative side.

Duncan (1957) proposed a modified version of the Kramer method, but the same criticism can be leveled at it, since the actual significance levels have never been shown to be conservative. The Duncan procedure is, in fact, less conservative than the Kramer, and consequently, might give grounds for greater concern.

Scheffé suggests that F projections are well adapted to what he calls *data-snooping*. Data-snooping is mucking around in the data to see if anything significant turns up. No specific contrasts or comparisons need be stipulated, or even half-consciously formulated, prior to the analysis. Data-snooping can be used either exclusively by itself, or in conjunction with tests of specific comparisons by allocating some of the error rate to the specific comparisons and the rest to data-snooping. To apply the F projections in this fashion, all that is required is to take the linear space $\mathcal{L}$ large enough to include any and all linear combinations which might be uncovered through snooping.

Although data-snooping with F projections is an intellectually nice idea, it does not seem to be of too great practical importance. With low α and large $\mathcal{L}$ (that is, large dimension d), the critical constant $(dF^{\alpha}_{d,\nu})^{\frac{1}{2}}$ will most likely be so large as to render the procedure ineffective because of insensitivity. There may be many combinations which are worth investigating further or speculating about, but which cannot surpass the all-encompassing critical value $(dF^{\alpha}_{d,\nu})^{\frac{1}{2}}$.

2.3 Comparison

The great generality of the Scheffé technique allows it to compete with each of the other four confidence interval techniques of this chapter in those situations where the other techniques are applicable. In a nutshell:

1. The studentized maximum modulus (Sec. 4), the many-one t statistics (Sec. 5), and the Tukey studentized range for pairwise mean comparisons (Sec. 1) are better for their specialized tasks.
2. For a small number of statements and low d the Bonferroni t statistics (Sec. 3) are better.

3. But the Scheffé method is always applicable and is best for general contrasts and general regression problems except where item 2 applies.

Scheffé (1953) gives a numerical comparison of Scheffé vs. Tukey in Tables 1, 2, and 3 of his article. Table 3b is partially reproduced in Scheffé (1959). The reader can refer to these tables for the actual numbers if he wishes, but the general flavor of the comparison is:

1. Tukey intervals are better for pairwise comparisons.
2. Scheffé intervals are better for contrasts involving more than two means.

In the problems for which they were designed the studentized maximum modulus and the many-one t statistics are better than the Scheffé method because they give precisely what is called for and nothing more. To cover these situations the Scheffé method requires large linear spaces, but does not use many of the possible F projections. The result is wastage and intervals which are too large for the projections used. Further discussion on this will be given in the sections on these techniques.

The comparison between the F projections and Bonferroni t statistics will be given in Sec. 3.3.

2.4 Derivation

The proof of the probability expression (37) can be given in two guises, one algebraic and one geometric. Both are the same proof. The only difference is how one likes to think of it. Both proofs will be given, and are embodied in the following two lemmas, which are just the Cauchy-Schwarz inequality.

Lemma 2 (algebraic) For $c > 0$, $\left| \sum_1^d a_i y_i \right| \leq c \left(\sum_1^d a_i^2 \right)^{\frac{1}{2}}$ for all $(a_1, \ldots, a_d)$ if and only if $\sum_1^d y_i^2 \leq c^2$.

Proof *If:* By the Cauchy-Schwarz inequality

$$\left| \sum_1^d a_i y_i \right| \leq \left(\sum_1^d y_i^2 \right)^{\frac{1}{2}} \left(\sum_1^d a_i^2 \right)^{\frac{1}{2}}.$$

The inequality $\left(\sum_1^d y_i^2 \right)^{\frac{1}{2}} \leq c$ establishes the assertion.

Only if: Choose $a_i = y_i$, $i = 1, \ldots, d$. Then, the inequality

$$\left| \sum_1^d a_i y_i \right| \leq c \left(\sum_1^d a_i^2 \right)^{\frac{1}{2}} \qquad \text{becomes} \qquad \left| \sum_1^d y_i^2 \right| \leq c \left(\sum_1^d y_i^2 \right)^{\frac{1}{2}}$$

which is the assertion. ||

Lemma 3 (*geometric*) The equations of the two hyperplanes perpendicular to the vector $\mathbf{a} = (a_1, \ldots, a_d)$ and tangent to the sphere of radius r, centered at $\mathbf{c} = (c_1, \ldots, c_d)$, are

$$\sum_1^d a_i y_i = \sum_1^d a_i c_i + r \left(\sum_1^d a_i^2\right)^{\frac{1}{2}} \tag{80a}$$

$$\sum_1^d a_i y_i = \sum_1^d a_i c_i - r \left(\sum_1^d a_i^2\right)^{\frac{1}{2}}. \tag{80b}$$

Proof The equation $\sum_1^d a_i y_i = c$ defines a $(d - 1)$-dimensional hyperplane in the space of $\mathbf{y} = (y_1, \ldots, y_d)$. The extension (or contraction) of $\mathbf{a}$ which lies in this plane is $\mathbf{a}^* = \left(c/\sum_1^d a_i^2\right) \mathbf{a}$. For any $\mathbf{y}$ in the plane

$$(\mathbf{y} - \mathbf{a}^*)^T \mathbf{a}^* = \frac{c}{\sum_1^d a_i^2} \sum_1^d y_i a_i - \left(\frac{c}{\sum_1^d a_i^2}\right)^2 \sum_1^d a_i^2 = \frac{c^2}{\sum_1^d a_i^2} - \frac{c^2}{\sum_1^d a_i^2} = 0$$

so the plane is perpendicular to $\mathbf{a}$. Thus, the planes (80a) and (80b) with $c = \sum_1^d a_i c_i + r \left(\sum_1^d a_i^2\right)^{\frac{1}{2}}, \sum_1^d a_i c_i - r \left(\sum_1^d a_i^2\right)^{\frac{1}{2}}$, respectively, are perpendicular to $\mathbf{a}$.

Let $\mathbf{s} = (s_1, \ldots, s_d)$ be a point in the sphere of radius r, centered at $\mathbf{c}$ (that is, $\sum_1^d (s_i - c_i)^2 \leq r^2$). By the Cauchy-Schwarz inequality

$$\left|\sum_1^d a_i(s_i - c_i)\right| \leq \left(\sum_1^d (s_i - c_i)^2\right)^{\frac{1}{2}} \left(\sum_1^d a_i^2\right)^{\frac{1}{2}} \leq r \left(\sum_1^d a_i^2\right)^{\frac{1}{2}}$$

so all points in the sphere lie between the two hyperplanes (80a) and (80b). The plane (80a) actually touches the sphere at the point $\mathbf{c} + \lambda \mathbf{a}$ where $\lambda = r/\left(\sum_1^d a_i^2\right)^{\frac{1}{2}}$ since

$$\sum_1^d a_i(c_i + \lambda a_i) = \sum_1^d a_i c_i + r \left(\sum_1^d a_i^2\right)^{\frac{1}{2}}$$

$$\sum_1^d [(c_i + \lambda a_i) - c_i]^2 = \lambda^2 \sum_1^d a_i^2 = r^2.$$

Similarly, the point $\mathbf{c} - \lambda \mathbf{a}$ belongs to the surface of the sphere and (80b). This establishes the tangency of the planes (80a) and (80b) to the sphere.

The tangency of the plane to the sphere could also be established by a

Lagrangian multiplier argument $\left[\text{i.e., minimize } \sum_1^d (y_i - c_i)^2 \text{ subject to } \sum_1^d a_i y_i = \sum_1^d a_i c_i + r\left(\sum_1^d a_i^2\right)^{\frac{1}{2}}\right]$. ||

The proof of (37) now follows easily via a change of basis.

Theorem 1 If $\mathbf{Y} \sim N(\mathbf{X}\boldsymbol{\beta},\sigma^2\mathbf{I})$ where $\mathbf{X}$ is an $n \times p$ matrix of rank p $(p < n)$, and $\mathcal{L}$ is a d-dimensional subspace of p-dimensional Euclidean space $(d \le p)$, then

$$P\{|l^T(\hat{\boldsymbol{\beta}} - \boldsymbol{\beta})| \le (dF^\alpha_{d,n-p})^{\frac{1}{2}} s(l^T(\mathbf{X}^T\mathbf{X})^{-1}l)^{\frac{1}{2}},\ \forall l \in \mathcal{L}\} = 1 - \alpha. \quad (81)$$

Proof Let $\mathbf{L}$ be a $d \times p$ matrix whose rows constitute a basis for $\mathcal{L}$. Let $\boldsymbol{\gamma} = \mathbf{L}\boldsymbol{\beta}$, and $\hat{\boldsymbol{\gamma}} = \mathbf{L}\hat{\boldsymbol{\beta}}$. For $l \in \mathcal{L}$ the linear combination $l^T\boldsymbol{\beta}$ (or $l^T\hat{\boldsymbol{\beta}}$) is equal to $\boldsymbol{\lambda}^T\boldsymbol{\gamma}$ (or $\boldsymbol{\lambda}^T\hat{\boldsymbol{\gamma}}$) for a unique $\boldsymbol{\lambda} = (\lambda_1, \ldots, \lambda_d)$, and conversely $\boldsymbol{\lambda}^T\boldsymbol{\gamma}$ (or $\boldsymbol{\lambda}^T\hat{\boldsymbol{\gamma}}$) for arbitrary $\boldsymbol{\lambda}$ is equal to $l^T\boldsymbol{\beta}$ (or $l^T\hat{\boldsymbol{\beta}}$) for a unique $l \in \mathcal{L}$. Hence, there is a 1-1 correspondence between $\mathcal{L}$ and the d-dimensional space of the $\boldsymbol{\lambda}$ created by $l^T = \boldsymbol{\lambda}^T\mathbf{L}$.

The random vector $\hat{\boldsymbol{\gamma}}$ is distributed as $N(\boldsymbol{\gamma},\sigma^2\mathbf{L}(\mathbf{X}^T\mathbf{X})^{-1}\mathbf{L}^T)$. The random linear combination $l^T\hat{\boldsymbol{\beta}} = \boldsymbol{\lambda}^T\hat{\boldsymbol{\gamma}}$ is distributed as $N(l^T\boldsymbol{\beta},\sigma^2 l^T(\mathbf{X}^T\mathbf{X})^{-1}l)$, or equivalently, $N(\boldsymbol{\lambda}^T\boldsymbol{\gamma},\sigma^2\boldsymbol{\lambda}^T\mathbf{L}(\mathbf{X}^T\mathbf{X})^{-1}\mathbf{L}^T\boldsymbol{\lambda})$. The $d \times d$ covariance matrix $\mathbf{L}(\mathbf{X}^T\mathbf{X})^{-1}\mathbf{L}^T$ is positive-definite so there exists a $d \times d$ nonsingular matrix $\mathbf{P}$ such that

$$\mathbf{P}(\mathbf{L}(\mathbf{X}^T\mathbf{X})^{-1}\mathbf{L}^T)\mathbf{P}^T = \mathbf{I}.$$

Define $\boldsymbol{\delta} = \mathbf{P}\boldsymbol{\gamma}$, and $\hat{\boldsymbol{\delta}} = \mathbf{P}\hat{\boldsymbol{\gamma}}$. Then, $\hat{\boldsymbol{\delta}} \sim N(\boldsymbol{\delta},\sigma^2\mathbf{I})$, i.e., $\hat{\delta}_1, \ldots, \hat{\delta}_d$ are independently normally distributed with means $\delta_1, \ldots, \delta_d$ and variance σ^2.

The argument can now split into either the algebraic or geometric vein.

Algebraic The ratio $\sum_1^d (\hat{\delta}_i - \delta_i)^2/ds^2$ has an F distribution because s^2 is independent of $\hat{\boldsymbol{\delta}}$ (by virtue of being independent of $\hat{\boldsymbol{\beta}}$). Since, by Lemma 2,

$$\sum_1^d (\hat{\delta}_i - \delta_i)^2 \le dF^\alpha_{d,n-p}s^2$$

if and only if

$$\left|\sum_1^d a_i(\hat{\delta}_i - \delta_i)\right| \le (dF^\alpha_{d,n-p})^{\frac{1}{2}} s \left(\sum_1^d a_i^2\right)^{\frac{1}{2}}$$

for all $\mathbf{a} = (a_1, \ldots, a_d)$, it follows that

$$P\{|\mathbf{a}^T(\hat{\boldsymbol{\delta}} - \boldsymbol{\delta})| \le (dF^\alpha_{d,n-p})^{\frac{1}{2}} s(\mathbf{a}^T\mathbf{a})^{\frac{1}{2}},\ \forall \mathbf{a}\} = 1 - \alpha. \quad (82a)$$

Geometric The $100(1 - \alpha)$ percent confidence sphere for $\boldsymbol{\delta}$ is given by

$$S_{\boldsymbol{\delta}} = \{\boldsymbol{\delta}: \sum_1^d (\delta_i - \hat{\delta}_i)^2 \le dF^\alpha_{d,n-p}s^2\}.$$

The sphere S_δ is centered at $\hat{\boldsymbol{\delta}}$ and has radius $(dF^\alpha_{d,n-p})^{\frac{1}{2}}s$. A point $\boldsymbol{\delta}$ is contained in S_δ if and only if it lies between all pairs of parallel hyperplanes tangent to the sphere. (This, hopefully, is a geometrically obvious fact.)

Except for a scale factor, each pair of parallel tangent planes is characterized by the vector $\mathbf{a} = (a_1, \ldots, a_d)$ perpendicular to them. As $\mathbf{a}$ is allowed to vary over d-dimensional space, all possible parallel, tangent planes are generated. Thus, by (80a) and (80b) of Lemma 3 and the if-and-only-if statement of the previous paragraph,

$$P\{|\mathbf{a}^T(\boldsymbol{\delta} - \hat{\boldsymbol{\delta}})| \leq (dF^\alpha_{d,n-p})^{\frac{1}{2}}s(\mathbf{a}^T\mathbf{a})^{\frac{1}{2}}, \forall \mathbf{a}\} = 1 - \alpha. \tag{82b}$$

The two veins of the proof now reunite (in the vena cava?). From the definition of $\boldsymbol{\delta}(\hat{\boldsymbol{\delta}})$ and $\mathbf{P}$,

$$\begin{aligned} \mathbf{a}^T\boldsymbol{\delta} &= \mathbf{a}^T\mathbf{P}\boldsymbol{\gamma} = \mathbf{a}^T\mathbf{P}\mathbf{L}\boldsymbol{\beta} \\ \mathbf{a}^T\hat{\boldsymbol{\delta}} &= \mathbf{a}^T\mathbf{P}\hat{\boldsymbol{\gamma}} = \mathbf{a}^T\mathbf{P}\mathbf{L}\hat{\boldsymbol{\beta}} \\ \mathbf{a}^T\mathbf{a} &= \mathbf{a}^T\mathbf{I}\mathbf{a} = \mathbf{a}^T\mathbf{P}\mathbf{L}(\mathbf{X}^T\mathbf{X})^{-1}\mathbf{L}^T\mathbf{P}^T\mathbf{a}. \end{aligned}$$

Since $\mathbf{P}$ is nonsingular, there is a 1-1 correspondence between $\mathbf{a}$ and $\boldsymbol{\lambda}$ through $\boldsymbol{\lambda}^T = \mathbf{a}^T\mathbf{P}$. Hence (82$a$) and (82$b$) can be written as

$$P\{|\boldsymbol{\lambda}^T\mathbf{L}(\hat{\boldsymbol{\beta}} - \boldsymbol{\beta})| \leq (dF^\alpha_{d,n-p})^{\frac{1}{2}}s(\boldsymbol{\lambda}^T\mathbf{L}(\mathbf{X}^T\mathbf{X})^{-1}\mathbf{L}^T\boldsymbol{\lambda}), \forall \boldsymbol{\lambda}\} = 1 - \alpha$$

or, because $\boldsymbol{l}^T = \boldsymbol{\lambda}^T\mathbf{L}$, as

$$P\{|\boldsymbol{l}^T(\hat{\boldsymbol{\beta}} - \boldsymbol{\beta})| \leq (dF^\alpha_{d,n-p})^{\frac{1}{2}}s(\boldsymbol{l}^T(\mathbf{X}^T\mathbf{X})^{-1}\boldsymbol{l}), \forall \boldsymbol{l} \in \mathcal{L}\} = 1 - \alpha. \qquad \|$$

The proofs of the two extensions of the Scheffé method are immediate. The Dwass (1955) extension was, in fact, proved when it was presented in the method section, since it only involves applying the triangle distance inequality to the points falling in the ellipsoidal confidence region.

The extension to random variables with a general covariance matrix follows from Theorem 1 via the reduction of the covariance matrix to the identity matrix by a nonsingular transformation. Suppose $\mathbf{Y} = \mathbf{X}\boldsymbol{\beta} + \mathbf{e}$ where $\mathbf{e} \sim N(\mathbf{0},\sigma^2\boldsymbol{\Sigma})$ and $\mathbf{0} = (0, \ldots, 0)$. Since $\boldsymbol{\Sigma}$ is assumed to be positive-definite, there exists a nonsingular matrix $\mathbf{P}$ such that $\mathbf{P}\boldsymbol{\Sigma}\mathbf{P}^T = \mathbf{I}$. Define $\mathbf{Y}^* = \mathbf{P}\mathbf{Y}$, $\mathbf{X}^* = \mathbf{P}\mathbf{X}$, and $\mathbf{e}^* = \mathbf{P}\mathbf{e}$. Then, $\mathbf{Y}^* = \mathbf{X}^*\boldsymbol{\beta} + \mathbf{e}^*$ where $\mathbf{e}^* \sim N(\mathbf{0},\sigma^2\mathbf{I})$, so Theorem 1 can be applied to $\mathbf{Y}^*$, $\mathbf{X}^*$, and $\boldsymbol{\beta}$. But $\mathbf{P}^T\mathbf{P} = \boldsymbol{\Sigma}^{-1}$, so

$$\begin{aligned} \mathbf{X}^{*T}\mathbf{X}^* &= \mathbf{X}^T\boldsymbol{\Sigma}^{-1}\mathbf{X} \\ \hat{\boldsymbol{\beta}}^* &= (\mathbf{X}^{*T}\mathbf{X}^*)^{-1}\mathbf{X}^{*T}\mathbf{Y}^* = (\mathbf{X}^T\boldsymbol{\Sigma}^{-1}\mathbf{X})^{-1}\mathbf{X}^T\boldsymbol{\Sigma}^{-1}\mathbf{Y} \\ s^2 &= \frac{\mathbf{Y}^{*T}[\mathbf{I} - \mathbf{X}^*(\mathbf{X}^{*T}\mathbf{X}^*)^{-1}\mathbf{X}^{*T}]\mathbf{Y}^*}{n - p} \\ &= \frac{\mathbf{Y}^T[\boldsymbol{\Sigma}^{-1} - \boldsymbol{\Sigma}^{-1}\mathbf{X}(\mathbf{X}^T\boldsymbol{\Sigma}^{-1}\mathbf{X})^{-1}\mathbf{X}^T\boldsymbol{\Sigma}^{-1}]\mathbf{Y}}{n - p}. \end{aligned} \tag{83}$$

Expression (58) follows immediately with these substitutions.

2.5 Distributions and tables

The only distribution involved in the Scheffé technique is the F distribution. This distribution is so well known, and so well discussed everywhere, and elsewhere, it will not be discussed here.

Tables of the F distribution can be found in most volumes of tables, such as Owen (1962), "Handbook of Statistical Tables;" Pearson and Hartley (1962), "Biometrika Tables for Statisticians," vol. I; and Fisher and Yates (1963), "Statistical Tables;" and in numerous textbooks such as Scheffé (1959), "The Analysis of Variance;" and Graybill (1961), "An Introduction to Linear Statistical Models," vol. I.

3 BONFERRONI t STATISTICS

This technique is an ancient statistical tool which depends solely on the simple probability inequality (1.13). The name of Bonferroni (or Boole) is attached to the probability inequality, but no name can be singled out to commemorate its first statistical application.

3.1 Method

Let $Y_1, \ldots, Y_k$ be normally distributed random variables with means $\mu_1, \ldots, \mu_k$ and variances $\sigma_1^2, \ldots, \sigma_k^2$, respectively. The Y_i's may or may not be independent. Let $s_1^2, \ldots, s_k^2$ be χ^2 estimators of $\sigma_1^2, \ldots, \sigma_k^2$ on $\nu_1, \ldots, \nu_k$ d.f., respectively; that is, $\nu_i s_i^2/\sigma_i^2 \sim \chi_{\nu_i}^2$, $i = 1, \ldots, k$. The s_i^2's may or may not be independent. However, it is assumed that Y_i is independent of s_i^2, $i = 1, \ldots, k$, so that

$$T_i = \frac{Y_i - \mu_i}{s_i} \qquad i = 1, \ldots, k \tag{84}$$

has a t distribution with ν_i d.f., $i = 1, \ldots, k$. Although it rarely occurs in practice, it is permissible for s_i^2 to be dependent with $Y_{i'}$, $i' \neq i$.

Let $t_{\nu_i}^{\alpha/2k}$, $i = 1, \ldots, k$, be the upper $\alpha/2k$ percentile points (or two-tailed α/k percentile points) of the t distribution with d.f. ν_i, $i = 1, \ldots, k$, respectively. Then, with probability greater than or equal to $1 - \alpha$, simultaneously,

$$\mu_i \in Y_i \pm t_{\nu_i}^{\alpha/2k} s_i \qquad \text{for } i = 1, \ldots, k. \tag{85}$$

For each component interval above, the significance level was set at α/k. If some of the intervals should be more sensitive or conservative than others, equal significance levels can be abandoned, and unequal allocation substituted. Any combination $\alpha_1, \ldots, \alpha_k$ for which $\alpha_1 + \cdots + \alpha_k = \alpha$ will produce the same bound α for the probability error rate. The expected error rate is exactly α.

Although scaling down the significance level when confronted with more than one statement is an old device, it is seldom taught to students and is rarely mentioned in textbooks and the literature.

3.2 Applications

The structure of the t statistics (84) required for most applications need not be as general as indicated in the method section. Customarily, $\sigma_i^2 = d_i\sigma^2$, $i = 1, \ldots, k$, where $d_1, \ldots, d_k$ are known constants, σ^2 is unknown, and $s_i^2 = d_i s^2$, $i = 1, \ldots, k$, where s^2 is a χ^2 estimate of σ^2 based on ν d.f. Thus, the numerators of the t statistics (84) consist of dependent or independent normal variables, and the denominators contain a common χ^2 variable with the appropriate normalizing constants.

This framework encompasses essentially all t tests or t intervals in (fixed effects) analysis-of-variance and linear regression problems. For example, if there are k row contrasts, $\sum_{i=1}^{r} c_{hi}\bar{Y}_{i..}$, $h = 1, \ldots, k$, of interest in an $r \times c$ two-way classification with n observations per cell, then the appropriate Bonferroni t intervals are

$$\sum_{i=1}^{r} c_{hi}\alpha_i \in \sum_{i=1}^{r} c_{hi}\bar{Y}_{i..} \pm t_{rc(n-1)}^{\alpha/2k} s \left(\frac{1}{cn} \sum_{i=1}^{r} c_{hi}^2\right)^{\frac{1}{2}} \qquad h = 1, \ldots, k. \tag{86}$$

The k contrasts might include the pairwise comparisons plus some general contrasts. Or, for a general linear regression problem, $\mathbf{Y} = \mathbf{X}\boldsymbol{\beta} + \mathbf{e}$, the p regression coefficients can be tested against hypothesized values $\beta_1^0, \ldots, \beta_p^0$ via

$$|\hat{\beta}_i - \beta_i^0| \leq t_{n-p}^{\alpha/2p} s(\mathbf{XX}^{ii})^{\frac{1}{2}} \qquad i = 1, \ldots, p \tag{87}$$

where $\mathbf{XX}^{ii}$ is the ith diagonal element of the inverse to the matrix $\mathbf{X}^T\mathbf{X}$ (that is, $\sigma^2\mathbf{XX}^{ii}$ is the variance of $\hat{\beta}_i$).

3.3 Comparison

The extreme generality of the Bonferroni t statistics makes them applicable to any of the problems discussed in Secs. 1 to 5.

For the special situations of Secs. 4 and 5 the techniques described therein will be better since they are exact procedures (i.e., the $P\{\mathfrak{F}\}$'s are exactly equal to α). Similarly, for pairwise mean comparisons, the studentized range technique will be better unless, perhaps, not all comparisons are of interest.

Where the Bonferroni t statistics stand a good chance of coming out the winner is in competition with the studentized range for general contrasts, the studentized augmented range, and the Scheffé F projections (for pairwise comparisons, general contrasts, and comparisons of sample values with hypothesized values). All of these procedures, including the Bonferroni t statistics, are wasteful in the sense that the family probability error rate is less than α instead of being equal to it. For the Bonferroni t statistics this arises because the Bonferroni inequality is indeed an inequality, and for the others because not all statements are made which are allowed under the probability error rate. For example, with the Scheffé F projections the probability is $1 - \alpha$ that *all* contrasts will be included in their respective intervals. However, it is only a *finite* number of contrasts that are ever examined in any application.

Examination of the Bonferroni vs. Scheffé match reveals that who is to be the victor depends on the type of application. For k statements the Bonferroni critical constant is $t_\nu^{\alpha/2k}$ where ν is the degrees of freedom of the common χ^2 variable in the denominator. Note that the constant depends directly on k and will increase as k increases. On the other hand, the Scheffé critical constant for k statements is $(dF_{d,\nu}^\alpha)^{\frac{1}{2}}$, where d is the dimension of the smallest linear space encompassing the k statements. Obviously, $d \leq k$.

The dimension d can equal k as, for instance, when the p regression coefficients $\hat{\beta}_1, \ldots, \hat{\beta}_p$ of a general multiple linear regression problem are tested against hypothesized values $\beta_1^0, \ldots, \beta_p^0$. Here, $d = k = p$. Alternatively, d can be considerably less than k if, for example, k is the number of row contrasts to be tested in an $r \times c$ two-way classification. In this case d is fixed at $r - 1$, but the number k of contrasts of interest can be arbitrarily large. Just the pairwise mean comparisons give $k = r(r - 1)/2$, and an unlimited number of general contrasts can be added for consideration.

Dunn (1959) has computed tables of $t_\nu^{\alpha/2k}$ and $(dF_{d,\nu}^\alpha)^{\frac{1}{2}}$ for $\alpha = .05$; $k = 1(1)\,10,15,20,50$; $d = 1(1)\,8$; and $\nu = 5,10,15,20,24,30,40,60,120, +\infty$. Inspection of the two tables readily shows that $t_\nu^{\alpha/2k} < (dF_{d,\nu}^\alpha)^{\frac{1}{2}}$ if d and k are at all similar in size. It is only when k is considerably bigger than d that the reverse inequality holds.

3.4 Derivation

The derivation follows immediately from the Bonferroni inequality (1.13). Since

$$P\left\{\left|\frac{Y_i - \mu_i}{s_i}\right| \leq t_{\nu_i}^{\alpha/2k}\right\} = 1 - \frac{\alpha}{k} \qquad i = 1, \ldots, k \tag{88}$$

the inequality (1.13) gives

$$P\left\{\bigcap_{i=1}^{k}\left[\left|\frac{Y_i - \mu_i}{s_i}\right| \le t_{\nu_i}^{\alpha/2k}\right]\right\} \ge 1 - k \cdot \frac{\alpha}{k} = 1 - \alpha. \tag{89}$$

3.5 Distributions and tables

The sole distribution needed in this section is the t distribution which does not require discussion. (Everyone is born knowing the t distribution.) Good tables are available in Owen (1962), "Handbook of Statistical Tables;" Pearson and Hartley (1962), "Biometrika Tables for Statisticians," vol. I; Fisher and Yates (1963), "Statistical Tables;" Graybill (1961), "An Introduction to Linear Statistical Models," vol. I; and elsewhere.

The only complication posed by the application of Bonferroni t statistics is the necessity for critical points $t_\nu^{\alpha/2k}$ at oddball percentiles $\alpha/2k$. This requirement can be met in several ways. Linear, or better yet, curvilinear, interpolation in the t tables cited above will frequently provide a critical point with accuracy sufficient for most applications. Or approximate critical points can be obtained from the approximation

$$t_\nu^\alpha \cong g^\alpha + \frac{1}{4\nu}\left[(g^\alpha)^3 + g^\alpha\right] \tag{90}$$

where g^α is the upper 100α percent point of the $N(0,1)$ distribution. This approximation is due to Peiser (1943).

The t statistic can be converted to a beta variable for which extensive tables are available [K. Pearson (1956), "Tables of the Incomplete Beta-Function"]. This is possible since $(t_\nu)^2 \sim F_{1,\nu}$ implies that

$$\frac{(t_\nu^2/\nu)}{1 + (t_\nu^2/\nu)} \sim B_{1,\nu} \tag{91}$$

where $B_{1,\nu}$ is a generic beta variable with d.f. 1 and ν (that is, the cdf of $B_{1,\nu}$ is an incomplete beta function with parameters $p = \frac{1}{2}$ and $q = \nu/2$).

Dunn (1959) gives a limited table of $t_\nu^{\alpha/2k}$ for $\alpha = .05$. A larger set of tables appears in Dunn (1961) where $t_\nu^{\alpha/2k}$ is given for $\alpha = .05, .01$; $k = 2(1)10(5)50, 100, 250$; $\nu = 5, 7, 10, 12, 15, 20, 24, 30, 40, 60, 120, \infty$. These are partially reproduced in Table II of Appendix B. Dunn and Massey (1965) also give $t_\nu^{\alpha/2k}$ for $\alpha = .01, .025, .05, .10\ (.1)\ .50$; $k = 2, 6, 10, 20$; $\nu = 4, 10, 30, \infty$.

4 STUDENTIZED MAXIMUM MODULUS

This technique was introduced by Tukey (1952b,1953b) and Roy and Bose (1953). It is the direct analog of the prototype in Sec. 3.2 of Chap. 1.

4.1 Method

Let $Y_1, \ldots, Y_k$ be independently, normally distributed random variables with means $\mu_1, \ldots, \mu_k$ and variances $d_1\sigma^2, \ldots, d_k\sigma^2$. The constants $d_1, \ldots, d_k$ are assumed known, but the parameters $\mu_1, \ldots, \mu_k, \sigma^2$ are unknown. Let s^2 be a χ^2 estimator of σ^2 on ν d.f. which is independent of $Y_1, \ldots, Y_k$. Then, with probability $1 - \alpha$,

$$\mu_i \in Y_i \pm |m|^{\alpha}_{k,\nu} \sqrt{d_i}\, s \qquad i = 1, \ldots, k \tag{92}$$

where $|m|^{\alpha}_{k,\nu}$ is the upper α point of the studentized maximum modulus distribution with parameters k,ν (that is, the distribution of $\max_i \{|Y_i - \mu_i|/\sqrt{d_i}\}/s)$. The numerator (divided by σ) of this statistic is just the maximum modulus $|M|_k$, the maximum absolute value of k independent, unit normal random variables, and the denominator (divided by σ) is $\sqrt{\chi^2_\nu/\nu}$.

Note in contradistinction to the Bonferroni t statistics that the variables $Y_1, \ldots, Y_k$ must be independent. In addition, there can be just a single common s in the denominator which must be independent of all numerators.

The maximum modulus technique can be extended to include confidence intervals for linear combinations of the means as well. The event (92), which has probability $1 - \alpha$, is equivalent to the event

$$\sum_1^k l_i\mu_i \in \sum_1^k l_iY_i \pm |m|^{\alpha}_{k,\nu} s \sum_1^k |l_i \sqrt{d_i}| \qquad \forall\, \boldsymbol{l} \in \mathcal{L}_{lc}. \tag{93}$$

Note that the intervals (92) are special cases of (93). Thus, the statistician can construct confidence intervals for any arbitrary linear combinations of interest without any change in the probability level.

4.2 Applications

The studentized maximum modulus can be applied wherever its assumptions are met, and the assumptions are quite simple and clear. A basic necessity is the independence of the numerators. Whenever independence is present, the studentized maximum modulus can be applied; without independence, it cannot.

For illustration, consider a general multiple linear regression problem. In testing $\hat{\beta}_1, \ldots, \hat{\beta}_p$ vs. $\beta_1^0, \ldots, \beta_p^0$, respectively, it is not customary for $\hat{\beta}_1, \ldots, \hat{\beta}_p$ to be independent. However, in the special case of polynomial curve-fitting with orthogonal polynomials, $\hat{\beta}_1, \ldots, \hat{\beta}_p$ will always be independent. [For a fuller discussion than will be given here of the statistical aspects of orthogonal polynomials in curve-fitting, the reader is referred to Anderson and Bancroft (1952).]

In fitting a pth-degree polynomial to the n pairs of points (x_1, Y_1), $\ldots$, (x_n, Y_n), an orthogonal polynomial parameterization may be convenient when the points $x_1, \ldots, x_n$ are equally spaced, that is, $x_i - x_{i'} = \Delta(i - i')$, $\Delta \neq 0$. In terms of the orthogonal polynomials the model is

$$Y_i = \beta_0 P_0(x_i) + \cdots + \beta_p P_p(x_i) + e_i \qquad i = 1, \ldots, n \tag{94}$$

where $P_j(x)$ is a jth-degree polynomial in x which is orthogonal to $P_{j'}(x)$ for $j' \neq j$. Orthogonality in this context means

$$\sum_i P_j(x_i) P_{j'}(x_i) = 0 \qquad j \neq j'. \tag{95}$$

The zero-, first-, and second-degree polynomials are

$$\begin{aligned} P_0(x_i) &\equiv 1 \\ P_1(x_i) &= x_i - \bar{x} \\ P_2(x_i) &= (x_i - \bar{x})^2 - \sum_h (x_h - \bar{x})^2; \end{aligned} \tag{96}$$

the higher-degree polynomials are obtained by a Gram-Schmidt orthogonalization procedure.

Because of the required orthogonality to the zero-degree polynomial $P_0(x) \equiv 1$, the higher-degree polynomials $P_j(x)$, $j \geq 1$, are usually more conveniently written in terms of $x - \bar{x}$ instead of x [see (96)]. Consequently, the model is often written as

$$Y_i = \beta_0 + \beta_1 Q_1(x_i - \bar{x}) + \cdots + \beta_p Q_p(x_i - \bar{x}) + e_i \tag{97}$$

where $Q_j(x_i - \bar{x}) = P_j(x_i)$.

The orthogonality property (95) permits the regression coefficient estimates to be computed separately:

$$\hat{\beta}_j = \frac{\sum_i Y_i Q_j(x_i - \bar{x})}{\sum_i Q_j^2(x_i - \bar{x})} \tag{98}$$

and
$$s^2 = \frac{\sum_i Y_i^2 - \hat{\beta}_0 \left(\sum_i Y_i\right) - \cdots - \hat{\beta}_p \left[\sum_i Y_i Q_p(x_i - \bar{x})\right]}{n - p - 1}. \tag{99}$$

These computations can be performed rapidly on a desk calculator with the aid of tables of orthogonal polynomial values. (Tables appear in Fisher and Yates (1963), "Statistical Tables," and elsewhere.)

The estimators $\hat{\beta}_0, \ldots, \hat{\beta}_p$ are *independently* [by (95)], normally distributed with means $\beta_0, \ldots, \beta_p$ and variances

$$\operatorname{Var}(\hat{\beta}_j) = \frac{\sigma^2}{\sum_i Q_j^2(x_i - \bar{x})} \qquad j = 0, 1, \ldots, p \tag{100}$$

respectively. Of course, $(n - p - 1)s^2/\sigma^2 \sim \chi^2_{n-p-1}$, and is independent of $\hat{\beta}_0, \ldots, \hat{\beta}_p$.

By virtue of the independence, normality, etc., of the $\hat{\beta}_j$'s, the studentized maximum modulus can be applied to all of them or any subset. For example, with probability $1 - \alpha$,

$$\beta_j \epsilon \hat{\beta}_j \pm |m|^{\alpha}_{p,n-p-1} s \left[\frac{1}{\sum_i Q_j^2(x_i - \bar{x})} \right]^{\frac{1}{2}} \qquad j = 1, \ldots, p. \tag{101}$$

The one crucial item to examine in switching to orthogonal polynomials from a straight polynomial fit

$$Y_i = \alpha_0 + \alpha_1 x_i + \alpha_2 x_i^2 + \cdots + \alpha_p x_i^p + e_i \tag{102}$$

is the physical interpretation and relevance of the orthogonalized regression coefficients β_j, $j = 0, 1, \ldots, p$. If they retain physical meaning, and tests of them are meaningful, then the orthogonalization of the model is a convenient tool. But if the interpretation of the regression coefficients is lost in the orthogonalization process, then the statistician is forced to stick with the nonorthogonal model.

4.3 Comparison

What there is to say can be said quickly. The competitors of the confidence intervals (92) are the projected F of Scheffé, the Bonferroni t statistics, and the extended studentized range. All of these are inexact due to not utilizing all the possible statements, or using probability bounds. Consequently, the studentized maximum modulus is better.

The extension (93) does not fare so well. The extended studentized range is better for contrasts, but the extended maximum modulus is better for linear combinations not involving differences, such as weighted averages. The Scheffé procedure will tend to do better for combinations of means with approximately equal (absolute) weights on the means, whereas the extended maximum modulus will give shorter intervals when one mean carries essentially all the weight. The author has not had any experience in comparing it with the Bonferroni t statistics, but very likely the same remarks apply here as in the Scheffé vs. Bonferroni case.

One temporary qualification to this evaluation is that the available tables of the studentized maximum modulus are quite limited, as will be seen in Sec. 4.5. Until this is remedied (which, hopefully, will be soon) the statistician may, in some situations, have to fall back on the Scheffé or Bonferroni statistics.†

As a testing device the studentized maximum modulus suffers from the same malady as the studentized range—overconservatism. A multiple stage analog will give greater sensitivity with the same protection for the null hypothesis. This will be discussed in more detail in Sec. 6.1.

† Footnote to second edition: See the new table of the studentized maximum modulus in the Addendum.

4.4 Derivation

The derivation of (92) lies entirely in the definition of the studentized maximum modulus distribution and the following double implication:

$$\left| \frac{Y_i - \mu_i}{\sqrt{d_i}\, s} \right| \leq c \qquad i = 1, \ldots, k \tag{103}$$

if and only if

$$\frac{\max\limits_{1 \leq i \leq k} \{|Y_i - \mu_i| / \sqrt{d_i}\}}{s} \leq c. \tag{104}$$

The extension (93) is a consequence of the following lemma.

Lemma 4

$$\max_{1 \leq i \leq k} \{|y_i|\} \leq c$$

if and only if

$$\left| \sum_1^k l_i y_i \right| \leq c \sum_1^k |l_i| \qquad \forall\, \boldsymbol{l} \in \mathcal{L}_{lc}.$$

Proof *If:* Choose $l_1 = 1, l_2 = \cdots = l_k = 0$. Then, by the inequality, $|y_1| \leq c$. Repeating this for $l_2, \ldots, l_k$ gives $|y_i| \leq c$, $i = 1, \ldots, k$, which is equivalent to $\max\{|y_i|\} \leq c$.

Only if:

$$\left| \sum_1^k l_i y_i \right| \leq \sum_1^k |l_i| \cdot |y_i| \leq \max_{1 \leq i \leq k} \{|y_i|\} \sum_1^k |l_i| \leq c \sum_1^k |l_i|. \qquad \|$$

The expression (93) follows from (92) and Lemma 4 with $(Y_i - \mu_i)/\sqrt{d_i}$ identified with y_i, and $l_i \sqrt{d_i}$ identified with l_i.

A geometric interpretation can be given to Lemma 4 and the extension (93) just as for the Scheffé procedure. The intervals (93) are the projections of the box-shaped confidence region (92) onto the one-dimensional lines spanned by the vectors $\boldsymbol{l}$. The pair of parallel planes, tangent to

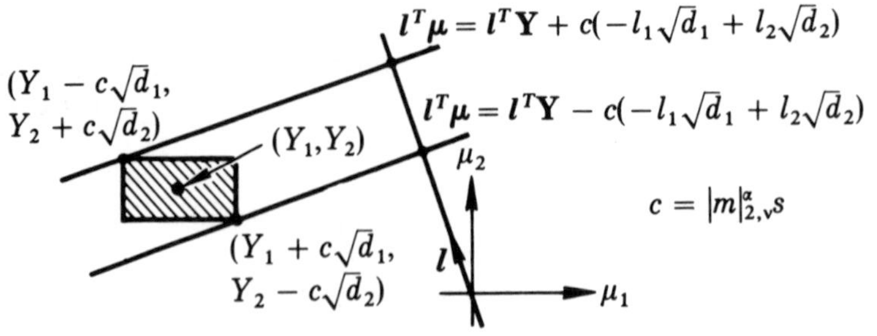

Figure 2

the box (92) and perpendicular to $\boldsymbol{l}$, intersect the line spanned by $\boldsymbol{l}$ at the distances $(\boldsymbol{l}^T\mathbf{Y} \pm |m|^{\alpha}_{k,\nu}s \sum_1^k |l_i \sqrt{d_i}|)/\|\boldsymbol{l}\|$ from the origin. These projections are illustrated graphically in Fig. 2 for the case $k = 2$ and $\|\boldsymbol{l}\| = 1$, where the scale factor $|m|^{\alpha}_{2,\nu}s$ is represented by c. Note that

$$-l_1\sqrt{d_1} + l_2\sqrt{d_2} = \sum_1^2 |l_i\sqrt{d_i}|$$

since l_1 is negative.

4.5 Distributions and tables

The cdf of the studentized maximum modulus is

$$P\left\{\frac{|M|_k}{\sqrt{\chi^2_\nu/\nu}} \le x\right\} = \int_0^{+\infty} P\left\{|M|_k \le xy \,\middle|\, \sqrt{\frac{\chi^2_\nu}{\nu}} = y\right\} dP\left\{\sqrt{\frac{\chi^2_\nu}{\nu}} \le y\right\}$$
$$= \int_0^{+\infty} \left[2\int_0^{xy} \varphi(u)\,du\right]^k \frac{\nu^{\nu/2}}{\Gamma\left(\frac{\nu}{2}\right)2^{(\nu/2)-1}} y^{\nu-1}e^{-\nu y^2/2}\,dy \tag{105}$$

for $x \ge 0$, where φ is the unit normal density (33). No simplification of the integral (105) that permits easy computation is possible.

The t statistics $(Y_i - \mu_i)/\sqrt{d_i}\,s$, $i = 1, \ldots, k$, have a multivariate density which is an analog of the t distribution (see Sec. 5.5). Numerical integration of this density does not appear to be a very practical method of obtaining critical points except for very small k.

Pillai and Ramachandran (1954) evaluated the integral (105) by expanding $\left[\int_0^x \varphi(u)\,du\right]^k$ in a power series in x and then integrating term by term. They give a table of $|m|^{\alpha}_{k,\nu}$ for the upper $\alpha = .05$ point; $k = 1(1)8$; $\nu = 5(5)\ 20, 24, 30, 40, 60, 120, +\infty$, and a slightly larger table for the lower $\alpha = .05$ point. Two decimal entries are given. The table of upper $\alpha = .05$ points is reproduced in Table III of Appendix B.

Dunn and Massey (1965) provide a table of $|m|^{\alpha}_{k,\nu}$ for $\alpha = .01, .025, .05, .10(.1).50$; $k = 2, 6, 10, 20$; $\nu = 4, 10, 30, \infty$.

Tukey (1953b) gives a method and tables for computing (approximately) additional points at $\alpha = .05$ and other α levels. Nemenyi (1953) has also computed additional points. These sources are not widely available.

Hopefully, more extensive, complete tables of the critical points of the studentized maximum modulus will appear in the not too distant future.†

† Footnote to second edition: See the new table of the studentized maximum modulus in the Addendum.

5 MANY-ONE t STATISTICS

This problem and technique were discussed by Dunnett (1955,1964), who computed the necessary tables to implement the technique. Earlier, Paulson (1952a) had considered the same problem but from a somewhat different point of view.

5.1 Method

Let $\{Y_{ij}; i = 0,1, \ldots, k, j = 1, \ldots, n\}$ be $k + 1$ independent samples of n independently, normally distributed random variables with variance σ^2 and means $E\{Y_{ij}\} = \mu_i$, $i = 0,1, \ldots, k$. The zero population ($i = 0$) will be referred to as the *control*, and the other k populations ($i \geq 1$) as *treatments*. The problem is to compare each treatment mean with the control mean in order to decide if any of the treatments differ from the control. For significance tests this constitutes testing

$$H_0\colon \quad \mu_i - \mu_0 = 0 \qquad i = 1, \ldots, k \tag{106}$$

versus the alternative that some of the $\mu_i - \mu_0$ differ from zero.

The class mean estimates are $\bar{Y}_{i\cdot}$, $i = 0, 1, \ldots, k$, and the variance σ^2 is estimated by

$$s^2 = \frac{1}{(k+1)(n-1)} \sum_{i,j} (Y_{ij} - \bar{Y}_{i\cdot})^2. \tag{107}$$

The natural statistic for comparing the ith treatment mean with the control is

$$T_i = \frac{(\bar{Y}_{i\cdot} - \bar{Y}_{0\cdot}) - (\mu_i - \mu_0)}{s\sqrt{2/n}} \qquad i = 1, \ldots, k. \tag{108}$$

Individually, each of these has a t distribution with $(k + 1)(n - 1)$ d.f. so that the critical point would be $t^{\alpha/2}_{(k+1)(n-1)}$. However, when the k comparisons are grouped into a family, the critical point must be increased to achieve a family probability error rate of α. The appropriate intervals are:

$$\mu_i - \mu_0 \in \bar{Y}_{i\cdot} - \bar{Y}_{0\cdot} \pm |d|^{\alpha}_{k,(k+1)(n-1)} s \sqrt{\frac{2}{n}} \qquad i = 1, \ldots, k \tag{109}$$

where $|d|^{\alpha}_{k,(k+1)(n-1)}$ is the upper 100α percent point of the distribution of

$$\frac{\max\limits_{1 \leq i \leq k} \{|Y_i - Y_0|/\sqrt{2}\}}{\sqrt{\chi^2_\nu/\nu}} \tag{110}$$

with $\nu = (k+1)(n-1)$. The variables $Y_0, Y_1, \ldots, Y_k$ in (110) are independent, unit normal deviates, and χ^2_ν is an independent χ^2 variable with ν d.f.

No title for the statistic (110) has been coined by its originators, and it is difficult to think of a suitable one. Statistic (110) shall therefore go nameless. Its distribution can be obtained from a multivariate analog of the t distribution, and this will be discussed in Sec. 5.5.

In some experiments the treatments are of interest only if they do better than the control, and not merely if they are different from it. This leads to one-sided confidence intervals or significance tests. *Better* could mean either larger or smaller expectations, depending on the context. For discovering expectations larger than the control the appropriate one-sided intervals are

$$\mu_i - \mu_0 \geq \bar{Y}_{i\cdot} - \bar{Y}_{0\cdot} - d^\alpha_{k,(k+1)(n-1)} s \sqrt{\frac{2}{n}} \qquad i = 1, \ldots, k \quad (111)$$

where $d^\alpha_{k,(k+1)(n-1)}$ is the upper 100α percent point of the distribution of (110) with the absolute value signs removed in the numerator and with $\nu = (k+1)(n-1)$.

The statements (109) can be extended to include comparisons of weighted averages of the μ_i, $i = 1, \ldots, k$, with μ_0 without any change in the confidence level. The appropriate intervals are

$$\left(\sum_1^k w_i\mu_i\right) - \mu_0 \in \left(\sum_1^k w_i\bar{Y}_{i\cdot}\right) - \bar{Y}_{0\cdot} \pm |d|^\alpha_{k,(k+1)(n-1)} s \sqrt{\frac{2}{n}} \qquad (112)$$

for any $\mathbf{w} = (w_1, \ldots, w_k)$ with $w_i \geq 0$, $\sum_1^k w_i = 1$. A similar extension can be made for the one-sided intervals (111).

There is no basic theoretical reason why the number of observations in each of the $k+1$ classes has to be the same. In fact, it would be more appealing to allow the control class to have m observations and each of the treatment classes n observations. Since each treatment mean is compared with the same control, the value of the control mean is more important than any single treatment value and should be known with greater accuracy. It therefore makes sense to have $m > n$, and Dunnett (1955) discusses the optimal allocation ratio m/n for a fixed total number of observations.

Unfortunately, tables are not available for $d^\alpha_{k,\nu}$ or $|d|^\alpha_{k,\nu}$ when $m \neq n$. When the need for greater accuracy in estimating μ_0 outweighs the nicety of exact critical points, the statistician can either try to approximate the correct many-one t critical point or retreat to the Scheffé and Bonferroni procedures.

5.2 Applications

Just as with the studentized maximum modulus, the range of application of the many-one t statistics is crisply demarcated. When k treatment means are to be compared with a single control, the many-one t statistics apply. Use of the existing tables is limited to the case of equal variances, which requires both equal sample size and homoscedasticity.

This type of experiment is reported to be not uncommon in the fields of pharmacology, toxicology, agronomy, horticulture, etc. However, the statistical technique is a good deal more specialized than the others discussed in Secs. 1 to 4. The author of this book cannot testify to having seen it used even once, but this must be purely circumstantial.

Dunnett (1955) discusses the use of this technique in parallel line bioassays of several treatment-control log-potencies. But here, as with unequal sample sizes, inequality of the correlations between the numerator statistics prevents the use of the tables, or allows at best an approximate significance level from the tables.

5.3 Comparison

For confidence intervals the many-one t statistics [(109) or (111)] offer an exact procedure where the tables are applicable, and their competitors (Scheffé, Bonferroni, etc.) suffer from inexactness. The form of the statistics is always the same, that is, $\sqrt{n}\,(\bar{Y}_{i.} - \bar{Y}_{0.})/\sqrt{2}\,s$. The only haggle is over the critical point, and the many-one t critical points will be smallest when they are available. The only qualification is the limited character of the tables.

For significance tests the many-one t statistics should be applied in a multiple stage fashion to avoid undue conservatism. The modification is similar to converting from the studentized range (Tukey) test to multiple range (Newman-Keuls-Duncan) tests, and will be described in Sec. 6.1.

5.4 Derivation

Trivially,

$$\left| \frac{(\bar{Y}_{i.} - \bar{Y}_{0.}) - (\mu_i - \mu_0)}{s\sqrt{2/n}} \right| \le c, \qquad i = 1, \ldots, k \tag{113}$$

if and only if

$$\frac{\max_{1 \le i \le k} \{|(\bar{Y}_{i.} - \bar{Y}_{0.}) - (\mu_i - \mu_0)|\}}{s\sqrt{2/n}} \le c. \tag{114}$$

The intervals (109) follow immediately. The nonabsolute form of (114) gives (111).

The extension (112) to weighted averages follows from Lemma 4 of Sec. 4.4 with the identifications $y_i = (\bar{Y}_{i.} - \bar{Y}_{0.}) - (\mu_i - \mu_0)$, $c = |d|^{\alpha}_{k,(k+1)(n-1)} s \sqrt{2/n}$, and $\mathbf{w} = \mathbf{l}$. Actually, the space $\{\mathbf{w}\colon w_i \geq 0, \sum_1^k w_i = 1\}$ is a subset of $\mathfrak{L}_{lc}$, but it contains the relevant points $[(1,0, \ldots, 0)$, etc.$]$ for the If-proof. Since $\mathfrak{L}_{lc}$ is more general than just weighted averages, (112) could be generalized, but it is doubtful whether any such generalization would be useful. The one-sided analog of (112) follows also from Lemma 4 with all absolute value signs dropped and the l_i restricted to be nonnegative (in the statement and proof of the lemma).

A geometric interpretation similar to Fig. 2 can be given for the extension (112) and its one-sided analog. The confidence intervals are the projections of the confidence regions on the w's.

5.5 Distributions and tables

Let $\mathbf{Z} = (Z_1, \ldots, Z_k)$ have a multivariate normal distribution with mean $\mathbf{0} = (0, \ldots, 0)$ and covariance matrix $\mathbf{\Sigma} = \sigma^2 \mathbf{P}$ where

$$\mathbf{P} = \begin{pmatrix} 1 & \rho_{12} & \cdots & \rho_{1k} \\ \rho_{21} & 1 & \cdots & \cdots \\ \cdots & \cdots & \cdots & \cdots \\ \rho_{k1} & \cdots & \cdots & 1 \end{pmatrix} \tag{115}$$

is the correlation matrix. The matrix (115) will be assumed positive-definite so $\mathbf{P}^{-1} = (\rho^{ij})$, the inverse to $\mathbf{P} = (\rho_{ij})$, is well-defined. Let $\nu s^2/\sigma^2$ be a χ^2 variable with ν d.f. which is independent of $\mathbf{Z}$.

The vector $\mathbf{T} = (T_1, \ldots, T_k) = (Z_1/s, \ldots, Z_k/s)$ constitutes a multivariate analog to the t distribution. It can be shown [see Dunnett and Sobel (1954,1955)] that the multivariate density of this random vector is

$$p(t_1, \ldots, t_k) = \frac{\Gamma\left(\frac{k+\nu}{2}\right)(\det \mathbf{P})^{-\frac{1}{2}}}{(\nu\pi)^{k/2}\Gamma(\nu/2)} \left(1 + \frac{1}{\nu} \mathbf{t}^T \mathbf{P}^{-1} \mathbf{t}\right)^{-(k+\nu)/2} \tag{116}$$

where $\mathbf{t} = (t_1, \ldots, t_k)$, and det $\mathbf{P}$ is the determinant of $\mathbf{P}$.

The distribution of the many-one t statistics is a special case of this multivariate t distribution. Let

$$Z_i = \frac{(\bar{Y}_{i.} - \bar{Y}_{0.}) - (\mu_i - \mu_0)}{\sqrt{\frac{1}{n_i} + \frac{1}{n_0}}} \qquad i = 1, \ldots, k \tag{117}$$

where $\bar{Y}_{i.}$ is the mean of n_i observations, $i = 0, 1, \ldots, k$. Thus, the

Z_i's have a multivariate normal distribution with means 0, variances σ^2, and correlations

$$\operatorname{Cor}(Z_i,Z_{i'}) = \rho_{ii'} = \frac{1}{\sqrt{\left(\frac{n_0}{n_i}+1\right)\left(\frac{n_0}{n_{i'}}+1\right)}} \tag{118}$$

so the many-one t statistics $(T_1, \ldots, T_k)$ have the density (116) with $\boldsymbol{\Sigma}$ given by (118). For those statistics considered in this section with $n_i = n, i = 0, 1, \ldots, k$, (118) reduces to the simple case $\rho_{ii'} = \frac{1}{2}, i \neq i'$.

The t statistics of the studentized maximum modulus (Sec. 4) have a multivariate density which is also a special case of (116). Here $Z_i = (Y_i - \mu_i)/\sqrt{d_i}$, $i = 1, \ldots, k$, where the Y_i's are the variables defined in Sec. 4.1. Thus, $\mathbf{Z} \sim N(\mathbf{0},\sigma^2\mathbf{I})$ so $\boldsymbol{\Sigma} = \mathbf{I}$ in this case since the Y_i are independent.

Dunnett and Sobel (1954) computed tables of $\iint_{-\infty}^{c} p(t_1,t_2)\,dt_1\,dt_2$ for the bivariate case with $\rho_{12} = \frac{1}{2}$ and $\rho_{12} = -\frac{1}{2}$. The former can be used to obtain quite accurate one-sided critical points $d^{\alpha}_{k,\nu}$ in the special case $k = 2$. For larger k Dunnett (1955) gives tables of $d^{\alpha}_{k,\nu}$ to two decimals for $k = 1(1)9$; $\alpha = .05, .01$; $\nu = 5(1)20, 24, 30, 40, 60, 120, +\infty$. These were computed by numerically integrating

$$\int_0^{+\infty} P\{Z_1 \leq cy, \ldots, Z_k \leq cy\} \frac{\nu^{\nu/2}}{\Gamma\left(\frac{\nu}{2}\right)2^{(\nu/2)-1}} y^{\nu-1}e^{-\nu y^2/2}\,dy \tag{119}$$

for selected values of c and ν with the aid of National Bureau of Standards tables for the multivariate normal probabilities $P\{Z_i \leq cy, i = 1, \ldots, k\}$. Inverse interpolation in c and ν then provided the tables which are reproduced as Table IV of Appendix B. Dunnett believes the accuracy to be within a unit in the second decimal.

Dunnett's (1955) tables for the two-sided critical points $|d|^{\alpha}_{k,\nu}$ do not give these points exactly, but rather upper bounds for them. Tables of the multivariate normal probabilities $P\{|Z_i| \leq cy, i = 1, \ldots, k\}$ are not available, so Dunnett resorted to a bound (explained below) utilizing the exact results for the bivariate case [see Dunnett and Sobel (1955)].

Let $Y, Y_0, Y_1, \ldots, Y_k$ be independent, unit normal, random variables and let s^2 be an independent χ^2_ν/ν variable with ν d.f. Then, the two-sided confidence probability of a point c is

$$\begin{aligned} P\left\{\left|\frac{Y_i - Y_0}{s}\right| \leq c, i = 1, \ldots, k\right\} &= \int_0^{+\infty}\int_{-\infty}^{+\infty} P\{|Y_i - y| \leq cu, \\ &\qquad i = 1, \ldots, k \mid Y_0 = y, s = u\} p_{Y_0}(y)p_s(u)\,dy\,du \\ &= \int_0^{+\infty}\int_{-\infty}^{+\infty} [P\{|Y - y| \leq cu\}]^k p_{Y_0}(y)p_s(u)\,dy\,du \end{aligned} \tag{120}$$

where p_{Y_0}, p_s are the density functions of Y_0, s, respectively. The second equality in (120) holds by the independence of $Y_1, \ldots, Y_k$ and the fact that the variable Y has the same distribution as each Y_i. A well-known [Loéve (1963, p. 156)] moment inequality for a random variable X is $(E|X|^p)^{1/p} \geq (E|X|^q)^{1/q}$ for $p \geq q$, or for a function of two variables $(E|f(X_1,X_2)|^p)^{1/p} \geq (E|f(X_1,X_2)|^q)^{1/q}$. Application of this inequality to the last integral in (120) with $p = k$, $q = 2$ gives

$$P\left\{\left|\frac{Y_i - Y_0}{s}\right| \leq c,\ i = 1, \ldots, k\right\}$$
$$\geq \left[\int_0^{+\infty} \int_{-\infty}^{+\infty} [P\{|Y - y| \leq cu\}]^2 p_{Y_0}(y) p_s(u)\, dy\, du\right]^{k/2}$$
$$= \left[P\left\{\left|\frac{Y_i - Y_0}{s}\right| \leq c,\ i = 1, 2\right\}\right]^{k/2}. \tag{121}$$

The probability $P\{|Y_i - Y_0| \leq cs, i = 1, 2\}$ involves just the bivariate t distribution, and can be evaluated with the aid of the tables in Dunnett and Sobel (1954). Dunnett's (1955) tables for the two-sided $|d|^\alpha_{k,\nu}$ were constructed by finding (via inverse interpolation) the value of c for which the last expression in (121) was equal to $1 - \alpha$. Because of the inequality sign appearing in (121) the table entries are somewhat larger than the true $|d|^\alpha_{k,\nu}$, but the discrepancy is on the conservative side. It can be readily shown that the inequality (121) is sharper than the corresponding Bonferroni inequality.

More recent tables by Dunnett (1964) give $|d|^\alpha_{k,\nu}$ exactly for $\alpha = .05, .01$; $k = 1(1)12, 15, 20$; $\nu = 5(1)20, 24, 30, 40, 60, 120, \infty$. These values of $|d|^\alpha_{k,\nu}$ were obtained by numerical integration and are reproduced in Table IV of Appendix B. Dunnett (1964) also provides an interpolation scheme for approximating $|d|^\alpha_{k,\nu}$ when $m \neq n$.

6. MULTIPLE RANGE TESTS (DUNCAN)

The first multiple range test was proposed by Newman (1939). The basic idea was either due to Student (W. S. Gosset) or was an outgrowth of some of his ideas. Later, in 1952, Keuls proposed the same test in a journal obscure to statisticians, *Euphytica*. Duncan (1947, 1951, 1952, 1955) advocated the use of multiple stage tests, but with less conservative p-mean significance levels than the Newman-Keuls test. Duncan studied the properties of multiple stage tests in much greater detail than the earlier authors, and, consequently, his name seems to be more commonly associated with them.

6.1 Method

The basic design and probability structure of the problem is the same as in Sec. 1.1, that is, the one-way classification. Only the form of the test has been changed.

Let $\{Y_{ij}; i = 1, \ldots, r, j = 1, \ldots, n\}$ be r independent samples of n independently, normally distributed random variables with common variance σ^2 and expectations $E\{Y_{ij}\} = \mu_i, i = 1, \ldots, r, j = 1, \ldots, n$. The null hypothesis

$$H_0: \quad \mu_1 = \mu_2 = \cdots = \mu_r \tag{122}$$

is to be tested against the alternative

$$H_1: \quad \mu_i \not\equiv \mu_{i'} \qquad \forall\, i,i'. \tag{123}$$

If H_0 is to be rejected, statements must be made about which means are thought to differ.

The basic credo of multiple range tests is: the difference between any two means in a set of r means is significant provided the range of each and every subset which contains the given means is significant according to an α_p-level studentized range test where p is the number of means in the subset concerned.

The α_p-level studentized range test is conducted by comparing the range (divided by $s/\sqrt{n}$) of the p means involved with the critical value $q_{p,r(n-1)}^{\alpha_p}$ of the studentized range distribution (cf. Sec. 1).

The difference between the Newman-Keuls and Duncan multiple range tests lies in the choice of α_p, $p = 2, 3, \ldots, r$. For a fixed α (for example, .05 or .01) their respective choices are

$$\begin{aligned} &\text{Newman-Keuls:} && \alpha_p = \alpha \\ &\text{Duncan:} && \alpha_p = 1 - (1 - \alpha)^{p-1}. \end{aligned} \tag{124}$$

The properties and rationale of these choices will be discussed later in this section and in Sec. 6.4.

To illustrate the mechanics of performing these multiple stage tests, consider the following numerical example in which the figures have been rounded to one decimal place for simplicity of exposition. Let 16.1, 17.0, 20.7, 21.1, and 26.5 be the five ordered means from a one-way classification with $r = 5$ classes and $n = 5$ observations per class. Let the standard error of a mean be $s/\sqrt{5} = 1.2$, which will have 20 d.f.

Table 1

p	2	3	4	5
Newman-Keuls	3.0	3.6	4.0	4.2
Duncan	3.0	3.1	3.2	3.3

The 5 percent critical points for measuring a group of p means, $p = 2$, 3, 4, 5, for the Newman-Keuls and Duncan procedures are obtained from Tables I and V of Appendix B, and are exhibited in Table 1.

Multiplication of the Table 1 entries by $s/\sqrt{5} = 1.2$ gives the appropriate constants (Table 2) for evaluating the range of p means.

Table 2

p	2	3	4	5
Newman-Keuls	3.5	4.3	4.7	5.1
Duncan	3.5	3.7	3.8	3.9

Comparison of the differences between sample means with the entries in Table 2 is most easily accomplished in a routine mechanical fashion. List the five ordered means horizontally on a sheet of paper. For each p, starting with the largest (5) and proceeding to the lowest (2), compare each group of p consecutive ordered means. If the group is nonsignificant (i.e., does not exceed the Table 2 entry), draw an unbroken line underneath this group of means. If the group is significant (i.e., exceeds the entry), draw nothing. When this is finished for all p, only those means which are not connected by a line underneath them are judged to be significantly different.

To save time some comparisons need not be calculated. Whenever a group of means is underlined, there is no point in comparing the range of any subgroup with Table 2. By the credo of multiple range tests no subgroup within this group can be significant, since it is contained in a larger nonsignificant group.

This routine is performed step-by-step in Table 3 for the five previously listed means. The Newman-Keuls significance levels are used.

Table 3

$p = 5$	$26.5 - 16.1 = 10.4 > 5.1$; draw no line.
$p = 4$	$21.1 - 16.1 = 5.0 > 4.7$; draw no line. $26.5 - 17.0 = 9.5 > 4.7$; draw no line.
$p = 3$	$20.7 - 16.1 = 4.6 > 4.3$; draw no line. $21.1 - 17.0 = 4.1 < 4.3$; underline. $26.5 - 20.7 = 5.8 > 4.3$; draw no line.
$p = 2$	$17.0 - 16.1 = .9 < 3.5$; underline. $20.7 - 17.0$, $21.1 - 20.7$ } omit because $\underline{17.0 \quad 20.7 \quad 21.1}$. $26.5 - 21.1 = 5.4 > 3.5$; draw no line.

The finished display for the underlined and nonunderlined means is illustrated in (125).

$$\begin{array}{ccccc} 16.1 & 17.0 & 20.7 & 21.1 & 26.5 \\ \cline{2-4} \\ \cline{1-2} \end{array} \tag{125}$$

The conclusions to be drawn from (125) are as follows: 26.5 differs (significantly) from the other four means, 16.1 differs from the highest three means (20.7, 21.1, 26.5), and no other differences can be regarded as significant. Thus, 26.5 seems to be separated from the group consisting of the four lower means, and this group of four is heterogeneous in that the higher two means seem to differ from the lowest.

The routine for the Duncan significance levels is similar. The other set of entries from Table 2 gives the same results except for the changes listed in Table 4.

Table 4

$p = 3$	$21.1 - 17.0 = 4.1 > 3.7$; draw no line.
$p = 2$	$20.7 - 17.0 = 3.7 > 3.5$; draw no line. $21.1 - 20.7 = .4$; underline.

The display (126) incorporates the changes of Table 4.

$$\underline{16.1 \qquad 17.0} \qquad \underline{20.7 \qquad 21.1} \qquad 26.5 \tag{126}$$

The conclusion to be drawn from (126) is as follows: the means divide themselves into three groups, {16.1,17.0}, {20.7, 21.1}, and {26.5}. The groups differ (significantly), and there are no differences within either of the lower groups.

Whether the Newman-Keuls, Duncan, or Tukey studentized range procedure is employed, the method described above of underlining nonsignificant groups of means is a convenient tool for reporting experimental results in the literature. It has been advocated by Duncan and others. Unfortunately, it has not found widespread acceptance and therefore requires an explanation of its use whenever it is applied.

The property which characterizes and distinguishes the Newman-Keuls and Duncan multiple range tests is their p-mean significance levels. Since mean differences are involved rather than mean vs. theoretical value comparisons, the definition of p-mean significance levels will have to be changed somewhat from that given in Sec. 4 of Chap. 1.

For the two means μ_1, μ_2, let $D(\mu_1 \neq \mu_2)$ denote the decision $\mu_1 \neq \mu_2$. For the three means μ_1, μ_2, μ_3, let $D(\mu_1 \neq \mu_2 \cup \mu_1 \neq \mu_3 \cup \mu_2 \neq \mu_3)$ denote the decision that $\mu_i \neq \mu_{i'}$, $i,i' = 1, 2, 3$, that is, at least one of the means differs from the other two and they may all be different. For a group of m means the *two-mean* and *three-mean significance levels* for the sets μ_1, μ_2, and μ_1, μ_2, μ_3 are, respectively,

$$\begin{aligned}\alpha(\mu_1,\mu_2) &= \sup_{\mu_3,\ldots,\mu_m} P\{D(\mu_1 \neq \mu_2)|\mu_1 = \mu_2; \\ &\qquad\qquad \mu_3, \ldots, \mu_m \text{ arbitrary}\} \\ \alpha(\mu_1,\mu_2,\mu_3) &= \sup_{\mu_4,\ldots,\mu_m} P\{D(\mu_1 \neq \mu_2 \cup \mu_1 \neq \mu_3 \cup \mu_2 \neq \mu_3)| \\ &\qquad\qquad \mu_1 = \mu_2 = \mu_3; \mu_4, \ldots, \mu_m \text{ arbitrary}\}.\end{aligned} \tag{127}$$

The p-mean significance levels for $p > 3$ and other combinations of means are defined analogously. Speaking generally, a p-mean significance level for a group of p means is the probability of falsely rejecting the hypothesis that the p means are all equal, this probability being maximized over the remaining $m - p$ means. As pointed out in Sec. 4 of Chap. 1, for a given p the $\binom{m}{p}$ different p-means significance levels for different mean combinations almost always have the same value due to symmetry in the problem and test procedure.

The difference in performance of the Newman-Keuls and Duncan tests is pinpointed by the p-mean significance levels. For Newman-Keuls the p-mean significance levels for any p are equal to α, whereas for Duncan the p-mean significance levels for a given p are $1 - (1 - \alpha)^{p-1}$, $p = 2, 3, \ldots, m$ (see Sec. 6.4).

In the numerical illustration the Newman-Keuls p-mean significance levels are all equal to .05. For the Duncan procedure, however, they shift to

$$\begin{aligned}\alpha(\mu_1,\mu_2) &= .05 \\ \alpha(\mu_1,\mu_2,\mu_3) &= .0975 \\ \alpha(\mu_1,\mu_2,\mu_3,\mu_4) &= .1426 \\ \alpha(\mu_1,\mu_2,\mu_3,\mu_4,\mu_5) &= .1855.\end{aligned} \tag{128}$$

For any set of equal means within the five, the chance of wrongly deciding that some of them differ is less than .05 under Newman-Keuls, but only less than .05, .0975, .1426, or .1855 for Duncan, depending upon whether the set contains two, three, four, or five means.

The rationale for the choice of the Duncan p-mean significance levels will be given in Sec. 6.4. The increasing liberalism of the higher-order levels creates greater sensitivity for the Duncan procedure, but for this author, at least, this liberalism misses the fundamental aim of simultaneous inference. The pro and con arguments will be aired in Sec. 6.3.

Although unmentioned in the literature, multiple stage tests for the studentized maximum modulus and many-one t statistics can be constructed just as for the studentized range. The basic idea is the same:

as some means are declared significant, the critical point for significance of the remaining means is decreased to conform with the size of the remaining group.

The credo for the studentized maximum modulus problem would be: a sample mean in a set of k means is significantly different from its hypothesized value provided the maximum modulus of each and every subset which contains the given mean is significant according to an α_p-level studentized maximum modulus test where p is the number of means in the subset concerned.

A similar credo would hold for the many-one t statistics.

The levels α_p can be selected in any fashion, but the choice most logically is between nonsimultaneity (i.e., controlling each t statistic separately), Duncan, Newman-Keuls, and Tukey (i.e., the original tests of Secs. 4 and 5). These levels are listed in order of increasing conservatism.

As with the multiple range test, the multiple maximum modulus test can be conveniently handled in steps. Let there be a total of k means to be compared with their hypothesized values.

1. Declare significant all means which differ from their hypothesized values by an amount exceeding $|m|_{k,\nu}^{\alpha_k}$ (see Sec. 4). If there are no significant means, stop. If there are $k_1 > 0$ significant means, proceed to stage 2.
2. Declare significant all remaining means which differ from their hypothesized values by an amount exceeding $|m|_{k-k_1,\nu}^{\alpha_{k-k_1}}$. If there are none, stop. If there are $k_2 > 0$ significant means, proceed to stage 3.
3. Declare significant all remaining means which differ from their hypothesized values by an amount exceeding $|m|_{k-k_1-k_2,\nu}^{\alpha_{k-k_1-k_2}}$. Et cetera.

A similar routine can be used on the many-one t statistics.

6.2 Applications

Application of the Newman-Keuls or Duncan multiple range tests is limited to testing pairwise mean differences in balanced one-way classifications. The use of the studentized range requires independent means, based on equal numbers of observations, and an independent estimate of the variance.

Multiple range tests can be applied to testing row differences in balanced two-way, three-way, etc., classifications. Except for changes in the degrees of freedom of the error variance, these higher-way classifications correspond to one-way classifications as far as testing row effects are concerned.

It will be mentioned again that multiple range tests can only be used as *tests of hypotheses;* they are not adaptable to confidence region construction. The multistage character of these tests has no counterpart in the confidence domain.

Kramer (1956,1957) suggested the use of multiple range tests for unbalanced designs, analyses of covariance, etc. The correct variance is used for each separate statistic, but the test critical points are the studentized range points whose validity requires the assumption of independent, equivariance means. Consequently, all known distributional properties, significance levels, etc., go down the drain. Duncan (1957) attempted to bolster this extension by heuristically arguing and conjecturing that the actual significance levels are on the conservative side. No mathematical proof or numerical substantiation was provided, however. Until this is done, this author will remain skeptical of the use of these inexact tests.

Nemenyi (1963) has substantiated this extension in a specialized form which is not applicable here.

6.3 Comparison

The comparison between the Newman-Keuls and Duncan multiple range tests has essentially been made in the method section. Both employ the studentized range statistic, but allot different critical values in order to achieve their respective p-mean significance levels:

$$\begin{array}{ll} \text{Newman-Keuls:} & \alpha(\mu_1, \ldots, \mu_p) = \alpha \\ \text{Duncan:} & \alpha(\mu_1, \ldots, \mu_p) = 1 - (1 - \alpha)^{p-1}. \end{array} \qquad (129)$$

The Duncan levels are allowed to increase with the size of the group in the systematic manner (129) because:

1. This will give increased power to the test while affording greater protection than that provided by $r(r - 1)/2$ separate nonsimultaneous tests.
2. This gives levels consistent with a series of independent tests (see Sec. 6.4).
3. This resembles a Bayes solution for a related problem with independent hypotheses and additive loss functions [see Duncan (1955)].

Although the Duncan levels do not rise as rapidly as the nonsimultaneous separate test levels, they do increase with a fair amount of speed and soon exceed $\frac{1}{2}$. For example, at $\alpha = .05$ the 10-mean level is .37, and the 20-mean level is .62. To this author this violates the spirit of what simultaneous inference is all about, namely, to protect a multiparameter null hypothesis against any false declarations due to the large number of declarations required.

The loss functions for what the author of this book considers to be a simultaneous problem are not additive, since the statistician wants to prevent *any* false statements under nullity, and one or several incorrect statements should be regarded with just about the same amount of disfavor. Simultaneous problems involve situations where it is desired or required to afford high protection to the whole null hypothesis. The prior probability of the null hypothesis being true would either be an irrelevant concept, or would be known to be fairly high (and not the product of independent probabilities for the component hypotheses).

The Newman-Keuls levels provide a high degree of protection for the entire null hypothesis (viz., α), and this is the multiple range test this author favors. Moreover, it does not suffer from the overconservatism of the Tukey test caused by utilizing just a single critical value.

Perhaps the Tukey procedure should be illustrated to dispel any conceivable confusion. Consider the numerical example of Sec. 6.1. The Tukey test uses just the single percentile point of the studentized range $q_{5,20}^{.05} = 4.2$, which, multiplied by $s/\sqrt{5}$, gives the critical value 5.1. The only mean differences which exceed this are 26.5 compared with 16.1, 17.0, 20.7, 21.1. The data thus divide themselves into two groups, {16.1, 17.0, 20.7, 21.1} and {26.5}; the nonsignificant (underlined) groups of means are displayed in (130).

$$\underline{16.1 \qquad 17.0 \qquad 20.7 \qquad 21.1} \qquad 26.5 \tag{130}$$

For this example the Tukey five-mean significance level is equal to .05, but the four-, three-, and two-mean levels decrease from this to approximately .033, .019, and .007, respectively. In general, for a total group of m means, the Tukey p-mean significance levels are given by

$$P\{Q_{p,\nu} > q_{m,\nu}^{\alpha}\} \tag{131}$$

where $Q_{p,\nu}$ is a generic studentized range variable with parameters p,ν. The levels (131) decrease from α as p decreases.

A decrease in the p-mean significance levels below α is unnecessarily conservative. There seems to be no good reason to penalize the comparisons within subgroups to this extent. The Newman-Keuls test relaxes this stringency by keeping all the levels uniformly at α.

6.4 Derivation

There is precious little to be derived for multiple range tests. The studentized range has been discussed in detail in Sec. 1.

From a collection of m means a subset of p means has a chance of being declared heterogeneous only if at all prior stages the decision favors significance. If all strictly larger subgroups containing the p means are judged significant, then the significance of the subset will not be masked by the values of extraneous means. Continuance to the given subset of means (i.e., no intervening nonsignificance) will be automatically guaranteed if the means not included in the subset are set equal to $\pm\infty$. Thus,

$$\begin{aligned}\alpha(\mu_1, \ldots, \mu_p) &= P\{D(\mu_1 \neq \mu_2 \cup \cdots \cup \mu_{p-1} \neq \mu_p) | \mu_1 \\ &\qquad = \cdots = \mu_p; \mu_{p+1} = \pm\infty, \ldots, \mu_m = \pm\infty\} \\ &= P\{Q_{p,\nu} > q_{h,\nu}^{\alpha_p}\} \end{aligned} \tag{132}$$

where $h = p$ ($\alpha_p = \alpha$, $1 - (1 - \alpha)^{p-1}$, respectively) for the Newman-Keuls and Duncan tests, $h = m$ ($\alpha_p = \alpha$) for the Tukey test, and $h = 2$ ($\alpha_p = \alpha$) for nonsimultaneous tests. In (132) $Q_{p,\nu}$ is a generic studentized range variable. This is admittedly just a heuristic argument, but the mathematical stuffings could be provided.

The rationale behind the choice of the Duncan p-mean significance levels is simple enough. For the $\binom{p}{2}$ different pairwise comparisons between $\mu_1, \ldots, \mu_p$, there are $p - 1$ linearly independent linear combinations of the means $\mu_1, \ldots, \mu_p$. Or, in other words, for p means the space of contrasts has dimension $p - 1$. When a statistician uses $p - 1$ independent α-level tests for $p - 1$ separate hypotheses, his overall probability error rate for the joint null hypothesis is then $1 - (1 - \alpha)^{p-1}$. Therefore, Duncan argues, since testing pairwise comparisons corresponds to $p - 1$ independent (except for the common divisor s) tests, the appropriate significance level is $1 - (1 - \alpha)^{p-1}$.

This argument hinges on the premise that for independent tests the statistician will use a level α test on each. In rebuttal, this author would handle independent tests in this fashion, provided they were unrelated, but *if* the hypotheses were related, and *if* the combined hypotheses should have greater protection, he would not conduct the separate tests at the level α, irrespective of whether or not the test statistics were independent. The degree of protection should be based on the relatedness of the null hypotheses and their need for protection, and not on the probabilistic dependence (or independence) of the test statistics. Dependence of the test statistics should play an incidental, minor role, not a major one. Thus, to this author, independence (unrelatedness) of the hypotheses is the primary concept, whereas for Duncan it is independence of the test statistics.

6.5 Distributions and tables

The only distribution involved in multiple range tests is the studentized range which was thoroughly discussed in Sec. 1.

The Newman-Keuls test only requires the critical values of the studentized range distribution at standard percentile points, so the tables listed in Sec. 1.5 and given in Table I of Appendix B should provide the necessary points.

The Duncan multiple range test, however, requires unorthodox (beatnik?) percentile points, and therefore demands additional tables. The original tables were published by Duncan (1955); the computations were based on the tables of Pearson and Hartley (1943) and the work of Beyer (1953). The tables include the critical points to two decimal places for $\alpha = .05, .01; p = 2(1)10(2)20, 50, 100; \nu = 1(1)20(2)30, 40, 60, 100, +\infty$. In locating these and later tables the reader should be advised that in the literature Duncan's multiple range test is titled "A New Multiple Range Test."

Later and more accurate tables have been published by Harter (1960b), who found that Duncan's original tables contained a number of inaccuracies. Harter's tables are based on the extensive tables of Harter and Clemm (1959) and Harter, Clemm, and Guthrie (1959). The accuracy of Harter's tables is estimated to be within a unit for the last digit of the four significant digits reported. The critical values are given for $\alpha = .10$, .05, .01, .005, .001; $p = 2(1)20(2)40(10)100$; $\nu = 1(1)20, 24, 30, 40, 60, 120, +\infty$. The tables for $\alpha = .05, .01$ are reproduced in the back of this book in Table V.

7 LEAST SIGNIFICANT DIFFERENCE TEST (FISHER)

This test was proposed by Fisher (1935) for locating those treatment effects which were creating a significant analysis of variance. It is a grandfather among simultaneous tests, but has not been put to pasture yet. Among its offspring are the multiple range and multiple F tests.

7.1 Method

The least significant difference (LSD) test can be applied to any (fixed effects) experimental design or multiple linear regression model. The test has two stages:

Stage 1 Test the null hypothesis (whatever it be) by the appropriate α-level F test.

(*a*) If the F value is nonsignificant, decide in favor of the null hypothesis.

(*b*) If the F value is significant, proceed to stage 2.

Stage 2 Test each single comparison (contained within the null hypothesis) by the appropriate α-level t test.

(*a*) If the t value is nonsignificant, decide the comparison is not significantly different from what the null hypothesis dictates.

(*b*) If the t value is significant, judge the comparison to be significant.

To remove any ambiguity, this general principle can best be amplified by examples. For a one-way classification with r populations, consider the null hypothesis

$$H_0: \quad \mu_1 = \mu_2 = \cdot \cdot \cdot = \mu_r. \tag{133}$$

Stage 1 of the LSD test compares the F statistic

$$F = \frac{N - r}{r - 1} \cdot \frac{\sum_1^r n_i(\bar{Y}_{i.} - \bar{Y}_{..})^2}{\sum_i \sum_j (Y_{ij} - \bar{Y}_{i.})^2} \tag{134}$$

$\left(\text{where } N = \sum_1^r n_i\right)$ with the critical F percentile point $F^{\alpha}_{r-1,N-r}$. Nonsignificance and curtailment of further testing occurs when $F \leq F^{\alpha}_{r-1,N-r}$; significance and advancement to stage 2 occurs when $F > F^{\alpha}_{r-1,N-r}$. At stage 2 any pairwise mean comparison of interest to the experimenter and statistician can be tested by comparing the relevant t statistic

$$T_{ii'} = \frac{\bar{Y}_{i.} - \bar{Y}_{i'.}}{s\sqrt{\dfrac{1}{n_i} + \dfrac{1}{n_{i'}}}} \tag{135}$$

with the critical t percentile point $t^{\alpha/2}_{N-r}$. Significance of the difference depends on whether $|T_{ii'}|$ exceeds or falls short of $t^{\alpha/2}_{N-r}$.

All $\binom{r}{2}$ pairs of means can be tested at stage 2, and more general contrasts as well, if stage 1 is significant. It can happen that the F test gives a significant value for the first stage, but none of the pairwise differences at the second stage are significant. This will happen when the sample point falls outside the elliptical F region, but inside the polyhedral region formed from the $\binom{r}{2}$ t tests. The analogous situation for the case of means vs. theoretical values is illustrated by the four crosshatched regions in Fig. 1.12.

For a multiple linear regression model ($\mathbf{Y} = \mathbf{X}\boldsymbol{\beta} + \mathbf{e}, \mathbf{X}\ n \times p$, full rank) the null hypothesis H_0: $\boldsymbol{\beta} = \mathbf{0}$ is tested by the F statistic

$$F = \frac{1}{p} \frac{\hat{\boldsymbol{\beta}}^T(\mathbf{X}^T\mathbf{X})\hat{\boldsymbol{\beta}}}{s^2} \tag{136}$$

which has an F distribution with parameters $p,n\text{-}p$. If this first stage indicates significance, then each individual regression coefficient could be tested by its own t statistic, that is, for β_i,

$$T_i = \frac{\hat{\beta}_i - 0}{s\sqrt{\mathbf{XX}^{ii}}} \tag{137}$$

where $\mathbf{XX}^{ii}$ is the ith diagonal element of $(\mathbf{X}^T\mathbf{X})^{-1}$.

As with the previous example it can happen that stages 1 and 2 are contradictory. Stage 1 might produce a significant value, but none of the regression coefficients $\hat{\beta}_1, \ldots, \hat{\beta}_p$ are significant by t tests. In this event, the statistician will just have to calm himself and his clients, and take heart in the realization of what has happened.

The title *least significant difference* (LSD) stems from the critical t percentile ($t_\nu^{\alpha/2}$ with ν d.f.) used in stage 2. This is the smallest or *least* value a difference must exceed in order to be declared a *significant difference,* and is the appropriate critical value when the difference is considered singly. If the difference is a member of a family, then the critical value will most likely be increased, or a prior stage of testing interposed. Tukey applies the term *wholly significant difference* (WSD) to the critical value of the studentized range ($q_{r,\nu}^\alpha$) which, when exceeded, guarantees significance for any difference, whether regarded singly or as a family member of $\binom{r}{2}$ mean differences. *Wholly,* of course, refers to being significant for the whole group of mean differences. The WSD is sometimes labeled the *honestly significant difference* (HSD).

7.2 Applications

The spectrum of applications of the LSD test is quite broad, since it applies to any linear hypothesis for any normal, linear model. The likelihood ratio tests are the F test for the whole hypothesis, and the t test for any component hypothesis, which makes the LSD test a natural one to use. It could be applied to any of the six testing problems discussed in the preceding sections of this chapter, and may be applied where these specialized techniques are inapplicable because of imbalance.

7.3 Comparison

The great virtue of the LSD test is its convenience and simplicity. With the current general availability of electronic digital computers, there is little hand numerical work required in the application of this test. Canned programs for the analysis of variance and regression frequently include in their outputs the value of the F statistic required in stage 1, and many of the t statistic values required in stage 2. Some programs even include the critical F and t percentile points, and perform the tests for you. (While the machines are grinding away, the statisticians can do arithmetical exercises to keep fit for the day the electricity goes off.)

Even without computer programs the ease of the test is apparent. The required statistics are the standard ones. Only two critical values (F^α and t^α) are needed from tables, in contrast to the number required for the multiple range tests. Furthermore, F and t tables are available practically everywhere, and are mentally engraved in some statisticians.

As compared with the multiple range tests, the LSD test has far greater versatility. It is not handcuffed when the one-way design is unbalanced, and is applicable to more general problems. The LSD test also has an historical edge on the multiple range tests. Some consumers (nonstatisticians) of simultaneous statistical tests have been weaned on the LSD test. For them to change now may not be emotionally possible.

The LSD test has sensitivity as good or better than any competitor. The stage 1 F test has been known over the years to have good power, and once the second stage is reached, all simultaneous conservatism is dropped. The tests are nonsimultaneous, individual t tests, and behave as such.

This abandonment of conservatism at the second stage creates poor, low-order significance levels. The stage 1 F test of m means (whether for differences or theoretical values) keeps the m-mean significance level (the family probability error rate) at α.† However, the p-mean significance levels for p lower than m return (increase) to those for nonsimultaneous tests, since the free, unspecified means can be selected so as to ensure a significant F value with probability arbitrarily close to one.

The preliminary F test guards against falsely rejecting the null hypothesis when the null hypothesis is true. However, when, in fact, the null hypothesis is false and likely to be rejected, the second stage of the LSD gives no increased protection to that part (if any) of the null hypothesis which still remains true.

7.4 Derivation

There is nothing to be derived.

† The level is α provided some type of nonnull decision is reached for sample points significant at stage 1, but not at stage 2 (i.e., points in the shaded regions of Fig. 1.7). If these points are included in the null region, then the m-mean significance level is somewhat lower than α.

7.5 Distributions and tables

The only tables required in conducting the LSD test are the F and t tables. Sources of F tables were discussed in Sec. 2.5, and t tables in Sec. 3.5.

8 OTHER TECHNIQUES

The techniques which are grouped together in this section are felt by the author to be worthwhile, but not quite as important, perhaps, as the seven principal techniques already discussed. This is because they are either not quite in the mainstream of simultaneous inference, or have been somewhat superseded in usage by other methods. They should not be overlooked, but the discussion will be briefer and less detailed than for the preceding techniques. The division into subsections (Method through Distributions and Tables) will therefore be abandoned, and just a single subsection will be devoted to each technique.

Similar sections with the same title and purpose will appear in Chaps. 3 to 5.

8.1 Tukey's gap-straggler-variance test

The gap-straggler-variance test was proposed by Tukey (1949) in the early days of modern simultaneous inference. This technique does not seem to have been extensively applied, perhaps because it is a bit cumbersome, and perhaps because its specific properties are unknown. Simpler, speedier techniques have supplanted it.

The aim of this test is to locate differences between means in a (balanced) one-way classification. Such differences manifest themselves in three ways:

1. "There is an unduly wide gap between adjacent variety means when arranged in order of size.
2. One variety mean straggles too much from the grand mean.
3. The variety means taken together are too variable."

Thus, the detection of mean differences boils down to testing for "(1) excessive gaps, (2) stragglers, (3) excess variability."

These three types of heterogeneity are tested as follows:

1. Excessive gaps: the natural criterion is the studentized maximum gap between ordered mean values. Unfortunately, this distribution does

not have any simple form which is amenable to calculation, and therefore a substitute must be found.

Tukey proposed the use of two-sided percentage points of the t distribution as critical points. He gave evidence (Monte Carlo and normal integration) to suggest that these percentage points will be on the conservative side.

2. Stragglers: the test for stragglers employs the studentized extreme deviate from the sample mean. This criterion is the same as that applied in the detection of outliers and will be discussed in detail in Sec. 1 of Chap. 6.

Only one table of this statistic [Nair (1948)] existed at the time of Tukey's article. Tukey suggested an alternative, approximate procedure (requiring only the normal distribution) which might have been preferable speedwise and which could be used in the absence of Nair's tables. Since then, bigger, better, and more readily available tables of the extreme studentized deviate have appeared, so the necessity of Tukey's approximation seems to be eliminated. Table look-up now seems to be faster than computing Tukey's approximation.

3. Excess variability: guess what?—the F test.

These tests are applied in the following order.

Stage 1 Divide the original set of ordered means into subgroups by regarding any significant gap between two means as a separation between subgroups. All further tests apply to each subgroup separately.

Stage 2 By the straggler test separate from a subgroup of three or more means any straggler means. If a straggler is found, reapply the test to the remaining means (subgroup minus one) to see if another mean is now straggling. Continue this process until there are no new stragglers.

Consider all straggler means separated from the right of the subgroup to be a new subgroup. Similarly, on the left. If either of these new subgroups contains three or more means, apply the straggler test to this new subgroup as at the beginning of stage 2.

Continue this process until no new stragglers or subgroups are created.

Stage 3 Apply the F test to each subgroup. Tukey is not specific as to what statements should be made in the event of significant variability other than a general pronouncement of heterogeneity.

That (in a nutshell?) is the Tukey gap-straggler-variance test. The procedure is simple enough, but when compared with the multiple range tests, it seems a little lengthy and Rube Goldbergish. The operating characteristics, significance levels, etc., of the *whole* test are unknown and unformulated. Tukey's intuition (which is usually a pretty good computer) told him that the frequency of false positives should be in the range 1.2α to 1.6α, where α is the significance level for each individual test. Since its innovation, some defects have been reported in the procedure, and today it is regarded somewhat as a relic.

For a discussion the reader is referred to Tukey (1953b), if a copy is available to him.

8.2 Shortcut methods

In actually performing the Tukey studentized range procedure the only operation which requires any time whatsoever is the calculation of s^2 (the residual variance) for the estimate of σ^2. A quicker estimate of σ^2 is the range, whose efficiency, distributional properties, etc., are not too bad. For a quick-and-dirty statistic, Tukey and the Princeton statistical research group proposed using a range estimate in the denominator of the studentized range. This converts the studentized range into a ratio of ranges, the numerator from between samples, the denominator within samples.

For a balanced one-way classification this proposal is straightforwardly executed. For each class calculate the mean $\bar{Y}_{i.}$ and the range R_i (unadjusted by any constants). The total range TR is the sum of the ranges, that is, $\sum_1^r R_i$. Then, the statistic which replaces the studentized range $Q_{r,r(n-1)}$ (see Sec. 1.1 of Chap. 1) is

$$\tilde{Q}_{r,n} = \frac{\max_{i,i'} \{|\bar{Y}_{i.} - \bar{Y}_{i'.}|\}}{TR/n} = \frac{\text{range } \{Y_{1.}, \ldots, Y_{r.}\}}{TR} \tag{138}$$

where $Y_{i.} = \sum_{j=1}^{n} Y_{ij}$.

Critical percentile points for the statistic $\tilde{Q}_{r,n}$ have been computed by Tukey and the Princeton group. In published form the 5 percent and 1 percent tables can be found in Tukey (1951,1953a), Federer (1955), and Kurtz, et al. (1965a). In unpublished form they can be found in Tukey (1953b,1954), Link and Wallace (1952), and other Princeton reports. The newest and best tables are Kurtz, et al. (1965a).

The shortcut idea can be extended to the two-way classification as well. In a two-way classification with equal multiple observations per cell, each cell can be regarded as a class, and the whole design treated as a one-way classification with $r \times c$ classes. The aforementioned tables for the one-way classification can be used to make comparisons between cells or combinations of cells. In a two-way classification with just a single observation per cell, the within sample ranges must be computed from differences between row (or column) values. Because of the mandatory choice between rows and columns for forming differences, there is a lack of uniqueness in the method. For the author of this book there is a lack of appeal for the two-way shortcut. Required tables can, however, be found in Tukey (1951;1953a,b;1954), Link and Wallace (1952), and Kurtz, et al. (1965a) for those who want to use them.

8.3 Multiple F tests

The spirit of multiple F tests is the same as that of multiple range tests, the only difference being that F tests are used instead of range tests.

The basic credo of multiple range tests can be restated for multiple F tests: the difference between any two means in a set of r means is significant provided the (between sample) variance of each and every subset which contains the given means is significant according to an α_p-level F test where p is the number of means in the subset concerned.

As with the range tests, the α_p levels can be chosen in any fashion. Duncan recommends $\alpha_p = 1 - (1 - \alpha)^{p-1}$. They could also be set at $\alpha_p \equiv \alpha$ in analogy with Newman-Keuls, or they could be anything the consumer wishes. For the Scheffé procedure the levels are very conservative, decreasing below α as p decreases.

A disadvantage of multiple F tests is the amount of computation involved. The time required to compute the numerator of the appropriate F statistic is considerably greater than the time needed to mentally calculate the corresponding range. The prospect of having to perform a number of F tests to analyze a single, one-way classification is somewhat unappealing. Proponents of multiple F tests argue that it really does not take that much more time to compute a variance than a range (especially for machines), and, frequently, only very few of the total number of possible tests actually have to be carried out. This happens because testing is stopped by nonsignificant values, and some variances are quite obviously significant without ever being calculated. Nonetheless, the fact remains that the appeal of simplicity and ease is gone.

On the positive side of the ledger for multiple F tests can be listed the indifference of F tests to balance in a design. Whereas the studentized range is restricted to balanced one-way classifications, the F test can be applied to any (balanced or unbalanced) one-way classification.

The p-mean significance levels of multiple F tests are not necessarily

α_p, which is the case for multiple range tests. This happens because there are sample points which, for instance, have a significant F value for all r means, but no significant lower-order F values. If these points are included in the null region along with those points having a nonsignificant F value, the actual r-mean significance level is less than α_r.

This situation is illustrated in Fig. 1.12 for the modulus (rather than difference) problem with $\alpha_1 = \alpha_2 = .05$. Since only two means are involved and $t^2 \sim F$, the multiple F test is identical to the LSD test. Points in the four corner crosshatched regions have a significant χ^2 value for the pair, but individually each coordinate is nonsignificant. The total null region for the multiple F test consists of the circle plus the four crosshatched regions. The union is a circle-star region with probability exceeding $1 - \alpha_2$.

For the Duncan levels $\alpha_1 = .05$, $\alpha_2 = .0975$ the multiple F test has a circle with radius 2.158. Although this circle is somewhat smaller, the null region is still a circle-star, and the same problem (of significance levels unequal to α_p) exists.

To cope with this, Duncan proposed a variant of the multiple F test which he called the *multiple comparisons test.* Simply stated for Fig. 1.12, the variant is to declare significant any linear combination of coordinates of points in the four crosshatched regions which has a significant t value and is of interest. Graphically, these linear combinations are represented by pairs of parallel lines which lie at a distance 1.96 from the origin and do not bracket the point (Y_1, Y_2) between them.

For differences this criterion for significance can be formalized as follows: a contrast is significantly different from zero provided the variance of each and every subset which contains all the means involved in the contrast (with nonzero coefficients) is significant according to an α_p-level F test (where p is the number of means in the subset concerned), *and* the contrast differs significantly from zero according to an α_1-level t test.

For greater detail the reader is referred to Duncan (1955) and Federer (1955, p. 32). The original work on multiple F tests and the multiple comparisons test appeared in Duncan (1947,1951,1952), but these sources are less readily available.

8.4 Two-sample confidence intervals of predetermined length

Stein (1945) created a two-sample method of obtaining a test of a univariate normal mean μ whose power was independent of the variance σ^2, or equivalently, of obtaining a confidence interval for μ of predetermined length.

The method for confidence intervals is described as follows. From an initial sample of independent normal variables $Y_1, \ldots, Y_n$ compute the sample variance s^2. For a confidence interval on μ of length less than $2l$, choose $c = (l/t_{n-1}^{\alpha/2})^2$, and define

$$N = \max\left\{\left\langle\frac{s^2}{c}\right\rangle + 1,\, n\right\}.\dagger \tag{139}$$

Take the additional observations $Y_{n+1}, \ldots, Y_N$ (if $N > n$). For the total sample the random variable

$$T_{n-1} = \frac{\left(\frac{1}{N}\sum_{1}^{N} Y_i\right) - \mu}{s/\sqrt{N}} \tag{140}$$

has a t distribution with $n - 1$ d.f. and can be used to construct a confidence interval on μ.

Note that the s in (140) is the same s as in (139) and is calculated from just the initial sample. Since $\sqrt{N} \geq st_{n-1}^{\alpha/2}/l$ by (139), the confidence interval based on (140) has total length less than $2l$.

To obtain a confidence interval of length *exactly* $2l$, the total sample size N and the average value of the Y_i's should be defined slightly differently. For details, proofs, and a discussion of the corresponding tests, the reader is referred to the original source.

Healy (1956) extended these ideas to simultaneous confidence intervals. The proofs hinge on the same conditional probability argument originally used by Stein. Healy handled the confidence interval problems covered by the studentized maximum modulus, the studentized range, and the F statistic.

Studentized maximum modulus Let $\{Y_{ij}; i = 1, \ldots, k, j = 1, \ldots, n_i\}$ be k independent samples of independent normal variables with unknown means $\mu_1, \ldots, \mu_k$. Let s^2 be the χ^2 estimator of the unknown σ^2 with $N - k$ $\left(N = \sum_{1}^{k} n_i\right)$ d.f. obtained from the k samples.

If the experimenter wishes the interval for μ_i to have length less than $2l_i$, $i = 1, \ldots, k$, determine the constants $c_i = (l_i/|m|_{k,N-k}^{\alpha})^2$, $i = 1, \ldots, k$, where $|m|_{k,N-k}^{\alpha}$ is the upper 100α percent point of the studentized maximum modulus distribution with parameters $k, N - k$. Enlarge the samples to the total sizes

$$N_i = \max\left\{\left\langle\frac{s^2}{c_i}\right\rangle + 1,\, n_i\right\} \qquad i = 1, \ldots, k \tag{141}$$

for the k populations, respectively. Then, with probability $1 - \alpha$,

$$\mu_i \in \left(\frac{1}{N_i}\sum_{j=1}^{N_i} Y_{ij}\right) \pm \frac{|m|_{k,N-k}^{\alpha}s}{\sqrt{N_i}} \qquad i = 1, \ldots, k \tag{142}$$

where s is the original s computed from the initial samples.

† $\langle a \rangle$ denotes the greatest integer less than a.

Studentized range Consider the same initial samples as above with $n_1 = \cdots = n_k = n$. If the client desires confidence intervals on the pairwise differences of length less than $2l$, calculate the constant $c = (l/q^{\alpha}_{k,k(n-1)})^2$, where $q^{\alpha}_{k,k(n-1)}$ is the upper studentized range α percentile point, and expand each sample to the size

$$N = \max\left\{\left\langle\frac{s^2}{c}\right\rangle + 1, n\right\}. \tag{143}$$

Then, with probability $1 - \alpha$,

$$\mu_i - \mu_{i'} \in \bar{Y}_{i\cdot} - \bar{Y}_{i'\cdot} \pm \frac{q^{\alpha}_{k,k(n-1)}s}{\sqrt{N}} \qquad i,i' = 1, \ldots, k \tag{144}$$

where $\bar{Y}_{i\cdot} = \left(\sum_{j=1}^{N} Y_{ij}/N\right)$.

Without any change in the probability level $1 - \alpha$, intervals on more general contrasts could be appended to (144), but their lengths would not necessarily be less than $2l$ since the factor $\frac{1}{2}\sum_1^k |c_i|$ would have to be included on the right (see Sec. 1.1).

F statistic For the balanced one-way classification, simultaneous confidence intervals of prescribed length can be constructed for all linear combinations $\sum_1^k l_i\mu_i$ through the F statistic.

Let the initial samples be $\{Y_{ij};\, i = 1, \ldots, k,\, j = 1, \ldots, n\}$, and suppose the goal is to have an interval of length less than $2l$ for any linear combination with $\sum_1^k l_i^2 = 1$. Then, for $c = l^2/kF^{\alpha}_{k,k(n-1)}$, expand the total sample size in each class to

$$N = \max\left\{\left\langle\frac{s^2}{c}\right\rangle + 1, n\right\}. \tag{145}$$

This will give

$$\sum_1^k l_i\mu_i \in \sum_1^k l_i\bar{Y}_{i\cdot} \pm (kF^{\alpha}_{k,k(n-1)})^{\frac{1}{2}}s\left(\sum_1^k l_i^2\right)^{\frac{1}{2}} \qquad \forall\, \boldsymbol{l} \in \mathfrak{L}_{lc} \tag{146}$$

with probability $1 - \alpha$.

For a normalized linear combination $\boldsymbol{l}$ with $\sum_1^k l_i^2 = 1$, the interval in (146) has length less than $2l$ by virtue of (145). Note that, just as in the two preceding applications, the means $(\bar{Y}_{i\cdot})$ are computed from the total sample, but the standard deviation (s) is computed from just the initial sample.

8.5 An improved Bonferroni inequality

A proof of the following lemma (in less general form) can be found in Kimball (1951).

Lemma 5 Let $g_i(x)$, $i = 1, \ldots, k$, be nonnegative, integrable, increasing functions of x. Then, for any cdf $F(x)$,

$$E\left\{\prod_1^k g_i(X)\right\} = \int_{-\infty}^{+\infty} \prod_1^k g_i(x)\, dF(x)$$
$$\geq \prod_1^k \int_{-\infty}^{+\infty} g_i(x)\, dF(x) = \prod_1^k E\{g_i(X)\}. \tag{147}$$

Kimball shows how (147) can be used to improve on the Bonferroni inequality (1.13) in certain statistical problems. The inequality (147) is, in essence, an extension of the independence equality (1.12) to these dependent cases.

This inequality is useful in problems where k ratios, $U_i = V_i/W$, $i = 1, \ldots, k$, are to be bracketed. It is assumed the numerators U_i are independent, the denominators are the same positive statistic W, and the numerators and denominators are independent. For the intervals $I_i = (a_i,b_i)$ with $-\infty \leq a_i \leq 0$, $0 \leq b_i \leq +\infty$, $i = 1, \ldots, k$,

$$P\{U_i \in I_i, i = 1, \ldots, k\} = \int_0^{+\infty} P\{U_i \in I_i, \forall i | W = w\}\, dF_W(w)$$
$$= \int_0^{+\infty} \prod_1^k P\{a_i w < V_i < b_i w\}\, dF_W(w)$$
$$\geq \prod_1^k \int_0^{+\infty} P\{a_i w < V_i < b_i w\}\, dF_W(w)$$
$$= \prod_1^k P\{U_i \in I_i\}. \tag{148}$$

In (148) the inequality follows from (147), since the probabilities $P\{a_i w < V_i < b_i w\}$ are nonnegative, bounded, increasing functions of w.

The inequality (148) is strictly better than the corresponding Bonferroni inequality, that is,

$$P\{U_i \in I_i, i = 1, \ldots, k\} \geq 1 - \sum_1^k P\{U_i \notin I_i\} \tag{149}$$

because of the easily verified algebraic inequality

$$\prod_1^k (1 - \alpha_i) > 1 - \sum_1^k \alpha_i \qquad 0 < \alpha_i < 1,\ i = 1, \ldots, k(k > 1). \tag{150}$$

Applications of (148) include, for instance, an (fixed effects) analysis of variance in which a number of independent numerator sums of squares are to be compared with the same error sum of squares. In a balanced two-way classification with row (R), column (C), and interaction (I) effects, this gives

$$P\left\{\frac{MSR}{MSE} \leq F^{\alpha_1}, \frac{MSC}{MSE} \leq F^{\alpha_2}, \frac{MSI}{MSE} \leq F^{\alpha_3}\right\} \geq (1 - \alpha_1)(1 - \alpha_2)(1 - \alpha_3) \tag{151}$$

where the F^{α_i}, $i = 1, 2, 3$, are based on the appropriate degrees of freedom.

Another application involves a set of t statistics with independent numerators but common denominators. If $Y_1, \ldots, Y_k$ are independent, $N(\mu_i,\sigma^2)$ variables, and s^2 is an independent χ^2 estimator of σ^2 with ν d.f., then

$$P\{\mu_i \in Y_i \pm st_\nu^{\alpha_i/2}, i = 1, \ldots, k\} \geq \prod_1^k (1 - \alpha_i). \tag{152}$$

If $\alpha_i \equiv \alpha$, then an exact probability can be obtained from the studentized maximum modulus distribution. However, in circumstances where tables of the studentized maximum modulus are unavailable, or unequal error rate allocation is desired, the inequality (152) gives a sharper result than the Bonferroni inequality.

Dunnett and Sobel (1955) used the inequality (147) in computing some tables for their multivariate analog of the t distribution.

9 POWER

The statistics utilized in this chapter in the construction of confidence intervals were the F and t statistics, the studentized range, the studentized maximum modulus, and the many-one t statistics. The question of the power of these statistics naturally arises when they are viewed as tests of their respective null hypotheses.

The corresponding multistage significance tests employ the same statistics, but vary the critical point parameters at the different stages. Because of their hierarchical structure, the power for these tests could be defined at different levels of the multiple stages. However, it seems fruitless to pursue this in detail, since the simplest problem of determining the probability of rejecting the overall null hypothesis under an alternative has been solved only for the F and t statistics.

This last sentence essentially summarizes our knowledge on power for multiple comparisons techniques. The power functions of the F and t statistics are pretty well known today. Those for the studentized range, the studentized maximum modulus, and the many-one t statistics are unknown. This is due to the obstacle that the power functions of these latter statistics are not functions of single noncentrality parameters which are simple functions of the theoretical means, as in the case of the F statistic.

A central F statistic arises in practice from a ratio of two independent quadratic forms, $Q_1 = \mathbf{Y}^T\mathbf{A}_1\mathbf{Y}$ and $Q_2 = \mathbf{Y}^T\mathbf{A}_2\mathbf{Y}$, which have central $\sigma^2\,\chi^2$ distributions.[1] The F statistic is $F = r_2Q_1/r_1Q_2$ where r_1, r_2 are the d.f. or ranks of Q_1, Q_2, respectively. In normal, linear models the denominator Q_2/σ^2 retains a central χ^2 distribution under any choice of the mean value $\boldsymbol{\mu}$ of $\mathbf{Y}$ permitted by the basic model. Under alternative hypotheses the numerator Q_1/σ^2, however, assumes a noncentral χ^2 distribution with parameters r_1 and $\boldsymbol{\mu}^T\mathbf{A}_1\boldsymbol{\mu}/\sigma^2$. This means that the power function of the F statistic is a function of r_1, r_2, α, and $\boldsymbol{\mu}^T\mathbf{A}_1\boldsymbol{\mu}/\sigma^2$.

The list of symbols which have been used for the noncentrality parameter $\boldsymbol{\mu}^T\mathbf{A}_1\boldsymbol{\mu}/\sigma^2$ seems about as large as the Greek alphabet, lowercase plus capital. No one of the multitude has succeeded to universal acclaim. Also, some authors like to multiply the ratio $\boldsymbol{\mu}^T\mathbf{A}_1\boldsymbol{\mu}/\sigma^2$ by various constants ($\frac{1}{2}$ being a favorite) in the definition of the noncentrality parameter. Throughout this monograph the choice will be $\delta^2 = \boldsymbol{\mu}^T\mathbf{A}_1\boldsymbol{\mu}/\sigma^2$ with the numerator $\boldsymbol{\mu}^T\mathbf{A}_1\boldsymbol{\mu}$ sometimes labeled Δ^2.

For a t statistic the form of δ^2 simplifies. If $t = Y/s$ where Y is $N(\mu,\sigma^2)$ and $\nu s^2/\sigma^2$ is an independent χ^2 variable with ν d.f., then the noncentrality parameter of the distribution of t is $\delta = |\mu|/\sigma$.

The reader is advised to check very carefully the definition of the noncentrality parameter in any table or book he is trying to use in conjunction with this monograph.

The overall power function of the Scheffé F projections is readily obtainable from existing tables. As a test of a linear null hypothesis the Scheffé technique is identical to the standard F test. A confidence interval which does not contain the hypothesized value of its linear combination will exist if and only if the F statistic exceeds its null critical point.

In the notation of Sec. 2, the noncentrality parameter of

$$F = \frac{1}{d}\,\frac{(\mathbf{L}\hat{\boldsymbol{\beta}} - \boldsymbol{\gamma}^0)^T(\mathbf{L}(\mathbf{X}^T\mathbf{X})^{-1}\mathbf{L}^T)^{-1}(\mathbf{L}\hat{\boldsymbol{\beta}} - \boldsymbol{\gamma}^0)}{s^2} \tag{153}$$

which tests the null hypothesis

$$H_0\colon\ \mathbf{L}\boldsymbol{\beta} = \boldsymbol{\gamma}^0 \tag{154}$$

is $$\delta^2 = \frac{(\boldsymbol{\gamma} - \boldsymbol{\gamma}^0)^T(\mathbf{L}(\mathbf{X}^T\mathbf{X})^{-1}\mathbf{L}^T)^{-1}(\boldsymbol{\gamma} - \boldsymbol{\gamma}^0)}{\sigma^2} \qquad \text{where } \boldsymbol{\gamma} = \mathbf{L}\boldsymbol{\beta}. \tag{155}$$

[1] For conditions on the symmetric matrices $\mathbf{A}_1$ and $\mathbf{A}_2$ and the variables $\mathbf{Y}$ which guarantee these properties, the reader is referred to Graybill (1961, chap. 4) or Graybill and Marsaglia (1957).

For a choice of $\boldsymbol{\beta}$, or $\boldsymbol{\gamma}$, under the alternative, the power is

$$1 - \beta = P\{F > F^{\alpha}_{d,n-p}\},$$

where F has a noncentral F distribution with parameters d, $n - p$, δ. This can be obtained easily from the Pearson and Hartley (1951) charts or the Fox (1956) charts. Tables are also available in Tang (1938) and Lehmer (1944). When $d = 1$ (that is, the t test) the tables of Neyman and Tokarska (1936) can also be used. Interpolation, either visual or numerical, is likely to be necessary in any of these charts or tables due to the discreteness of r_1, r_2, α, and δ or β.

These charts and tables conveniently appear in places other than their original sources. For instance, Scheffé (1959) contains the Pearson-Hartley and Fox charts, Owen (1962) contains the Fox charts, and Kempthorne (1952) contains the Tang tables.

In an experimental design where δ^2 involves constants depending on the alternative and the adjustable sample size, it is possible to determine the sample size to achieve a specified power $1 - \beta$ for a prescribed alternative $\boldsymbol{\beta}$ or $\boldsymbol{\gamma}$. Usually the sample size will be a factor in δ^2 and r_2 (error d.f.) which necessitates a trial-and-error search in the tables rather than a direct look-up. In the special case of the t test, tables are available [Owen (1962)] for a direct look-up.

It should be borne in mind that even though the sample size is determined to give power $1 - \beta$ for the alternative $\boldsymbol{\gamma}$, the probability is less than $1 - \beta$ of selecting as significant at least one of the coordinates of $\hat{\boldsymbol{\gamma}}$. The null hypothesis can be rejected by the Scheffé technique without any of the estimates $\hat{\gamma}_i$, $i = 1, \ldots, d$, being significantly different from γ_i^0, $i = 1, \ldots, d$, respectively. This happens when the confidence ellipsoid does not contain $\boldsymbol{\gamma}^0$ but the circumscribed rectangle on the ellipse does contain $\boldsymbol{\gamma}^0$.

If the statistician wishes to determine the sample size that gives a prescribed probability $1 - \beta_i$ of finding $\hat{\gamma}_i$ significantly different from γ_i^0 (by the Scheffé technique) when γ_i is the true parameter value, he must resort to the existing tables of the noncentral t distribution. These are the Resnikoff and Lieberman (1957) tables and the Johnson and Welch (1940) tables. Tables specifically designed for the t test cannot be used, even though the relevant statistic is a noncentral t statistic, because the test critical point is $(dF^{\alpha}_{d,n-p})^{\frac{1}{2}}$ rather than $t^{\alpha/2}_{n-p}$.

Use of either the Resnikoff-Lieberman or Johnson-Welch tables is laborious because the tables were not designed for this application. The required sample size can only be found through trial selections of sizes. Furthermore, because of the discrete choice of the parameter values for which the power is calculated, the answer will be approximate at best. In some cases, the only answer on the correct size may be crude bounds.

Let $Z \sim N(0,\sigma^2)$, and s^2 be an independent $\sigma^2\chi^2$ variable divided by its d.f. ν. Let

$$T_\delta = \frac{Z + \Delta}{s} = \frac{(Z/\sigma) + \delta}{(s/\sigma)} \qquad \text{where } \delta = \frac{\Delta}{\sigma}. \tag{156}$$

The Resnikoff-Lieberman tables give $P\{T_\delta \leq \sqrt{\nu}\, x\}$ for appropriate two-decimal values of x; $\nu = 2(1)24(5)49$; and $p = .25, .15, .10, .065, .04, .025, .01, .004, .0025$, and $.001$, where δ is determined by p and ν through

$$\begin{aligned} \delta &= \sqrt{\nu + 1}\, K_p \\ p &= \frac{1}{\sqrt{2\pi}} \int_{K_p}^{+\infty} e^{-x^2/2}\, dx. \end{aligned} \tag{157}$$

This means that for a trial selection of the sample size the statistician must compute δ (which is a function of the sample size, the alternative, and the variance of the estimate) and find that column of p for which $\sqrt{n - p + 1}\, K_p$ most closely approximates δ (or those two columns which bound δ). If the power entry for $x = (dF^\alpha_{d,n-p})^{\frac{1}{2}}/\sqrt{n - p}$ is adequate, this sample size will suffice. Otherwise, the statistician must try another sample size, and continue until he finds an adequate one. For $n - p$ above 50 the authors feel that the t distribution can be conveniently replaced by the normal.

For selected values of t_0, ν, and β, the Johnson-Welch tables provide an entry from which the value of δ giving

$$P\{T_\delta \leq t_0\} = \beta \tag{158}$$

can be computed. If this δ is larger than the one for which the statistician is trying to achieve power $1 - \beta$, then the power at the prescribed δ is smaller than $1 - \beta$ and the sample size must be increased. The Johnson-Welch tables for $\beta = .10, .05, .01$ appear in Owen (1962).

Note that both these tables give one-sided probabilities, whereas the Scheffé interval is two-sided. Provided δ is at all reasonably sized, the amount of probability in the other tail will be so small as to be of no practical import. It is also unlikely that the statistician will want to include in his power the probability of rejecting the null hypothesis in the wrong direction.

Both the Resnikoff-Lieberman and Johnson-Welch tables are better adapted for use in the reverse direction. Given the sample size the Resnikoff-Lieberman tables quickly give the power as a function of the noncentrality parameter, and the Johnson-Welch tables give the noncentrality parameter as a function of the power.

Tables for the overall power functions (i.e., the probabilities of rejecting the null hypotheses) of the studentized maximum modulus, the studentized range, and the many-one t statistics do not exist. The studentized maximum modulus for just two means is a simple case in point. Let $Z_1 \sim N(\mu_1,\sigma^2)$, $Z_2 \sim N(\mu_2,\sigma^2)$, $s^2 \sim \sigma^2\chi^2_\nu/\nu$, and let Z_1, Z_2, and s be independent. Then, the power function is

$$\begin{aligned}1 - \beta &= P\left\{\frac{\max\{|Z_1|,|Z_2|\}}{s} > |m|^{\alpha}_{2,\nu}\right\}\\ &= P\left\{\frac{\max\left\{\left|\dfrac{(Z_1-\mu_1)}{\sigma}+\dfrac{\mu_1}{\sigma}\right|,\left|\dfrac{(Z_2-\mu_2)}{\sigma}+\dfrac{\mu_2}{\sigma}\right|\right\}}{(s/\sigma)} > |m|^{\alpha}_{2,\nu}\right\}.\end{aligned} \tag{159}$$

The above probability is expressible either as

$$\begin{aligned}1 - \int_0^\infty [\Phi(y|m|^{\alpha}_{2,\nu} - \delta_1) - \Phi(-y|m|^{\alpha}_{2,\nu} - \delta_1)][\Phi(y|m|^{\alpha}_{2,\nu} - \delta_2)\\ - \Phi(-y|m|^{\alpha}_{2,\nu} - \delta_2)]\,\frac{\nu^{(\nu/2)-1}}{\Gamma\left(\dfrac{\nu}{2}\right)2^{(\nu/2)-1}}\,y^{\nu-1}e^{-\nu y^2/2}\,dy\end{aligned} \tag{160}$$

where Φ is given in (33) and $\delta_1 = |\mu_1|/\sigma$, $\delta_2 = |\mu_2|/\sigma$; or as

$$1 - \frac{1}{2\pi}\iint_R \left[1 + \frac{x^2+y^2}{\nu}\right]^{-(\nu+2)/2} dx\,dy \tag{161}$$

where the region of integration is

$$R = \{\delta_1 - |m|^{\alpha}_{2,\nu} \le x \le \delta_1 + |m|^{\alpha}_{2,\nu},\ \delta_2 - |m|^{\alpha}_{2,\nu} \le y \le \delta_2 + |m|^{\alpha}_{2,\nu}\}.$$

The density appearing in (161) is the bivariate Student's t density discussed in Sec. 5.

The contour lines of the power function (159) to (161) do not seem to be expressible in terms of a simple, single-valued function of (δ_1,δ_2) as can be done for the F test (viz., $\delta_1^2 + \delta_2^2$). The difficulty is evident in (161) where it is necessary to integrate the spherically symmetric bivariate t density over an off-center square region. Consequently, the required amount of power tabulation is greatly increased, since entries would have to be computed for the pair (δ_1,δ_2) rather than a single parameter. As the number of means increases, the corresponding amount of tabulation increases.

The marginal power functions can be handled just as in the Scheffé case. Whether for the studentized range, the studentized maximum modulus, or the many-one t statistics, the marginal statistics are noncentral t statistics. Only the critical points change. For the studentized maximum modulus of two means the marginal power of the first mean is

$$1 - \beta_1 = P\left\{\frac{|Z_1|}{s} > |m|^{\alpha}_{2,\nu}\right\}$$
$$= P\left\{\frac{\left|\left(\frac{Z_1 - \mu_1}{\sigma}\right) + \delta_1\right|}{(s/\sigma)} > |m|^{\alpha}_{2,\nu}\right\}. \tag{162}$$

This can be attacked with the aid of the Johnson-Welch or Resnikoff-Lieberman tables as discussed previously for the Scheffé technique. The critical constant $(dF^{\alpha}_{d,n-p})^{\frac{1}{2}}$ is merely replaced by $|m|^{\alpha}_{2,\nu}$.

Harter (1957) has tabulated a strange hybrid power function. His noncentrality parameter is defined to be $\delta' = \Delta/s$ with the sample standard deviation s replacing the theoretical σ in the denominator. The quantity δ' is held constant in the calculations, but s is treated as a random variable in the sample statistics. This permits the author to use the t distribution in his computations. However, it is not clear whether the resulting power and sample sizes are good for anything.

Tukey (1953b) gives methods for approximately determining the sample size required to have the confidence intervals for pairwise mean differences less than a specified length in one-way and two-way classifications. The method depends on a prior estimate or guess as to the size of the error variance, and guarantees only with a specified probability (.5 or .9) that the length of the interval will be less than the prescribed amount. This reference, unfortunately, is not widely available.

10 ROBUSTNESS

The story on robustness is very much the same as that on power. A great deal is known for the F and t statistics, but essentially nothing is known for the other statistics useful in simultaneous inference.

All linear models dealt with in this chapter include in their basic structure the following three assumptions:

1. The observational errors are normally distributed (normality).
2. The observational errors are independent (independence).
3. The observational errors all have the same variance (homoscedasticity).

Whenever any one or more of these assumptions is violated the statistical analysis may be in peril. (For regression analysis the assumed linear model for the means may be inappropriate as well, but this type of model breakdown is usually not categorized under the heading of robustness.)

The study of the sensitivity of statistical techniques to departures from these underlying assumptions dates back to E. S. Pearson (1929, 1931) or possibly earlier. Since the F test or F statistic were, and still are, the mainstays of statistical inference, it is natural that much of the work has been concerned with these. However, the F statistic has virtually monopolized the studies in robustness to the exclusion of the other techniques.

From all the effort that has been exerted in this field, one image emerges: in tests of location the F statistic is robust [i.e., the F statistic is not sensitive to departures from the assumptions (1) to (3)]. Unless the assumptions are grossly violated, or the experimental design is badly unbalanced, the F test (or F statistic) will have, to a close degree, the properties claimed for it under the assumptions.

This image has been created from many widely scattered papers by a variety of authors. No attempt will be made here to give a synopsis of all the results leading to this impression. Such an undertaking would lengthen this monograph by a considerable fraction of its present size. An excellent exposition on robustness is given by Scheffé (1959, chap. 10). Scheffé also gives some calculations and results which have not appeared elsewhere. The reader is referred to Scheffé if he wishes to investigate this question in more detail.

No image or impression of robustness has evolved for the studentized range, the studentized maximum modulus, and the many-one t statistics. In fact, if one had to hazard a guess, one would guess that they were more sensitive to the assumptions than the F statistic. Each depends on an extreme statistic, i.e., the largest of a set of variables or the difference between the largest and the smallest of the set. The distribution of an extreme statistic is a good deal more sensitive to the form of the tails of the distribution (nonnormality), the largest variance (heteroscedasticity), and the interdependence between variables (dependence), than the distribution of a sum of squares. In a sum the oddball variable has a chance of being averaged out by the others, but in an extreme statistic the oddball variable is the statistic.

This should give the user a greater feeling of insecurity with these other statistics than with the F. Whether queasiness over the broad use of these statistics is justified has not been evaluated. Practically no work on the robustness of these statistics has appeared in the literature. The nonrobustness of these statistics is probably not catastrophic, but it is likely to be worse than for the F statistic.

Tukey (1953b) has some words and calculations on the use of the studentized range vs. the F distribution in the face of nonnormality and heteroscedasticity.

CHAPTER 3

Regression Techniques

The previous chapter covered the techniques available for simultaneously testing, or bracketing in confidence intervals, a set of regression coefficients. This chapter treats the other three problems often associated with regression analysis: banding the regression surface, prediction, and discrimination.

These problems call primarily for confidence procedures, so the distinction maintained in Chaps. 1 and 2 between confidence regions and significance tests does not exist in this chapter. For each problem the technique customarily applied to it is discussed. Because the application of each technique is limited to the motivating problem, there is no need to include sections on applications. Also, for each of the principal methods the only distributions required are the F and t distributions, so all sections on distributions and tables are omitted.

The basic model which will be assumed throughout this chapter is the general multiple linear regression model:

$$\begin{gathered} \mathbf{Y} = \mathbf{X}\boldsymbol{\beta} + \mathbf{e} \\ \mathbf{e} \sim N(\mathbf{0},\sigma^2\mathbf{I}) \\ \mathbf{X}\ n \times p \text{ has rank } p \qquad (p < n). \end{gathered} \tag{1}$$

The matrix $\mathbf{X}$ of independent variables will be regarded as containing all quantitative variables except possibly for the first column, which qualitatively may insert a general mean value. Variables other than the first may be qualitative, however. In this case their actual ranges of values will be finite, discrete subsets of those considered here.

The estimators of the unknown parameters $\boldsymbol{\beta}$ and σ^2 will be the standard ones: $\hat{\boldsymbol{\beta}} = (\mathbf{X}^T\mathbf{X})^{-1}\mathbf{X}^T\mathbf{Y}$ and $s^2 = \mathbf{Y}^T(\mathbf{I} - \mathbf{X}(\mathbf{X}^T\mathbf{X})^{-1}\mathbf{X}^T)\mathbf{Y}/(n - p)$. Their distributional properties are well known, and are summarized in (2.36).

1 REGRESSION SURFACE CONFIDENCE BANDS

For fixed $\boldsymbol{\beta}$ the functions

$$f(x_1, \ldots, x_p) = \beta_1 x_1 + \cdots + \beta_p x_p \tag{2a}$$

or, with a general mean,

$$f(x_2, \ldots, x_p) = \beta_1 + \beta_2 x_2 + \cdots + \beta_p x_p \tag{2b}$$

create surfaces over the p and $(p - 1)$-dimensional spaces of $(x_1, \ldots, x_p)$ and $(x_2, \ldots, x_p)$, respectively. In regression problems these are customarily referred to as *regression surfaces* or *response surfaces*. In the case of *simple linear regression* ($p = 2$; $x_{i1} \equiv 1$) the regression surface becomes the familiar straight line

$$f(x) = \alpha + \beta x. \tag{3}$$

The problem of confidence-banding the regression surface is to construct two functions $\bar{f}$ and $\underline{f}$, based on the sample data, which lie entirely above and below, respectively, the unknown true regression surface f with probability $1 - \alpha$.† That is, $\bar{f}$ and $\underline{f}$ should satisfy

$$P\{\bar{f}(x_1, \ldots, x_p) \geq f(x_1, \ldots, x_p) \geq \underline{f}(x_1, \ldots, x_p), \forall\, x_1, \ldots, x_p\} = 1 - \alpha \tag{4}$$

or a similar expression for the surface (2b). As compared with single upper and lower confidence limits on a single parameter, this problem requires upper and lower confidence limits on the value of the regression function for all values of the independent variables.

Practical problems which can be formulated in this manner are those in which knowledge of the behavior of the regression surface is required over a range of values of the independent variables. A variable's range which is of interest to an experimenter will seldom (never?) be from $-\infty$ to $+\infty$,

† The region between the upper and lower confidence functions is called the *confidence band* for the regression surface, and sometimes the functions themselves are referred to as *confidence bands*. This double usage of the word *band* for both surface and region is somewhat ambiguous.

but mathematical considerations make it easier to solve the infinite problem rather than the finite. The band over that part of the regression surface which is not of interest, or is physically meaningless, is ignored. The net effect is that the confidence in the band on the regression surface over the limited range somewhat exceeds the prescribed $1 - \alpha$.

Working and Hotelling (1929) completely solved this problem for simple linear regression. Their approach was not from a general viewpoint of simultaneous inference, nor was their proof as translucent as those presented today. Yet the idea was the same, and the extension to multiple linear regression is immediate through Scheffé projections.

1.1 Method

For the regression surface (2a) the $100(1 - \alpha)$ percent confidence band is

$$\sum_1^p \beta_i x_i \in \sum_1^p \hat{\beta}_i x_i \pm (pF^\alpha_{p,n-p})^{\frac{1}{2}} s(\mathbf{x}^T(\mathbf{X}^T\mathbf{X})^{-1}\mathbf{x})^{\frac{1}{2}} \qquad \forall\, \mathbf{x} = (x_1, \ldots, x_p). \tag{5a}$$

Correspondingly, for (2b) the $100(1 - \alpha)$ percent band is

$$\beta_1 + \sum_2^p \beta_i x_i \in \hat{\beta}_1 + \sum_2^p \hat{\beta}_i x_i \pm (pF^\alpha_{p,n-p})^{\frac{1}{2}} s(\mathbf{x}^T(\mathbf{X}^T\mathbf{X})^{-1}\mathbf{x})^{\frac{1}{2}} \qquad \forall\, \mathbf{x} = (1, x_2, \ldots, x_p). \tag{5b}$$

Note carefully that the critical constant $(pF^\alpha_{p,n-p})^{\frac{1}{2}}$ in (5b) is the same as in (5a). It does not drop to $[(p-1)F^\alpha_{p-1,n-p}]^{\frac{1}{2}}$ even though only $p - 1$ variables $x_2, \ldots, x_p$ are allowed to vary.

Had the experimenter requested just a confidence interval on the value of the regression function at a particular point $(x_1, \ldots, x_p)$, instead of a band on the entire regression surface, the critical constant would have been $t^{\alpha/2}_{n-p}$ instead of $(pF^\alpha_{p,n-p})^{\frac{1}{2}}$. This other is what is frequently presented in elementary statistics textbooks.

For the special case of simple straight line regression the band (5b) simplifies to:

$$\alpha + \beta x \in a + bx \pm (2F^\alpha_{2,n-2})^{\frac{1}{2}}\, s \left[\frac{1}{n} + \frac{(x - \bar{x})^2}{\sum_1^n (x_i - \bar{x})^2}\right]^{\frac{1}{2}} \tag{6}$$

for all x. The estimates a, b, s^2 are the customary ones [see (2.71)]. This is the original band of Working and Hotelling, and gives the familiar hyperbolic curves about the straight line $a + bx$. Figure 1 (on page 112) depicts the confidence band graphically.

1.2 Comparison

This Working-Hotelling-Scheffé confidence band on the regression surface has no competition in general. Gafarian (1964) has provided an alternative procedure for certain special problems. If the experimenter is interested in banding a straight regression line ($p = 2$; $x_{i1} \equiv 1$) over a finite interval $I = [x_*, x^*]$, he can have a confidence band of uniform width with Gafarian's method. The alternative probability statement is

$$P\{\alpha + \beta x \in a + bx \pm \delta s, \ \forall\, x \in [x_*, x^*]\} = 1 - \alpha \tag{7}$$

where δ is a constant independent of x.

The width of the Gafarian band is a constant 2δ, whereas the bandwidth for Working-Hotelling is variable. The latter has its minimum width at $x = \bar{x}$, and expands out to infinity as x diverges from $\bar{x}$. Whether one does or does not want a band of uniform width should depend on the problem.

Tables for the Gafarian band are available only when n is even and $\bar{x} = (x_* + x^*)/2$. These conditions should not be too restrictive, however. For greater detail the reader is referred to Sec. 4.1.

1.3 Derivation

The proof of (5a) has already been given in Sec. 2.4 of Chap. 2. It is just Theorem 2.1 with $\mathfrak{L}$ taken to be $\mathfrak{L}_{lc}$, the space of all linear combinations of the coordinates of $\boldsymbol{\beta}$. This space obviously has dimension p.

The statements (5b) and (5a) are equivalent, so the validity of (5b) is a consequence of (5a). The equivalence of (5a) and (5b) is easily estab-

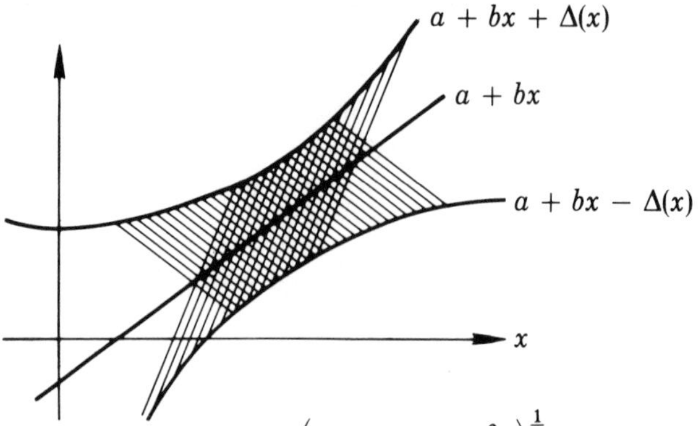

$$\Delta(x) = (2F^{\alpha}_{2,n-2})^{\frac{1}{2}}\, s\left(\frac{1}{n} + \frac{(x-\bar{x})^2}{\Sigma(x_i - \bar{x})^2}\right)^{\frac{1}{2}}$$

Figure 1

lished. The statement (5a) clearly implies (5b) by choosing $x_1 = 1$. Conversely, the statement (5b) implies (5a) by choosing $x_i = x_i'/x_1'$, $i = 2, \ldots, p$, in (5b), and then multiplying both sides by x_1'. This gives (5a) with $\mathbf{x} = (x_1, \ldots, x_p)$ replaced notationally by $\mathbf{x}' = (x_1', \ldots, x_p')$ *except* in the case of a zero in the first coordinate. This exceptional case is included, however, as a limiting result by letting $x_1' \to 0$.

The equivalence of (5a) and (5b) is due to the fact that the intervals in (5a) are invariant (in meaning and validity) under a scale change in one variable. For those who desire a more formal proof, Theorem 2.1 could be restated with $l_1 = 1$ throughout ($\mathcal{L}$ = p-dimensional space) and Lemma 2.2 replaced by the slightly modified version given here as Lemma 1.

Lemma 1 For $c > 0$,

$$\left|\sum_1^d a_i y_i\right| \leq c\left(\sum_1^d a_i^2\right)^{\frac{1}{2}} \qquad \text{for all } a_2, \ldots, a_d$$

if and only if

$$\sum_1^d y_i^2 \leq c^2 \qquad (\text{provided } a_1 y_1 \neq 0).$$

Proof *If:* Same as Lemma 2.2 (Cauchy-Schwarz inequality).
Only if: Choose $a_i = a_1 y_i / y_1, i = 2, \ldots, d$. Then

$$\left|\sum_1^d a_i y_i\right| \leq c\left(\sum_1^d a_i^2\right)^{\frac{1}{2}}$$

becomes

$$\left|a_1 y_1 + \left(\frac{a_1}{y_1}\right)\sum_2^d y_i^2\right| \leq c\left[a_1^2 + \left(\frac{a_1}{y_1}\right)^2 \sum_2^d y_i^2\right]^{\frac{1}{2}}.$$

Multiplication of both sides of the inequality by $|y_1/a_1|$ gives

$$\left|\sum_1^d y_i^2\right| \leq c\left(\sum_1^d y_i^2\right)^{\frac{1}{2}}. \qquad \|$$

Lemma 1 can be used to extend the Scheffé F projections to affine subspaces, i.e., linear subspaces translated away from the origin along a fixed vector. If $\mathcal{L}_a$ is an affine subspace, then points $\mathbf{a} \in \mathcal{L}_a$ have the representation $\mathbf{a} = \boldsymbol{l} + \mathbf{a}_0$ where $\boldsymbol{l} \in \mathcal{L}$ is a $(d-1)$-dimensional subspace, and $\mathbf{a}_0$ is a fixed vector. To establish Theorem 2.1 for $\mathcal{L}_a$ take the first row of $\mathbf{L}$ to be $\mathbf{a}_0$, and the remaining $d - 1$ rows to be any basis for $\mathcal{L}$. The argument then proceeds verbatim except that $\lambda_1 = 1$ is held fixed and Lemma 1 replaces Lemma 2.2.

Besides the Working-Hotelling application ($\mathcal{L}_a = \{(1,x)\}$) there are other applications of affine subspaces. H. Chernoff has mentioned an instance in which a simultaneous confidence band was needed for the derivative of a quadratic regression curve. If the regression curve is $f(x) = \alpha + \beta x + \gamma x^2$, then its derivative is $f'(x) = \beta + 2\gamma x$. The affine subspace generating this line is $\mathfrak{L}_a = \{(0,1,2x)\}$ so the appropriate critical constant in (2.81) is $(2F^{\alpha}_{2,n-3})^{\frac{1}{2}}$. A similar application arises in multiple linear regression when a simultaneous band is desired for the regression surface with only some of the independent variables allowed to vary. For example, a band could be sought for $\beta_1 x_1^0 + \cdots + \beta_k x_k^0 + \beta_{k+1}x_{k+1} + \cdots + \beta_p x_p$ for all $x_{k+1}, \ldots, x_p$ at a fixed point $(x_1^0, \ldots, x_k^0)$. There may also be some application when the linear combinations represent mixtures: $\mathbf{l} = (l_1, \ldots, l_p)$ where $\sum_1^p l_i = 1$ [see, for example, Scheffé (1958, 1963)].

2 PREDICTION

From a sample regression analysis the experimenter (and statistician) may wish to predict the values of k future observations on the dependent variable at k different settings or readings on the independent variables. Let $Y_1^0, \ldots, Y_k^0$ be the (future) random variables at the settings $\mathbf{x}_i^0 = (x_{i1}^0, \ldots, x_{ip}^0)$, $i = 1, \ldots, k$, respectively. The natural, and customarily used, predictors of the values of these random variables are the variables $\hat{Y}_i^0 = \mathbf{x}_i^{0T}\hat{\boldsymbol{\beta}}$, $i = 1, \ldots, k$.

The experimenter may also wish to bracket the values of the future observations with some degree of certainty. Intervals of this type are called *prediction intervals*. For simultaneous prediction intervals the statistician must conjure up k constants Δ_i^0, each dependent on the corresponding $\mathbf{x}_i^0$, such that

$$P\{Y_i^0 \in \hat{Y}_i^0 \pm \Delta_i^0\text{s}, i = 1, \ldots, k\} = 1 - \alpha. \tag{8}$$

This problem for a single future observation is frequently discussed in elementary statistics texts [e.g., Mood and Graybill (1963)].

There is a definite distinction between prediction intervals and *tolerance intervals* which the reader should be aware of. The former bracket a single future observation; the latter bracket a specified proportion P of the (normal) distribution centered at $\mathbf{x}_i^{0T}\boldsymbol{\beta}$ with variance σ^2. If a large number of future observations were taken at the setting $\mathbf{x}_i^0$, then approximately $100P$ percent of them would fall in the tolerance interval (with probability $1 - \alpha$ for the original sample $\mathbf{Y}$).

This section deals with prediction intervals, and not tolerance intervals, since the author guesses the demand for the former is greater. For a treatment of tolerance intervals the reader is referred in the nonsimultaneous case to Wallis (1951), and in the simultaneous case to Sec. 4.2 or Lieberman and Miller (1963).

2.1 Method

A Scheffé-type argument can be applied to the construction of simultaneous prediction intervals. This approach was expounded by Lieberman (1961). The result is: with probability greater than $1 - \alpha$,

$$Y_i^0 \in \hat{Y}_i^0 \pm (kF_{k,n-p}^{\alpha})^{\frac{1}{2}} s(1 + \mathbf{x}_i^{0T}(\mathbf{X}^T\mathbf{X})^{-1}\mathbf{x}_i^0)^{\frac{1}{2}} \qquad i = 1, \ldots, k \tag{9}$$

where $\hat{Y}_i^0 = \mathbf{x}_i^{0T}\hat{\boldsymbol{\beta}}$ is the predicted value.

Had each interval been treated separately under the nonsimultaneous theory, the critical constant would have been $t_{n-p}^{\alpha/2}$ instead of $(kF_{k,n-p}^{\alpha})^{\frac{1}{2}}$.

In the special case of straight line regression the expressions for the intervals (9) simplify to

$$Y_i^0 \in a + bx_i^0 \pm (kF_{k,n-2}^{\alpha})^{\frac{1}{2}} s \left[1 + \frac{1}{n} + \frac{(x_i^0 - \bar{x})^2}{\sum_1^n (x_j - \bar{x})^2} \right]^{\frac{1}{2}} \qquad i = 1, \ldots, k. \tag{10}$$

The confidence attached to the intervals (9) or (10) is strictly greater than $1 - \alpha$ because not all the statements are used which are permitted. For the same probability $1 - \alpha$ one can, in fact, bracket the values of all linear combinations of the Y_i^0, that is, $\sum_1^k l_i Y_i^0$, $\forall\, \mathbf{l} \in \mathcal{L}_{lc}$. But it is doubtful that any such combination would ever be of practical interest. The intervals (9) give the box-shaped region in k-dimensional space which circumscribes the elliptical region generated by all the linear combinations.

The chief competitor to this Scheffé-type technique is the corresponding Bonferroni procedure. The Bonferroni inequality (1.13) gives

$$Y_i^0 \in \hat{Y}_i^0 \pm t_{n-p}^{\alpha/2k} s(1 + \mathbf{x}_i^{0T}(\mathbf{X}^T\mathbf{X})^{-1}\mathbf{x}_i^0)^{\frac{1}{2}} \qquad i = 1, \ldots, k \tag{11}$$

with probability greater than $1 - \alpha$ since each statistic

$$\frac{Y_i^0 - \hat{Y}_i^0}{s\sqrt{1 + \mathbf{x}_i^{0T}(\mathbf{X}^T\mathbf{X})^{-1}\mathbf{x}_i^0}} \tag{12}$$

has a t distribution with $n - p$ d.f.

2.2 Comparison

The intervals (9) and (11) are identical except for the critical constants $(kF^{\alpha}_{k,n-p})^{\frac{1}{2}}$ and $t^{\alpha/2k}_{n-p}$, respectively. The statistician is at liberty to compute both constants and select whichever is smaller for his particular k, $n - p$, and α. Inspection of the Dunn (1959) tables reveals that for $\alpha = .05$, $t^{\alpha/2k}_{\nu} < (kF^{\alpha}_{k,\nu})^{\frac{1}{2}}$.

As k increases the critical constants $(kF^{\alpha}_{k,\nu})^{\frac{1}{2}}$ and $t^{\alpha/2k}_{\nu}$ increase without limit. For k large it may be advantageous to switch to the simultaneous tolerance intervals of Lieberman and Miller (1963). This will also be necessary if the precise number of predictions to be made is unknown and cannot be closely bounded on the upper side. For a discussion of the tolerance interval approach the reader is referred to Sec. 4.2.

2.3 Derivation

The Bonferroni intervals require no additional proof. To establish the validity of the Scheffé-type procedure let $Z_i = Y^0_i - \tilde{Y}^0_i$, $i = 1, \ldots, k$. Viewed as random functions of the normal deviates $Y_1, \ldots, Y_n$, $Y^0_1, \ldots, Y^0_k$, the variables $Z_1, \ldots, Z_k$ have a multivariate normal distribution with mean vector $\mathbf{0} = (0, \ldots, 0)$ and covariance matrix $\sigma^2\mathbf{\Sigma}$. The diagonal elements of $\mathbf{\Sigma}$ are

$$1 + \mathbf{x}^{0T}_i(\mathbf{X}^T\mathbf{X})^{-1}\mathbf{x}^0_i \qquad i = 1, \ldots, k \tag{13}$$

that is, the sums of the variances of Y^0_i and $\tilde{Y}^0_i$. The off-diagonal elements of $\mathbf{\Sigma}$ could be evaluated as well, but they are not explicitly required in the proof.

By the extension of Theorem 2.1 to dependent variables

$$P\{|\boldsymbol{l}^T\mathbf{Z} - \boldsymbol{l}^T\mathbf{0}| \leq (kF^{\alpha}_{k,n-p})^{\frac{1}{2}}s(\boldsymbol{l}^T\mathbf{\Sigma}\boldsymbol{l})^{\frac{1}{2}}, \forall\, \boldsymbol{l} \in \mathcal{L}_{lc}\} = 1 - \alpha. \tag{14}$$

The space $\mathcal{L}_{lc}$ in (14) is the space of all linear combinations, but the only combinations which are used for the intervals (9) are $\boldsymbol{l}_1 = (1,0, \ldots, 0)$, $\boldsymbol{l}_2 = (0,1,0, \ldots, 0), \ldots, \boldsymbol{l}_k = (0, \ldots, 0,1)$. When these multiply the matrix $\mathbf{\Sigma}$ in the term $\boldsymbol{l}^T\mathbf{\Sigma}\boldsymbol{l}$, they give the diagonal elements of the matrix $\mathbf{\Sigma}$, namely, the variances of the Z_i's given by (13). This then produces the intervals (9).

The intervals (9) form the box-shaped region circumscribing the ellipsoid $(\mathbf{Y}^0 - \tilde{\mathbf{Y}}^0)^T\mathbf{\Sigma}^{-1}(\mathbf{Y}^0 - \tilde{\mathbf{Y}}^0) \leq s^2kF^{\alpha}_{k,n-p}$, so the intervals (9) have probability greater than $1 - \alpha$. If all the linear combinations $\sum_1^k l_i(Y^0_i - \tilde{Y}^0_i)$ were utilized, the intersected region would be the ellipse, and the confidence would be exactly $1 - \alpha$.

3 DISCRIMINATION

The discrimination problem is the reverse of prediction. It is customarily only discussed for straight line regression, and the discussion in this section will be limited to this case as well. The statistician has n pairs of values $(x_1, Y_1), \ldots, (x_n, Y_n)$ from which to estimate the regression line $\alpha + \beta x$. He also has k additional observations $Y_1^0, \ldots, Y_k^0$ for which the corresponding independent variable values $x_1^0, \ldots, x_k^0$ are unknown. The problem is to estimate the unknown quantities $x_1^0, \ldots, x_k^0$ and bracket them in confidence intervals.

The prediction problem was the reverse. The values $x_1^0, \ldots, x_k^0$ were known, and the objective was to predict $Y_1^0, \ldots, Y_k^0$ and bracket them in intervals.

3.1 Method

In the absence of any statistical fluctuation the equality $Y_i^0 = a + bx_i^0$ would hold so the standard estimator of x_i^0 is $\hat{x}_i^0 = (Y_i^0 - a)/b$.

As with prediction intervals the two competitors are the Bonferroni and Scheffé intervals.

Bonferroni Since

$$Y_i^0 - a - bx_i^0 \sim N\left(0, \sigma^2 \left\{1 + (1/n) + \left[(x_i^0 - \bar{x})^2 / \sum_j (x_j - \bar{x})^2\right]\right\}\right),$$

$$P\left\{|Y_i^0 - a - bx_i^0| \leq t_{n-2}^{\alpha/2k} s \left[1 + \frac{1}{n} + \frac{(x_i^0 - \bar{x})^2}{\sum_1^n (x_j - \bar{x})^2}\right]^{\frac{1}{2}}, \quad i = 1, \ldots, k\right\} \geq 1 - \alpha \quad (15)$$

or, equivalently,

$$P\left\{(Y_i^0 - a - bx_i^0)^2 \leq F_{1,n-2}^{\alpha/k} s^2 \left[1 + \frac{1}{n} + \frac{(x_i^0 - \bar{x})^2}{\sum_1^n (x_j - \bar{x})^2}\right], \quad i = 1, \ldots, k\right\} \geq 1 - \alpha. \quad (16)$$

Thus, all points x which satisfy the quadratic inequality

$$(Y_i^0 - a - bx)^2 \le F_{1,n-2}^{\alpha/k} s^2 \left[1 + \frac{1}{n} + \frac{(x - \bar{x})^2}{\sum_1^n (x_j - \bar{x})^2} \right] \tag{17}$$

form a confidence region for the unknown x_i^0, and the k regions for $x_1^0, \ldots, x_k^0$ have a combined probability greater than $1 - \alpha$ for the simultaneous inclusion of these unknowns.

In the nonsimultaneous problem where the critical constant is $F_{1,n-2}^{\alpha}$ this result is referred to either as Fieller's (1940) or Paulson's (1942) theorem.

The region generated by (17) experiences the same difficulties suffered in the univariate case. One wants a finite interval for a confidence region, but one may get the entire real line, or two semi-infinite lines. Why, and when these three cases occur is illustrated in Figs. 2, 3, and 4. In each figure

$$\begin{aligned} \bar{f}(x) &= a + bx + (F_{1,n-2}^{\alpha/k})^{\frac{1}{2}} s \left[1 + \frac{1}{n} + \frac{(x - \bar{x})^2}{\sum_1^n (x_j - \bar{x})^2} \right]^{\frac{1}{2}} \\ \underline{f}(x) &= a + bx - (F_{1,n-2}^{\alpha/k})^{\frac{1}{2}} s \left[1 + \frac{1}{n} + \frac{(x - \bar{x})^2}{\sum_1^n (x_j - \bar{x})^2} \right]^{\frac{1}{2}} \end{aligned} \tag{18}$$

and the solid part of the horizontal line with ordinate at Y^0 is the confidence region for x^0. The desirable case (Fig. 2) will occur if and only if $b^2 > F_{1,n-2}^{\alpha/k} s^2 / \sum_1^n (x_j - \bar{x})^2$.

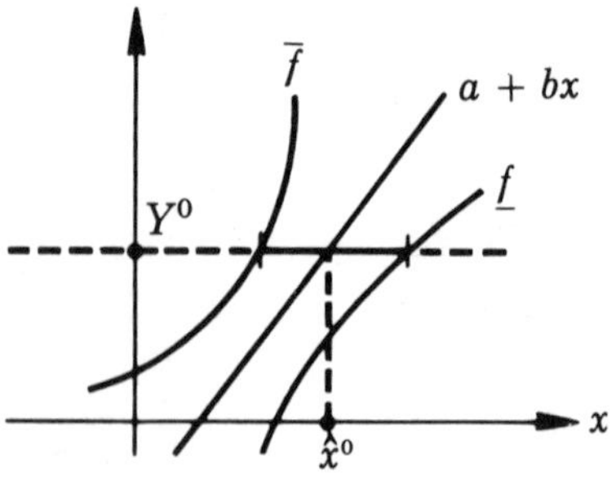

Figure 2

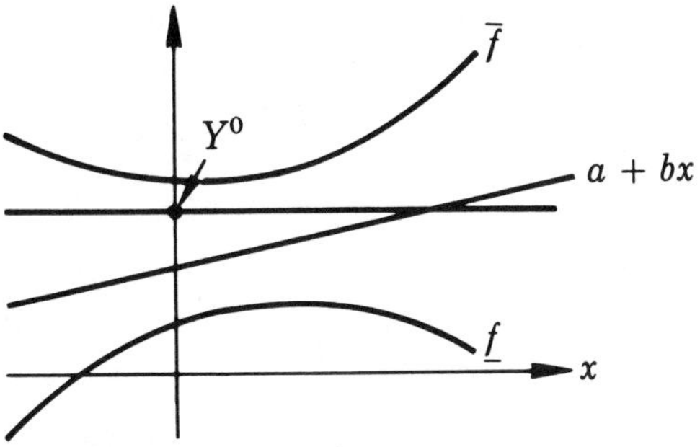

Figure 3

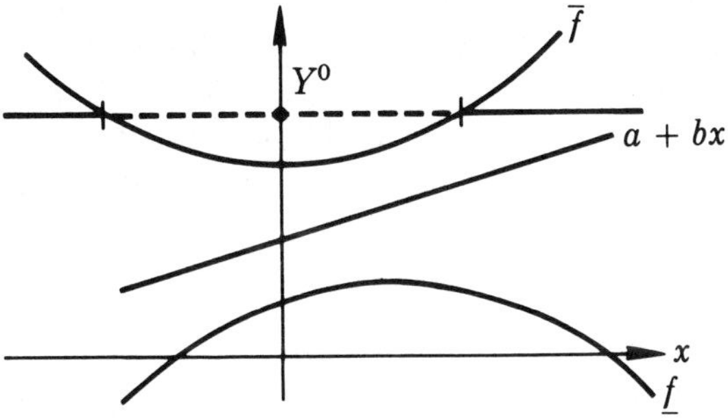

Figure 4

When the expression (17) with equality has two real roots, they are

$$x = \bar{x} - \frac{(Y_i^0 - \bar{Y})}{b\left[\dfrac{F^{\alpha/k}_{1,n-2}s^2}{b^2\sum_1^n (x_j - \bar{x})^2} - 1\right]}$$

$$\pm \frac{\sqrt{(Y_i^0 - \bar{Y})^2 - \left[\dfrac{F^{\alpha/k}_{1,n-2}s^2}{b^2\sum_1^n (x_j - \bar{x})^2} - 1\right]\left[F^{\alpha/k}_{1,n-2}s^2\left(1 + \dfrac{1}{n}\right) - (Y_i^0 - \bar{Y})^2\right]}}{b\left[\dfrac{F^{\alpha/k}_{1,n-2}s^2}{b^2\sum_1^n (x_j - \bar{x})^2} - 1\right]}. \tag{19}$$

Whether the confidence region is a finite interval or the union of two semi-infinite lines will either be obvious or can be determined by checking whether or not the estimate $\hat{x}_i^0$ lies between the two roots.

When the expression (17) with equality has no real roots, the confidence region for x_i^0 is the entire real line.

Scheffé The Scheffé intervals stem from the inequality

$$P\left\{|Y_i^0 - a - bx_i^0| \leq (kF_{k,n-2}^{\alpha})^{\frac{1}{2}}s\left[1 + \frac{1}{n} + \frac{(x_i^0 - \bar{x})^2}{\sum_1^n (x_j - \bar{x})^2}\right]^{\frac{1}{2}}, \quad i = 1, \ldots, k\right\} \geq 1 - \alpha. \quad (20)$$

The only alteration in the Scheffé approach from the Bonferroni lies in the critical constant: $kF_{k,n-2}^{\alpha}$ vs. $F_{1,n-2}^{\alpha/k}$. The previous discussion for Bonferroni discrimination intervals can be applied verbatim to the Scheffé intervals after changing the critical constant to $kF_{k,n-2}^{\alpha}$.

Mandel (1958) applied the Scheffé ideas to the discrimination problem. He took a slightly different attack from that presented in this section, and obtained the same intervals except for the critical constant. His constant, $(k+2)F_{k+2,n-2}^{\alpha}$, is larger than the constant, $kF_{k,n-2}^{\alpha}$, of this section.

3.2 Comparison

Whichever method has the smaller critical constant will give the better results. The critical constant determines the width of the bands (18) about the line $a + bx$. The smaller the constant, the tighter the bands will be about the sample regression line. Figures 2, 3, and 4 show that tighter bands will produce tighter confidence intervals.

An analytical comparison of $F_{1,n-2}^{\alpha/k}$ vs. $kF_{k,n-2}^{\alpha}$ has not been carried out. As mentioned in Sec. 2.2, $F_{1,\nu}^{\alpha/k} < kF_{k,\nu}^{\alpha}$ for the only tables [$\alpha = .05$; Dunn (1959)] constructed to date.

There may be problems in which the number of discriminations to be performed is unknown. For example, this will be the case in constructing a standard curve for a bioassay from which an unlimited number of assays are to be made in the future. The techniques of this section are not applicable under these circumstances. The statistician will have to resort to the method of Sec. 4.3.

When k is large, the Bonferroni and Scheffé critical constants may be so large as to render the discrimination intervals useless. The method of Sec. 4.3 may also be of some aid in this event.

3.3 Derivation

The probability inequality (15) for the Bonferroni method follows from the Bonferroni inequality (1.13).

The Scheffé inequality (20) was proved in Sec. 2.3.

4 OTHER TECHNIQUES

4.1 Linear confidence bands

Gafarian (1964) showed how to construct a confidence band of uniform width over a finite interval for simple straight line regression. He also computed the critical constants necessary to implement this technique in certain special cases.

Suppose the experimenter desires a confidence band of the form depicted in Fig. 5. The corresponding probability statement is

$$P\{|(a - \alpha) + (b - \beta)x| \leq \delta s, \forall\, x \in [x_*,x^*]\} = 1 - \alpha. \qquad (21)$$

It is convenient to have a and b independent, so reparameterize (21) as

$$P\{|(a - \alpha) + (b - \beta)(x - \bar{x})| \leq \delta s, \forall\, x \in [x_*,x^*]\} = 1 - \alpha \qquad (22)$$

where now $a = \bar{Y}$ and b is the usual estimator. This can be rewritten as

$$P\left\{\left|\frac{T_1}{\sqrt{n}} + \frac{T_2}{\sqrt{n}\, v}(x - \bar{x})\right| \leq \delta, \forall\, x \in [x_*,x^*]\right\} = 1 - \alpha \qquad (23)$$

where $T_1 = \sqrt{n}\,(a - \alpha)/s$, $T_2 = \sqrt{n}\,v(b - \beta)/s$, and

$$v^2 = \frac{\left[\sum_1^n (x_i - \bar{x})^2\right]}{n}.$$

The variables T_1 and T_2 are independent, unit normal variables divided by a common standard deviation estimator s. Hence, they have a bivariate density which is the bivariate analog of the t distribution dis-

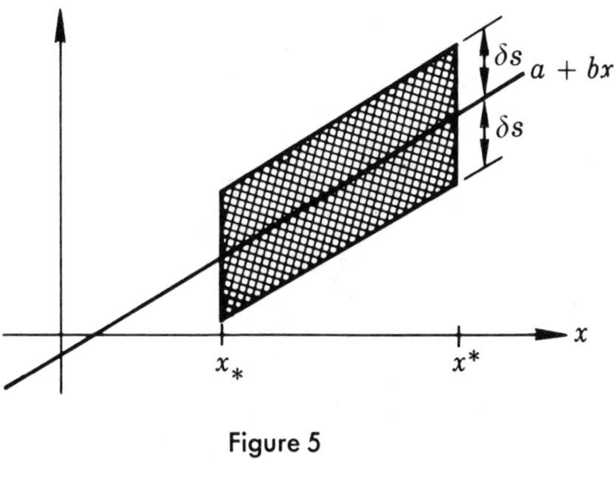

Figure 5

cussed in Sec. 5.5 of Chap. 2. The density is given by

$$p(t_1,t_2) = \frac{1}{2\pi\left(1 + \dfrac{t_1^2 + t_2^2}{n-2}\right)^{n/2}}. \tag{24}$$

The region in the (t_1,t_2)-plane which satisfies the inequality in (23), that is,

$$\left\{(t_1,t_2): \left|\frac{t_1}{\sqrt{n}} + \frac{t_2}{\sqrt{n}\,v}(x - \bar{x})\right| \le \delta,\ \forall\, x \in [x_*,x^*]\right\} \tag{25}$$

is a parallelogram as depicted in Fig. 6. Thus, to determine the appropriate δ, the density (24) must be integrated over the region (25). Whichever δ gives the value $1 - \alpha$ to this integral is the appropriate one.

In the special case that $\bar{x} = (x_* + x^*)/2$ (that is, when the mean of the independent variable coincides with the center of the interval over which the regression line is to be bracketed), the top and bottom vertices of the parallelogram in Fig. 6 fall on the t_2 axis, and the integration problem greatly simplifies. Due to the spherical symmetry of (24) and the symmetry of the region of integration, it is only necessary to integrate (24) over the triangle in the positive quadrant and multiply the integral by 4.

Gafarian further simplifies his integration problem by restricting n to even integers, specifically, $n = 4(2)20, 30, 50, \infty$. For each value of $c = 2v/(x^* - x_*) = 1(.1)2(.2)3(.4)5, 6, 8, 10, 20, \infty$, a table in n and $d = \sqrt{n}\delta = 1(.05)2.5(.1)4(.2)5(.5)7(1)10(5)20(10)50$ gives the value of the integral (i.e., the confidence coefficient) to three decimals. By switching axes the table can also be used for values of c less than 1.

The conditions ($\bar{x} = (x_* + x^*)/2$ and n even) imposed by Gafarian to

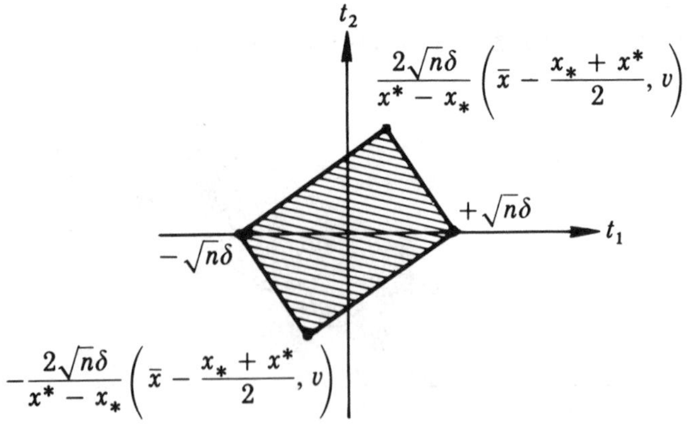

Figure 6

obtain the tables do not seem very restrictive. Unless the experiment is poorly designed the mean $\bar{x}$ will be somewhere near the middle of the interval $I = [x_*, x^*]$. The interval of interest I is usually sufficiently indefinitely defined to permit shrinking, expanding, or shifting to achieve $(x_* + x^*)/2 = \bar{x}$. Interpolation in the tables will give approximate results for odd n.

Gafarian compares his confidence band with the customary Working-Hotelling band restricted to the interval $I = [x_*, x^*]$. Typically, the Working-Hotelling band will be narrower in the middle near $\bar{x}$, but wider near the tails, x_* and x^*. Other than on grounds of taste or expediency, the most sensible way to compare the two methods is in terms of the area of their respective confidence bands. But with this measure neither method dominates the other. From numerical results in Gafarian (1964) the Working-Hotelling technique seems to be better for long intervals I, and the Gafarian technique seems better for short intervals. For greater detail the reader is referred to the original source.

Gafarian also discusses the optimal allocation of the independent variable.

4.2 Tolerance intervals

When the number (k) of prediction intervals to be handled simultaneously is large, the critical constant $kF^{\alpha}_{k,n-2}$ may be so large as to render the intervals useless. In other cases, the total number of predictions to be made may be unknown or may possibly be subject to chance. For these two situations, as well as when the experimenter actually wants a tolerance interval instead of a prediction interval, the simultaneous tolerance intervals of Lieberman and Miller (1963) may be useful.

At the value x of the independent variable the true interval which contains the symmetric, central $100P$ percent of the normal distribution centered at $\alpha + \beta x$ is

$$I_x(P) = (\alpha + \beta x - N(P)\sigma,\ \alpha + \beta x + N(P)\sigma) \tag{26}$$

where $N(P)$ is the two-sided P percentile point of the unit normal distribution, that is,

$$P = \frac{1}{\sqrt{2\pi}} \int_{-N(P)}^{+N(P)} e^{-y^2/2}\, dy. \tag{27}$$

An interval, $T_x(P)$, which contains the interval $I_x(P)$ with probability (at least) $1 - \alpha$, is a tolerance interval.[1] Families of intervals which

[1] This differs slightly from the usual definition. Usually, $T_x(P)$ is only required to cover $100P$ percent of the underlying (normal) distribution; it does not have to be the symmetric, central portion [cf. (31)].

contain their corresponding $I_x(P)$ for all x (or for all x and all P) with probability (at least) $1 - \alpha$ are called simultaneous tolerance intervals.

The reader should note that the probability $1 - \alpha$ refers to the original sample $(x_1, Y_1), \ldots, (x_n, Y_n)$; that is, under repeated sampling, $100(1 - \alpha)$ percent of all the original samples will have *all* subsequent statements on tolerance intervals correct. The probability P refers to the proportion of the normal distribution captured in $I_x(P)$. For a correct tolerance interval $T_x(P)$ [that is, one for which $I_x(P) \subset T_x(P)$] at least $100P$ percent of all future observations at this x value will fall in the tolerance interval.

A quick and easy family of simultaneous tolerance intervals can be patched together with the aid of the Bonferroni inequality. The interval $I_x(P)$ given in (26) can be covered if the center $\alpha + \beta x$ is bounded in an interval and its width $2N(P)\sigma$ is bounded above. But the center is bounded in the Working-Hotelling band,

$$\alpha + \beta x \in a + bx \pm (2F^{\alpha/2}_{2,n-2})^{\frac{1}{2}} s \left[\frac{1}{n} + \frac{(x - \bar{x})^2}{\sum_1^n (x_i - \bar{x})^2} \right]^{\frac{1}{2}} \tag{28}$$

with probability $1 - (\alpha/2)$. The unknown scale factor σ appearing in the width is bounded above by

$$\sigma \leq \left(\frac{n - 2}{{}^{\alpha/2}\chi^2_{n-2}} \right)^{\frac{1}{2}} s \tag{29}$$

where ${}^{\alpha/2}\chi^2_{n-2}$ is the lower $(\alpha/2)$ percentile point of a χ^2 distribution with $n - 2$ d.f. Putting (28) and (29) together with the Bonferroni inequality gives

$$I_x(P) \subset a + bx \pm s \left\{ (2F^{\alpha/2}_{2,n-2})^{\frac{1}{2}} \left[\frac{1}{n} + \frac{(x - \bar{x})^2}{\sum_1^n (x_i - \bar{x})^2} \right]^{\frac{1}{2}} + N(P) \left(\frac{n - 2}{{}^{\alpha/2}\chi^2_{n-2}} \right)^{\frac{1}{2}} \right\} \tag{30}$$

for *all* x and *all* P, with probability at least $1 - \alpha$.

Lieberman and Miller also give three other techniques which constitute alternatives to the one above. The first extends the ideas of Wald and Wolfowitz (1946) and Wallis (1951) to simultaneous intervals in the regression situation. With a specified probability $(1 - \alpha)$ these tolerance intervals contain $100P$ percent of the underlying normal distribution, but the $100P$ percent can be from any part of the distribution. It does not have to be the symmetric, central portion given by the interval

$I_x(P)$. All that is required is

$$\frac{1}{\sqrt{2\pi}\,\sigma}\int_{T_x(P)} e^{-(y-\alpha-\beta x)^2/2\sigma^2}\,dy \geq P \qquad \text{for all } x. \tag{31}$$

Also, P is fixed for any set of intervals. It cannot vary as in (30).

The second technique depends on Lemma 1 of this chapter and yields the following probability statement:

$$1-\alpha = P\left\{ I_x(P) \subset a + bx \pm c^{*\alpha} s \left[\frac{1}{n} + \frac{(x-\bar{x})^2}{\sum_1^n (x_i - \bar{x})^2} + N^2(P) \right]^{\frac{1}{2}}, \forall\, x, P \right\}. \tag{32}$$

The critical constant $c^{*\alpha}$ is defined by

$$P\left\{ \frac{Z_1^2 + Z_2^2 + 1}{\chi_{n-2}^2/(n-2)} \leq (c^{*\alpha})^2 \right\} = 1 - \alpha \tag{33}$$

where Z_1, Z_2 are independent, unit normal random variables, and χ_{n-2}^2 is an independent χ^2 variable with $n-2$ d.f. The ratio in (33) is almost distributed as $3F_{3,n-2}$; the difference is that a third variable in the numerator (Z_3^2) has been replaced by its expected value (1). A table of $c^{*\alpha}$ is given in Lieberman and Miller (1963) for $\alpha = .5, .3, .1, .05, .01$, and $.001$; and d.f. $= 1(1)30(5)50(10)100$.

Just as in (30) the tolerance intervals are variable with P. For the same original sample the experimenter can vary the proportion of the distribution he wishes to bracket depending upon the circumstances.

The idea behind the third technique is similar to (32), but P is kept fixed and not allowed to vary. Special critical constants are required, and a limited table is given in the article.

Numerical comparison of the four competing techniques reveals that no one is uniformly better than the others. Depending on α, n, $\bar{x}$, $\sum_1^n (x_i - \bar{x})^2$, and x, any one of the four can be best. The simplest and easiest to use is (30), and it seems to perform just as well as the more sophisticated methods.

4.3 Unlimited discrimination intervals

The same problem may arise for discrimination as motivated the construction of tolerance intervals in the preceding section. From the sample data a standard regression line $a + bx$ can be established, but the number of discriminations to be made from it may be unknown. This occurs

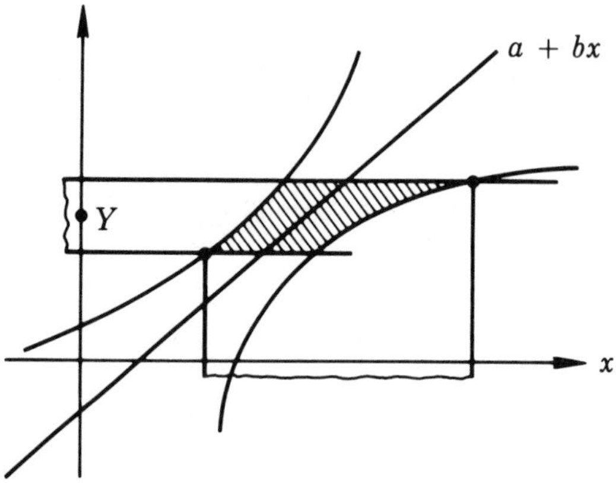

Figure 7

frequently in bioassay where a standard curve is constructed on which all future assays (discriminations) are to be run. The number of future assays is unknown and essentially unlimited. Consequently, the results of Sec. 3.1 are inapplicable.

In other cases the number of discriminations may be known, but the critical constant $kF^{\alpha}_{k,n-2}$ may be so large as to make the intervals ridiculously wide.

No fancy techniques have been proposed to handle this problem. However, a crude (but possibly satisfactory) procedure can be put together with the aid of the Bonferroni inequality. The idea is simple enough. A confidence band can be placed on the unknown regression line, and confidence intervals can be put on the expected values of the future observations. The confidence intervals for the $E\{Y\}$'s on the Y axis can be converted into discrimination intervals on the x axis by observing where the Y axis interval intersects the confidence band on the regression line. The idea is illustrated in Fig. 7. The wiggly interval on the Y axis is the confidence interval for $E\{Y\}$ corresponding to Y. The wiggly interval on the x axis is the appropriate discrimination interval because *if* $E\{Y\} = \alpha + \beta x$ is contained in the Y axis interval, and *if* the regression line is covered by its confidence band, then the point $(x, \alpha + \beta x)$ must lie in the intersection (shaded region), and x must be contained in the projected x axis interval.

This proposal is mathematically executed as follows. The confidence band on the regression line with confidence $1 - (\alpha/2)$ is

$$\alpha + \beta x \in a + bx \pm (2F^{\alpha/2}_{2,n-2})^{\frac{1}{2}} s \left[\frac{1}{n} + \frac{(x - \bar{x})^2}{\sum_{1}^{n} (x_i - \bar{x})^2} \right]^{\frac{1}{2}} \tag{34}$$

which holds for all x. If the variance σ^2 were known, than a $100P$ percent confidence interval for $E\{Y\}$ would be $E\{Y\} \in Y \pm N(P)\sigma$, where $N(P)$ is defined by (27). Although σ is unknown, it can be bounded above through the inequality

$$P\left\{\sigma \leq \left(\frac{n-2}{{}^{\alpha/2}\chi^2_{n-2}}\right)^{\frac{1}{2}} s\right\} = 1 - \frac{\alpha}{2} \tag{35}$$

where ${}^{\alpha/2}\chi^2_{n-2}$ is the lower $(\alpha/2)$ percentile point of the χ^2 distribution with $n-2$ d.f. Since σ is bounded above by the multiple of s in (35), it follows that *all* intervals $Y \pm N(P)\sigma$ must be contained in $Y \pm N(P)[(n-2)/{}^{\alpha/2}\chi^2_{n-2}]^{\frac{1}{2}}s$ with probability $1-(\alpha/2)$ (for the original sample). The probability that both (34) and (35) hold is at least $1-\alpha$ by the Bonferroni inequality.

The discrimination interval is obtained by intersecting the confidence interval on $E\{Y\}$ with the confidence band on $\alpha + \beta x$, and projecting the intersection onto the x axis. The nice case is exhibited in Fig. 7. Actually, the discrimination interval can be the entire real line or the union of two semi-infinite intervals, as well as the conventional interval of Fig. 7. These pathological cases will occur when the regression line is (relatively) too flat (i.e., when b is too near zero), and are analogous to what can happen for the simpler problem of just a single determination. The reader can visualize when these other cases occur by redrawing Fig. 7 with a much flatter slope and shifting the Y interval up and down.

In the nice case (viz., Fig. 7) let $D_Y(P) = (\underline{D}_Y(P), \bar{D}_Y(P))$ be the discrimination interval. If the sample regression slope is positive (that is, $b > 0$), the upper end point $\bar{D}_Y(P)$ is the solution of the equation

$$a + bx - (2F^{\alpha/2}_{2,n-2})^{\frac{1}{2}} s \left[\frac{1}{n} + \frac{(x-\bar{x})^2}{\sum_1^n (x_i - \bar{x})^2}\right]^{\frac{1}{2}} = Y + N(P)\left(\frac{n-2}{{}^{\alpha/2}\chi^2_{n-2}}\right)^{\frac{1}{2}} s \tag{36}$$

and $\underline{D}_Y(P)$ is the solution of the equation

$$a + bx + (2F^{\alpha/2}_{2,n-2})^{\frac{1}{2}} s \left[\frac{1}{n} + \frac{(x-\bar{x})^2}{\sum_1^n (x_i - \bar{x})^2}\right]^{\frac{1}{2}} = Y - N(P)\left(\frac{n-2}{{}^{\alpha/2}\chi^2_{n-2}}\right)^{\frac{1}{2}} s. \tag{37}$$

If b is negative, then $\bar{D}_Y(P)$ is the solution of (37), and $\underline{D}_Y(P)$ is the solution of (36). Equations (36) and (37) are readily solved by transposing and squaring terms to obtain quadratic equations in x.

The interpretation of these intervals should be emphasized. Every time a standard regression line is constructed for performing discriminations there is a chance that for all discriminations run on it at least

$100P$ percent of the discrimination intervals will contain the true x's. The chance of this is at least $1 - \alpha$. Thus, for at least $100(1 - \alpha)$ percent of all standard regression lines that are created, at least $100P$ percent of the discrimination intervals for each line will enclose the correct x's.

For the other fraction (less than 100α percent) the percentage of discrimination intervals enclosing their true x's may be greater or less than $100P$ percent for each line. This will depend on which values of x (and Y) arise. For example, if σ is correctly bounded above, but the confidence band only covers part of the regression line, more than $100P$ percent of the discrimination intervals will still contain their true x's when the x's arise predominantly in the region where the confidence band covers the regression line.

It is frequently the case in practice that only the estimated value $\hat{x} = (Y - b)/a$ is read from the standard regression line. The corresponding discrimination interval is either overlooked or dispensed with. This is okay provided that over the range of x values which actually will occur the discrimination intervals will be sufficiently narrow to make the difference between any two values in the interval of no physical or practical significance.

The technique of this section will aid the statistician in figuring how accurately the standard line has to be determined in order for the discrimination intervals to be superfluous as indicated above. Often the initial sample size (n) will be so large that the line $\alpha + \beta x$ is known for all practical purposes (i.e., the confidence band is extremely narrow). In this case the only worry is the inherent size of σ which directly affects the width of the confidence interval on $E\{Y\}$ and hence indirectly the width of the discrimination interval. If σ is too large, repeated determinations (observations) will have to be made for the same x value to reduce σ.

CHAPTER 4

Nonparametric Techniques

This chapter covers those simultaneous techniques whose test statistics have a null distribution which does not depend on the form of the underlying distribution generating the sample observations. A few restrictions are usually placed on the underlying sample distribution, but they are minor in character.

These nonparametric techniques are more universally applicable than the normal techniques of Chap. 2. They are appropriately applied when there is danger of the normal theory procedures leading to fallacious results. This can occur when there is evidence of sufficient nonnormality to make the use of a normal technique a foolhardy venture, even though the normal technique may be reasonably robust. Or, due to the risks involved, great caution in the inference may require the significance levels to be as specified whatever the underlying sample distribution.

Throughout this chapter all sample observations will be assumed to be distributed according to probability densities. This prevents any sample point having positive probability, and thus, theoretically eliminates the possibility of two different observations being equal to the same value (i.e., ties). In practice, of course, ties do occur. Sometimes they can be resolved by finer recording of the observations (i.e., examining more decimals). But other times there is no way of breaking the tie. Provided the occurrence of unbreakable ties is not extensive, the techniques of this chapter can still be applied by treating the ties fairly from the point of view of all the populations (e.g., by assigning the average rank of the tied observations to each of them). The true critical point of the test statistic involving the ties cannot be very different from the tabulated point when the number of ties is quite small. When tying is extensive, however, some alternative method of analysis will have to be found.

The methods of this chapter will all be aimed at studying location parameters. That is, if one population has the density $p(x)$, the method will examine whether another population has a shifted density $p(x - \theta)$ for $\theta \neq 0$. The parameter θ is referred to as the shift or location parameter. This is the type of problem considered so far in Chaps. 1 and 2. The techniques of this chapter are also reasonable procedures to use against more general alternatives such as stochastic shifting [that is, $F_1(x) < F_2(x), \forall x$]. However, the author likes to think of the problems from the location parameter point of view and then, as an afterthought, examine for what wider class of alternatives the test remains reasonable.

Each method is most conveniently introduced from the hypothesis-testing point of view. However, to each significance test there corresponds a confidence interval procedure which will also be discussed. No relative importance is implied in the ordering of hypothesis-testing prior to confidence intervals. The corresponding confidence interval is sometimes tedious to construct numerically, even though its theoretical basis is simple. Fortunately, there is often an easy graphical method of constructing the confidence interval.

The significance tests will be discussed from the single stage (*à la* Tukey) point of view. Any of the techniques can be directly adapted to multistage testing (*à la* Newman-Keuls-Duncan) by shifting the critical points at each stage. There is, of course, no analog to this in the confidence realm.

Nonparametric simultaneous techniques naturally divide themselves into two groups. There are those which utilize the maximum (over all possible pairs of two samples) value of the usual nonparametric statistic for two samples. The major ones of this type are treated in Secs. 1 to 5. The other type treats all the observations together as a whole (viz., the rank of an observation is determined by its ordering with respect to all observations in all the samples). Two principal methods of this type are discussed in Secs. 6 and 7. Section 8 contains additional miscellaneous procedures with shorter discussions for each.

1 MANY-ONE SIGN STATISTICS (STEEL)

This technique was the first nonparametric multiple comparisons procedure. It was introduced by Steel (1959a), and is a nonparametric procedure for the problem posed in Sec. 5 of Chap. 2.

1.1 Method

Let $\{Y_{ij};\ i = 0, 1, \ldots, k,\ j = 1, \ldots, n\}$ be n independent samples of size $k + 1$ with one observation in each sample from each of $k + 1$ different populations. The $k + 1$ observations in each sample are assumed independent. The zero population ($i = 0$) is the *control* population, and the other k populations are *treatment* populations. Each sample $j = 1, \ldots, n$ can be thought of as a block of $k + 1$ observations, one from each control and treatment population. Interest centers on whether all treatment populations in a block have the same density function as the control population in that block, or whether some of the treatment densities have been shifted away from the control.

The model which corresponds to the classical analysis-of-variance structure is

$$p_{Y_{ij}}(x) = p(x - \theta_i - \beta_j) \qquad i = 0, 1, \ldots, k \qquad j = 1, \ldots, n \tag{1}$$

where $p_{Y_{ij}}$ denotes the density function of Y_{ij}, $p(x)$ is a fixed density function, and $\theta_0 = \beta_1 = 0$. The parameter θ_i is the shift for population i, and β_j is the shift for block j. For the classical linear model the density $p(x)$ is taken to be the normal density.

The null hypothesis is that the treatment populations do not differ from the control; that is,

$$H_0\colon\ \theta_i \equiv 0 \qquad i = 1, \ldots, k. \tag{2}$$

This hypothesis can also be stated in terms of the differences between the treatment observations and the control observations. Let $D_{ij} = Y_{ij} - Y_{0j}$, $i = 1, \ldots, k$, $j = 1, \ldots, n$. Under the model (1),

$$p_{D_{ij}}(x) = q(x - \theta_i) \qquad i = 1, \ldots, k \qquad j = 1, \ldots, n \tag{3}$$

where $q(x) = \int_{-\infty}^{+\infty} p(y)p(x + y)\, dy$. The parameter θ_i is the median of the density $p_{D_{ij}}(x)$, and the null hypothesis can be stated as

$$H_0\colon\ \text{median of } D_{ij} = 0 \qquad i = 1, \ldots, k \qquad j = 1, \ldots, n.† \tag{4}$$

The two-sided alternative hypothesis is simply

$$H_1\colon\ \theta_i \not\equiv 0 \qquad i = 1, \ldots, k. \tag{5}$$

If the θ_i are not identically zero, then the medians of D_{ij} are not identically zero. The one-sided alternative hypothesis for positive shifts is

$$H_1\colon\ \theta_i > 0 \qquad \text{for some } i. \tag{6}$$

† The basic model and hypotheses can, in fact, be more general than just described, but it would be confusing at this point to delve further into the basic structure until the test statistic has been defined. A fuller discussion appears at the end of Sec. 1.5.

The corresponding hypothesis with "$<$" is the one-sided alternative for negative shifts.

The Steel many-one sign test consists in computing the usual one-sample sign statistic for each of the k sets of variables $\{D_{ij}, j = 1, \ldots, n\}$, and comparing the maximum value of these k statistics with the appropriate critical point. Specifically, let S_i^+ be the number of positive D_{ij}, $j = 1, \ldots, n$. The variable S_i^+ is equal to the sum $\sum_{j=1}^{n} D_{ij}^+$ where

$$D_{ij}^+ = \begin{cases} 1 & \text{if } D_{ij} > 0 \\ 0 & \text{if } D_{ij} < 0. \end{cases} \tag{7}$$

The one-sided many-one sign test for positive shifts is to reject the null hypothesis when

$$S^+ = \max\{S_1^+, \ldots, S_k^+\} \tag{8}$$

equals or exceeds the upper α percentile point s_+^α of its distribution under the null hypothesis. Note that s_+^α will be a function of α, k, and n.

The corresponding negative counting variables are

$$D_{ij}^- = \begin{cases} 1 & \text{if } D_{ij} < 0 \\ 0 & \text{if } D_{ij} > 0 \end{cases} \tag{9}$$

and $S_i^- = \sum_{j=1}^{n} D_{ij}^- = n - S_i^+$. The one-sided test for negative shifts rejects H_0 when

$$S^- = \max\{S_1^-, \ldots, S_k^-\} \tag{10}$$

equals or exceeds $s_-^\alpha (= s_+^\alpha)$.

The two-sided test which examines for medians θ_i unequal to zero in either direction is to reject H_0 when

$$S^\pm = \max\{S_1^\pm, \ldots, S_k^\pm\} \tag{11}$$

equals or exceeds $s_\pm^\alpha$, the appropriate critical point. The component variables $S_i^\pm$ are the maximum number of counts of one sign, either plus or minus; that is,

$$S_i^\pm = \max\{S_i^+, S_i^-\} = \max\{S_i^+, n - S_i^+\}. \tag{12}$$

Under the model (1) the critical constants $(s_+^\alpha, s_-^\alpha, s_\pm^\alpha)$ can be used to construct simultaneous confidence intervals for the shift parameters $\theta_1, \ldots, \theta_k$. A two-sided simultaneous confidence interval for θ_i is the interval $(D_{i(n-s_\pm^\alpha+1)}, D_{i(s_\pm^\alpha)})$ where $D_{i(1)} < \cdots < D_{i(n)}$ are the order statistics corresponding to $D_{i1}, \ldots, D_{in}$. When all k intervals are taken together, they have a combined probability approximately equal to $1 - \alpha$ of covering the k true shifts $\theta_1, \ldots, \theta_k$. The term *approxi-*

mately is used in the preceding sentence because many values in the existing tables of $s_{\pm}^{\alpha}$ are based on large sample approximations and due to the discreteness of $S_1^{\pm}, \ldots, S_k^{\pm}$ there is no value of $s_{\pm}^{\alpha}$ which actually produces a probability of exactly α in the upper tail (except in accidental cases). The existing tables are thought to be on the conservative side.

The values of $s_{\pm}^{\alpha}$ given in Table VI in Appendix B are all integers. If the reader computes some other $s_{\pm}^{\alpha}$ by an approximation, a nonintegral answer may arise. In constructing the interval $(D_{i(n-s_{\pm}^{\alpha}+1)}, D_{i(s_{\pm}^{\alpha})})$ the nonintegral answer $s_{\pm}^{\alpha}$ should be rounded up to the nearest integer for a conservative interval. For an approximate interval $s_{\pm}^{\alpha}$ can be rounded to the nearest integer, or an interpolated value between the two nearest order statistics can be used.

The one-sided simultaneous confidence intervals for θ_i are

$$(D_{i(n-s_{+}^{\alpha}+1)}, +\infty)$$

for positive shifts and $(-\infty, D_{i(s_{-}^{\alpha})})$ for negative shifts.

1.2 Applications

The range of application of this technique is strictly limited by its narrow design. A number of treatments (k) are to be compared with a standard control. The number of observations (n) must be the same for all treatments and the control, and furthermore, the observations must occur in n blocks of size $k+1$ with one observation on each treatment and the control in a block. If the observations do not naturally occur in blocks, the statistician is faced with either abandoning this technique or randomly grouping the observations into blocks. The latter idea is unpleasant because the statistician would always be haunted by the thought that the results might be due to the grouping instead of the data.

1.3 Comparison

The Steel many-one sign test frees the statistician of the normality assumption imposed by the Dunnett many-one t statistics (Sec. 5 of Chap. 2). The price for this includes the restriction that the observations come in blocks. Also included in the price is the inability of the statistician to put more observations into the control group than in each of the treatment groups. These could be serious drawbacks compared with the risk of the maximum many-one t statistic being somewhat nonrobust.

If the observations do not naturally occur in blocks, it would be foolhardy to randomly assign them to blocks, because the many-one rank test of Sec. 3 is a good nonparametric test for this problem when there are no block effects.

If the observations do indeed arise in blocks, one might also propose a signed-rank test. This is discussed in Sec. 5. However, the signed-rank test is not strictly a competitor because it fails to be truly distribution-free. Its null distribution depends somewhat on the underlying distribution. It may be possible to use the signed-rank procedure as an almost nonparametric alternative to the many-one sign test, but for the moment the known results are too meager to fairly compare them.

Comparison with the technique of Sec. 7 is postponed until this later section.

1.4 Derivation

No derivation is required for the test described above. All that needs be done is compute the critical points (see Sec. 1.5).

The derivation of the confidence intervals from the test structure proceeds as follows. Let $\boldsymbol{\theta} = (\theta_1, \ldots, \theta_k)$. If θ_i is the true amount of shift of treatment population i, then the random variables $Y_{ij}(\theta_i) = Y_{ij} - \theta_i$, $j = 1, \ldots, n$, have the densities $p(x - \beta_j)$, respectively. That is to say, the $Y_{ij}(\theta_i)$ are distributed as in (1) under the null hypothesis (2). This means the differences $D_{ij}(\theta_i) = Y_{ij}(\theta_i) - Y_{0j}$, and the related statistics $D_{ij}^+(\theta_i)$, $D_{ij}^-(\theta_i)$, $S_i^+(\theta_i)$, $S_i^-(\theta_i)$, $S_i^\pm(\theta_i)$, $S^+(\boldsymbol{\theta})$, $S^-(\boldsymbol{\theta})$, and $S^\pm(\boldsymbol{\theta})$, all are distributed as their corresponding test statistics under the null hypothesis. For instance,

$$P\{S^\pm(\boldsymbol{\theta}) = \max\,\{S_1^\pm(\theta_1), \ldots, S_k^\pm(\theta_k)\} < s_\pm^\alpha\} = 1 - \alpha \qquad (13)$$

if $\boldsymbol{\theta}$ is the true shift.

The simultaneous confidence interval for θ_i consists of all values of θ for which $S_i^\pm(\theta) < s_\pm^\alpha$. If the intervals for θ_i, $i = 1, \ldots, k$, are defined in this fashion, then by virtue of (13) the probability is $1 - \alpha$ that these intervals will contain the true θ_i, $i = 1, \ldots, k$, simultaneously.

Let the interval for θ_i be denoted by $(\underline{\theta}_i, \bar{\theta}_i)$. It consists of all θ such that $S_i^\pm(\theta) < s_\pm^\alpha$. Hence, since $D_{ij}(\theta) = D_{ij} - \theta$,

$$\bar{\theta}_i = \sup\,\{\theta\colon \text{number of } D_{ij} - \theta < 0, j = 1, \ldots, n, \text{ is less than } s_\pm^\alpha\}. \qquad (14)$$

But if the $D_{i1}, \ldots, D_{in}$ are arranged in order of magnitude (taking account of sign) $D_{i(1)} < \cdots < D_{i(n)}$, then the supremum in (14) is merely $D_{i(s_\pm^\alpha)}$. For if $\theta < D_{i(s_\pm^\alpha)}$, then subtracting it from all the D_{ij} will still leave $D_{i(s_\pm^\alpha)} - \theta > 0$, which means that less than $s_\pm^\alpha$ values $D_{ij}(\theta)$ are less than zero; and if $\theta > D_{i(s_\pm^\alpha)}$, then there are at least $s_\pm^\alpha$ values $D_{ij}(\theta)$ less than zero. Hence, $\bar{\theta}_i = D_{i(s_\pm^\alpha)}$. A similar argument proves $\underline{\theta}_i = D_{i(n-s_\pm^\alpha+1)}$.

Similar arguments establish the one-sided confidence intervals.

1.5 Distributions and tables

Under the assumptions that the Y_{ij} are independent and their densities have the form (1), the null hypothesis (2) implies that any arrangement of the $k + 1$ observations in a block is equally likely. The observations in block j could be thought of as being generated by a random arrangement over populations of $k + 1$ independent observations distributed according to $p(x - \beta_j)$. In other words, given the values of the order statistics $Y_{(0)j} < \cdots < Y_{(k)j}$, the conditional probability of any arrangement $Y_{0j}, \ldots, Y_{kj}$ of these values is $1/(k + 1)!$.

The values of the statistics D_{ij}^+ and D_{ij}^- for a fixed block j depend on the numerical sizes of $Y_{0j}, \ldots, Y_{kj}$ only through their ordering. Thus, the unconditional distributions of D_{ij}^+ and D_{ij}^- are the same as their conditional distributions conditioned on the values of the order statistics $Y_{(0)j} < \cdots < Y_{(k)j}$.

Consider the vector $\mathbf{D}_j^+ = (D_{1j}^+, \ldots, D_{kj}^+)$. All coordinate entries are 0 or 1, and there are 2^k different possible vectors. Each arrangement of the order statistics $Y_{(0)1} < \cdots < Y_{(k)j}$ gives rise to one of these possible vectors. Hence, the probability of a specific vector $\mathbf{d}^+$ is determined by the number of different arrangements which give this vector; that is,

$$P\{\mathbf{D}_j^+ = \mathbf{d}^+\} = \frac{\text{number of arrangements giving } \mathbf{d}^+}{(k + 1)!}. \tag{15}$$

For illustration, take $k = 3$. There are 3! arrangements giving the vector (1,1,1), namely, those arrangements in which Y_{0j} is the smallest observation, and the observations Y_{1j}, Y_{2j}, Y_{3j} are in any order. There are also 3! arrangements for (0,0,0). All the remaining vectors with one zero or two zeros each have 2! arrangements producing them. For the former (latter) the two highest (lowest) observations can be permuted, thereby leading to 2! permutations.

The vector $\mathbf{S}^+ = (S_1^+, \ldots, S_k^+)$ consists of the sign statistics for each of the k treatments, and is the sum of the within-block comparisons, i.e., $\mathbf{S}^+ = \sum_{j=1}^{n} \mathbf{D}_j^+$. Since the $\mathbf{D}_j^+$ are independent and identically distributed by assumption, the distribution of $\mathbf{S}^+$ is the n-fold convolution of the multivariate distribution (15).

Although it is conceptually simple, it is not particularly easy to compute the n-fold convolution of (15). Let $\mathbf{d}_1^+, \ldots, \mathbf{d}_{2^k}^+$ denote the 2^k different possible vectors obtainable from a single block, and $p_1, \ldots, p_{2^k}$ their corresponding probabilities given by (15). For n blocks let n_i be the number of blocks producing the sign vector $\mathbf{d}_i^+$, $i = 1, \ldots, 2^k$.

Obviously, $\sum_1^{2^k} n_i = n$. For any configuration of blocks $(n_1, n_2, \ldots, n_{2^k})$ the probability of its occurrence is

$$\prod_{i=1}^{2^k} p_i^{n_i} \tag{16}$$

so the probability attached to any specific vector $\mathbf{S}^+ = (S_1^+, \ldots, S_k^+)$ is the sum of the probabilities (16) for those block configurations $(n_1, \ldots, n_{2^k})$ yielding $\mathbf{S}^+$, that is, those $(n_1, \ldots, n_{2^k})$ satisfying

$$\sum_{i=1}^{2^k} n_i \mathbf{d}_i^+ = \mathbf{S}^+ \qquad \sum_{i=1}^{2^k} n_i = n. \tag{17}$$

The distribution of the maximum coordinate, $S^+ = \max \{S_1^+, \ldots, S_k^+\}$, is then obtainable from the multivariate distribution of $\mathbf{S}^+$.

Steel (1959a) has given an algorithm, based on lexicographic ordering of the $\mathbf{d}_i^+$, which mechanizes the computation of the distribution of $\mathbf{S}^+$. He uses this to compute the exact distribution of S^+ for $k = 2$, $n = 4(1)10$, and $k = 3$, $n = 4(1)7$.

The distribution of S^- is, of course, the same as that of S^+. The distribution for $S^\pm$ could be obtained in a similar fashion.

Except for very small k (2 or 3) and small $n(\leq 10)$ the enumerations required in constructing the distribution of S^+ or $S^\pm$ represent a sizable amount of work. Fortunately, approximations based on the asymptotic distributions of $\mathbf{S}^+$ and $\mathbf{S}^\pm$ appear to give good results.

Since $\mathbf{S}^+$ is a sum $\left(\sum_{j=1}^{n} \mathbf{D}_j^+\right)$ of independent, identically distributed random vectors, it is asymptotically (multivariate) normally distributed. Under the null hypothesis

$$\begin{aligned} E\{D_{ij}^+\} &= \tfrac{1}{2} \\ E\{(D_{ij}^+)^2\} &= \tfrac{1}{2} \\ E\{D_{ij}^+ D_{i'j}^+\} &= \frac{2!}{3!} = \frac{1}{3} \qquad \text{for } i \neq i'; \end{aligned} \tag{18}$$

so

$$\begin{aligned} E\{S_i^+\} &= \frac{n}{2} \\ \operatorname{Var}(S_i^+) &= \frac{n}{4} \\ \operatorname{Cor}(S_i^+, S_{i'}^+) &= \tfrac{1}{3} \qquad \text{for } i \neq i'.\dagger \end{aligned} \tag{19}$$

† Cor (X, Y) denotes the correlation coefficient between X and Y.

Thus, an approximate value for s_+^α is the smallest integer equalling or exceeding

$$\frac{n}{2} + \frac{1}{2} + m_{k(\rho)}^\alpha \frac{\sqrt{n}}{2} \tag{20}$$

where $m_{k(\rho)}^\alpha$ is the upper α percentile point of the distribution of the maximum of k equally correlated ($\rho = \frac{1}{3}$) unit normal random variables. The term $\frac{1}{2}$ appearing in (20) is the usual continuity correction. By always rounding upwards in (20) one hopes the critical points are conservative (i.e., the actual significance levels are less than or equal to α).

Gupta (1963) has tabled $m_{k(\rho)}^\alpha$ for $\rho = \frac{1}{3}$, and these tables were used to compute s_+^α in Table VI in Appendix B for $\alpha = .05, .01$; $k = 2(1)10$; $n = 6(1)20(5)50, 100$. A short version of these tables appears in Nemenyi (1963). Steel's original tables were based on Dunnett's tables for $m_{k(\rho)}^\alpha$ with $\rho = \frac{1}{2}$ since those for $\rho = \frac{1}{3}$ were not available. This approximation was quite good since $m_{k(\rho)}^\alpha$ varies only slightly as ρ changes from 0 to $\frac{1}{2}$.

Wherever an exact critical point was available, a comparison of the approximation (20) with the exact point was made. The approximation (20) was found to be excellent even for small n.

Approximate critical points $s_\pm^\alpha$ can be computed similarly. To reject the null hypothesis for large values of $S^\pm$ is exactly equivalent to rejecting it for large or small values of S^+; that is, the one-sided test using $S^\pm$ is the same as the two-sided test using S^+. This is because $S_i^+ + S_i^- \equiv n$ for all i. Consequently, the large sample distribution theory of S^+ suggests that $s_\pm^\alpha$ can be approximated by the smallest integer equalling or exceeding

$$\frac{n}{2} + \frac{1}{2} + |m|_{k(\rho)}^\alpha \frac{\sqrt{n}}{2} \tag{21}$$

where $|m|_{k(\rho)}^\alpha$ is the upper α percentile point of the distribution of the maximum absolute value of k equally correlated ($\rho = \frac{1}{3}$) unit normal random variables.

Unfortunately, tables of $|m|_{k(\rho)}^\alpha$ with $\rho = \frac{1}{3}$ are not readily available. Steel (1959a) derived tables based on Dunnett's (1955) tables with $\rho = \frac{1}{2}$. Since $|m|_{k(\rho)}^\alpha$ changes only slightly from $\rho = \frac{1}{2}$ to $\rho = 0$, the values for $\rho = \frac{1}{2}$ are a good approximation for $\rho = \frac{1}{3}$. Steel's tables are reproduced in Table VI of Appendix B. They give approximate $s_\pm^\alpha$ for $\alpha = .05, .01$; $k = 2(1)9$; $n = 6(1)20$. Nemenyi (1963) gives a short table of $s_\pm^\alpha$ for $k = 2$ or 3 based on the computations of $|m|_{k(\rho)}^\alpha$ for $\rho = \frac{1}{3}$ by Thigpen and David (1961).

The model (1) and the null hypothesis (2) are more specialized than is necessary. In deriving the null distributions of S^+ and $S^\pm$ above, all that was required was that the probability of any permutation $(Y_{0j}, \ldots, Y_{kj})$

of the order statistics in a block ($Y_{(0)j} < \cdot\cdot\cdot < Y_{(k)j}$) be $1/(k+1)!$. Everything followed from this. Obviously, situations more general than (1) and (2) will lead to the permutations in blocks being equally likely. For example, the densities could change between blocks other than by location parameters β_j, just so long as the density was the same for every population in a block. Also, the Y_{ij} need not be independent. If they were interchangeable or exchangeable, the same results would hold.

No results have been obtained for the overall power function of the test. For a single treatment i the marginal power is $P\{S_i^+ \geq s_+^\alpha\}$ or $P\{S_i^\pm \geq s_\pm^\alpha\}$. This can be obtained from binomial tables [e.g., "Tables of the Cumulative Binomial Distribution" (1955)] since S_i^+ has a binomial distribution with parameters n and $p = P\{Y_{ij} > Y_{0j}\}$. Under the null hypothesis $p = \frac{1}{2}$. Any alternative (not necessarily a rigid shift) leading to a value $p \neq \frac{1}{2}$ will have the same marginal power as any other alternative producing the same p. If the block densities differ other than by shifting, then the marginal power will be a function of $p_j = P\{Y_{ij} > Y_{0j}\}$, $j = 1, \ldots, n$.

2 k-SAMPLE SIGN STATISTICS

This procedure is discussed in Nemenyi (1963). [See also, Wormleighton (1959) and Nemenyi (1961).]

The structure of the problem is analogous to the preceding section except that, instead of just the comparisons between k treatments and a control, all the comparisons between k (treatment) populations are of interest.

2.1 Method

Let $\{Y_{ij}; i = 1, \ldots, k, j = 1, \ldots, n\}$ be n independent samples of k independent observations (one from each of k different populations). The density of the random variable Y_{ij} is

$$p_{Y_{ij}}(x) = p(x - \theta_i - \beta_j) \tag{22}$$

where p is a fixed density function, θ_i is the shift for population i, and β_j is the shift for block j. As they stand now, the shift parameters θ_i and β_j are not uniquely defined because p is not specified. Some constraints like $\theta_1 = \beta_1 = 0$ could be imposed, but it is not really necessary because the parametric quantities of interest are the differences $\theta_i - \theta_{i'}$, which are uniquely defined.

Let $D_{ii';j} = Y_{ij} - Y_{i'j}$. Then the differences $D_{ii';j}$ have the density

$$p_{D_{ii';j}}(x) = q(x - (\theta_i - \theta_{i'})) \qquad i,i' = 1, \ldots, k,\ i \neq i' \quad j = 1, \ldots, n \tag{23}$$

where $q(x) = \int_{-\infty}^{+\infty} p(y)p(x + y)\,dy$. The density q is symmetric about zero so $\theta_i - \theta_{i'}$ is the population median of $D_{ii';j}$.

The null hypothesis to be tested is

$$H_0\colon \quad \theta_i \equiv \theta_{i'} \qquad \forall\ i, i' \tag{24}$$

that is, all k populations are the same. The usual alternative is

$$H_1\colon \quad \theta_i \not\equiv \theta_{i'} \qquad \forall\ i, i'. \tag{25}$$

To perform the k-sample sign test compute the statistics

$$\begin{aligned} D^+_{ii';j} &= \begin{cases} 1 & \text{if } D_{ii';j} > 0 \\ 0 & \text{if } D_{ii';j} < 0 \end{cases} \\ S^+_{ii'} &= \sum_{j=1}^{n} D^+_{ii';j} \\ S_{ii'} &= \max \{S^+_{ii'}, n - S^+_{ii'}\} \\ S &= \max_{i,i'} \{S_{ii'}\}. \end{aligned} \tag{26}$$

The rejection region consists of those sample points leading to $S \geq s^\alpha$, where s^α is the upper α percentile point of the distribution of S under the null hypothesis. The critical point s^α will depend on α, k, and n.

Note the distinction between the final statistic S of this section and $S^\pm$ of the preceding section. The individual components of each are computed in the same way, but S takes the maximum over all possible pairs (i,i') with $i \neq i'$ whereas $S^\pm$ takes the maximum over just the pairs $(i,0)$.

The simultaneous two-sided confidence intervals for $\theta_i - \theta_{i'}$ are constructed the same way as in Sec. 1.1. The only change is that s^α replaces $s^\alpha_\pm$. Let $D_{ii';(1)} < \cdots < D_{ii';(n)}$ be the order statistics corresponding to the differences $D_{ii';1}, \ldots, D_{ii';n}$. The confidence interval for $\theta_i - \theta_{i'}$ is $(D_{ii';(n-s^\alpha+1)}, D_{ii';(s^\alpha)})$. When the $\binom{k}{2}$ confidence intervals are taken together, they have a probability (approximately) $1 - \alpha$ of containing the true differences $\theta_i - \theta_{i'}$. The adverb *approximately* arises from the discreteness of S and the necessity of using a large sample approximation for s^α in most cases. The approximation should be pretty good, however.

2.2 Applications

The k-sample sign analysis can be applied whenever k populations are to be compared for location and the observations are produced in blocks, i.e., in a two-way classification. The test examines whether any populations are shifted away from others. No assumptions are placed on the underlying density, and the densities may change from block to block by shifting or other means (see Sec. 2.5). The observations also need not be strictly independent (see Sec. 2.5).

2.3 Comparison

The direct parametric competitor to the k-sample sign test is Tukey's studentized range test (Sec. 1 of Chap. 2). Other possible competitors are the Scheffé F projections (Sec. 2 of Chap. 2) and the Bonferroni t tests (Sec. 3 of Chap. 2). Each of these involves the normality assumption which the sign test does not, but they are applicable to designs other than the two-way classification [e.g., one-way classification (no block effects)].

The Scheffé and Bonferroni tests are known to be reasonably robust. The Tukey test, which depends on an extreme deviate, may be a bit more sensitive. This might give some impetus to the use of the sign test instead of Tukey's in a two-way classification where there is danger of serious nonnormality.

For the Scheffé and Bonferroni tests the numbers of observations on each population need not be the same. They must be the same for the Tukey studentized range and for the sign test.

If the experimental design is a one-way classification rather than a two-way, it would be foolish to randomly group the observations into artificial blocks. The rank test of Sec. 4 is a good nonparametric test which covers this situation. No fallacious results could then be attributable to random pairing.

One might hope to have a signed-rank procedure which was applicable to the two-way classification of this section, and thus be a competitor to the sign test. But the k-sample signed-rank test is not strictly distribution-free. It may be possible to still use it as a nonparametric test, but the information is too meager at this point to make any firm statement. This will be discussed in Sec. 5.

Comparison with the nonparametric technique of Sec. 7 is delayed until this later section.

2.4 Derivation

The test requires no derivation. The computation of the critical point s^α will be covered in the next subsection.

The derivation of the confidence intervals for $\theta_i - \theta_{i'}$ from the test structure is essentially identical to the derivation for the many-one problem (Sec. 1.4). The only changes are to include all comparisons (i vs. i') inside the probability braces and to substitute s^α for $s^\alpha_\pm$.

2.5 Distributions and tables

For the densities (22) and independent Y_{ij}, any arrangement of the k observations within a block is equally likely under the null hypothesis (24). Given the values of the order statistics $Y_{(1)j} < \cdots < Y_{(k)j}$, the conditional probability of any permutation $Y_{1j}, \ldots, Y_{kj}$ is $1/k!$.

Let $\mathbf{D}_j^+ = (D^+_{12;j}, \ldots, D^+_{k-1,k;j})$ be the vector with $\binom{k}{2}$ coordinates giving the positive signs of the differences $Y_{ij} - Y_{i'j}$ for $i < i'$ within the jth block. The coordinate entries are 0 or 1, so there are $2^{\binom{k}{2}}$ different possible vectors with 0 or 1 entries. However, only $k!$ vectors will have positive probability. The other combinations of 0's and 1's lead to impossible (inconsistent) orderings of the Y_{ij}'s, and thus cannot arise in the physical problem. Thus, to determine the distribution of $\mathbf{D}_j^+$ it suffices to determine those $k!$ vectors $\mathbf{d}_j^+$ which lead to consistent orderings of $Y_{1j}, \ldots, Y_{kj}$, and to assign each the probability $1/k!$.

The block vectors $\mathbf{D}_1^+, \ldots, \mathbf{D}_n^+$ are independent by assumption. Thus, the probability attached to any realizable collection $\mathbf{d}_1^+, \ldots, \mathbf{d}_n^+$ is

$$P\{\mathbf{D}_1^+ = \mathbf{d}_1^+, \ldots, \mathbf{D}_n^+ = \mathbf{d}_n^+\} = \left(\frac{1}{k!}\right)^n. \tag{27}$$

From any set $\mathbf{d}_1^+, \ldots, \mathbf{d}_n^+$ of block vectors the sample value s of the random variable S can be computed by (26). Thus, the probability S equals s is $(1/k!)^n$ times the number of different sets $\mathbf{d}_1^+, \ldots, \mathbf{d}_n^+$ of realizable block vectors which lead to the value s.

A complete enumeration of the sets of block vectors producing the various specific values s of S is impossibly tedious if attempted with pencil and paper, unless k and n are extremely small (viz., $k = 3$ or 4, $n = 2$, 3, or 4). There seem to be no easy short cuts to the answer. To perform the enumerations on an electronic digital computer by brute force is feasible for somewhat higher values of k and n. Based on a program written by Howard Givner, Nemenyi (1963) gives the critical points for $k = 3$, $n = 7(1)16$, $\alpha = .05$ and $.01$.

The only way around the dilemma of having so few known exact critical points is through a large sample approximation just as for the many-one problem. The vector $\mathbf{S}^+ = \sum_{j=1}^{n} \mathbf{D}_j^+$ has asymptotically a multivariate normal distribution since it is a sum of independent, identically distributed random vectors. Under the null hypothesis

$$
\begin{aligned}
E\{D^+_{ii';j}\} &= \tfrac{1}{2} \qquad E\{(D^+_{ii';j})^2\} = \tfrac{1}{2} \\
E\{D^+_{ii';j}D^+_{hh';j}\} &= \tfrac{1}{2}\cdot\tfrac{1}{2} = \tfrac{1}{4} \qquad i \neq h, h' \qquad i' \neq h, h' \\
E\{D^+_{ih;j}D^+_{ih';j}\} &= E\{D^+_{hi;j}D^+_{h'i;j}\} = \frac{2!}{3!} = \frac{1}{3} \qquad h \neq h' \\
E\{D^+_{ih;j}D^+_{hi';j}\} &= \frac{1}{3!} = \frac{1}{6} \qquad i < h < i'
\end{aligned}
\tag{28}
$$

so

$$
\begin{aligned}
E\{S^+_{ii'}\} &= \frac{n}{2} \qquad \operatorname{Var}(S^+_{ii'}) = \frac{n}{4} \\
\operatorname{Cor}(S^+_{ii'},S^+_{hh'}) &= 0 \qquad i \neq h, h' \qquad i' \neq h, h' \\
\operatorname{Cor}(S^+_{ih},S^+_{ih'}) &= \operatorname{Cor}(S^+_{hi},S^+_{h'i}) = \tfrac{1}{3} \qquad h \neq h' \\
\operatorname{Cor}(S^+_{ih},S^+_{hi'}) &= -\tfrac{1}{3} \qquad i < h < i'.
\end{aligned}
\tag{29}
$$

To reject H_0 when S is large is exactly equivalent to rejecting H_0 when

$$\max_{i,i'} \left\{ \left| S^+_{ii'} - \frac{n}{2} \right| \right\} \tag{30}$$

is large, since both large values of $S^+_{ii'}$ and small values of $S^+_{ii'}$ (that is, large values of $n - S^+_{ii'}$) lead to large values of $S_{ii'}$. Consequently, a reasonable approximation to s^α is the smallest integer equalling or exceeding

$$\frac{n}{2} + \frac{1}{2} + |m|^\alpha_k \frac{\sqrt{n}}{2} \tag{31}$$

where $|m|^\alpha_k$ is the upper α percentile point of the distribution of the maximum absolute value of $\binom{k}{2}$ unit normal random variables with correlations given by (29).

Tables of $|m|^\alpha_k$ do not exist for the correlation structure of (29). If, instead of $\frac{1}{3}$ and $-\frac{1}{3}$ in (29), the correlations had been $\frac{1}{2}$ and $-\frac{1}{2}$, then $|m|^\alpha_k$ would have equalled $q^\alpha_{k,\infty}/\sqrt{2}$ where $q^\alpha_{k,\infty}$ is the upper α percentile point of the range of k independent, unit normal random variables.[1] This assertion can be checked straightforwardly by computing the correlations between all the differences $Y_i - Y_{i'}$ and $Y_h - Y_{h'}$ whose maximum absolute value gives the range of $Y_1, \ldots, Y_k$. This suggests using

$$\frac{n}{2} + \frac{1}{2} + \frac{q^\alpha_{k,\infty}\sqrt{n}}{2\sqrt{2}} \tag{32}$$

as an approximation to (31). Table VII in Appendix B gives s^α based on this approximation. The approximation seems to be good when checked against the known exact critical values of s^α.

[1] See Sec. 1 of Chap. 2 for the definition of $q^\alpha_{r,\nu}$ for $\nu \leq +\infty$.

Nemenyi (1963) proposes approximating (31) by a $(\frac{2}{3},\frac{1}{3})$ weighting of (32) and the value of $|m|_k^\alpha$ when all the correlations are zero; i.e.,

$$|m|_k^\alpha = N\left((1-\alpha)^{1/\binom{k}{2}}\right) \tag{33}$$

where $N(P)$ is defined by

$$P = \frac{1}{\sqrt{2\pi}} \int_{-N(P)}^{+N(P)} e^{-x^2/2}\, dx. \tag{34}$$

The weighting $(\frac{2}{3},\frac{1}{3})$ is motivated by its being the linear combination of the correlation coefficients $(\frac{1}{2},0)$ which gives the desired correlation $\frac{1}{3}$. There is close agreement between Nemenyi's tables and Table VII in Appendix B.

As in the many-one problem the distributional structure of the Y_{ij} was presented in specialized form for pedagogical convenience. *Any* set of assumptions on the Y_{ij} which leads to independent blocks and to each permutation of the order statistics $Y_{(1)j} < \cdots < Y_{(k)j}$ within a block having the conditional probability $1/k!$ will produce the same null distribution for S. Thus the model and the null hypothesis could be considerably generalized. The densities could vary wildly from block to block, and independence within a block could be replaced by interchangeability.

No results have been obtained on the overall power function of the k-sample sign test. The marginal power functions can be obtained from binomial tables as in the many-one situation (see Sec. 1.5).

3 MANY-ONE RANK STATISTICS (STEEL)

This was the first simultaneous technique based on ranking the observations. It was introduced by Steel (1959b).

3.1 Method

Let $\{Y_{ij};\ i = 0,1, \ldots, k,\ j = 1, \ldots, n_i\}$ be $k + 1$ independent samples consisting of $n_0, n_1, \ldots, n_k$ independent observations, respectively. The zero population ($i = 0$) is the *control* population, and the other k populations are *treatment* populations. Although it is conceptually feasible to allow $n_0, n_1, \ldots, n_k$ to be completely general, computation of the necessary critical constants is currently possible only if the sample sizes are specialized. It will be assumed that $n_0 = m$, and $n_i \equiv n$,

$i = 1, \ldots, k$. This requires all treatment samples to be the same size, but the control sample can be different (presumably larger).

The observations are assumed to have the densities

$$p_{Y_{ij}}(x) = p(x - \theta_i) \tag{35}$$

where $p(x)$ is a fixed (unknown) density function and $\theta_0 = 0$. Note that in comparison with (1) no block parameters appear. The parameter θ_i is the shift of the treatment population i away from the control.

Interest centers on whether the treatment and control populations are all identical,

$$H_0\colon \quad \theta_i = 0 \qquad i = 1, \ldots, k \tag{36}$$

or whether some of the treatment populations differ from the control, either in any direction,

$$H_1\colon \quad \theta_i \not\equiv 0 \qquad i = 1, \ldots, k \tag{37}$$

or in the positive direction,

$$H_1\colon \quad \theta_i > 0 \qquad \text{for some } i. \tag{38}$$

The alternative (38) with "$<$" is for negative shifts.

The Steel many-one rank test compares the largest two-sample Wilcoxon rank statistic for a treatment vs. control with the appropriate critical constant.

For the ith treatment sample $(Y_{i1}, \ldots, Y_{in})$ and the control sample $(Y_{01}, \ldots, Y_{0m})$, combine the two samples into one sample of size $n + m$ and arrange the observations in order of size irrespective of which populations they are from. Assign the ranks $1, 2, \ldots, n + m$ to the ordered observations (1 to the smallest, 2 to the next smallest, etc.). Let $R_{i1}, \ldots, R_{in}$ be the ranks assigned to the treatment observations $Y_{i1}, \ldots, Y_{in}$, respectively. ($R_{01}, \ldots, R_{0m}$ are the ranks of the control observations, but they are not needed.) The rank sum of the treatment observations is $R_i = \sum_{j=1}^{n} R_{ij}$.

To test for positive shifts [i.e., for (38)] the Steel rank test rejects H_0 when

$$R = \max\,\{R_1, \ldots, R_k\} \geq r^\alpha \tag{39}$$

where the critical point r^α is a function of k, n, m, and α. Any population i with $R_i \geq r^\alpha$ is inferred to be shifted positively away from the control.

For the two-sided alternative the critical region consists of those sample points with

$$R^* = \max\,\{R_1^*, \ldots, R_k^*\} \geq r_*^\alpha \tag{40}$$

where

$$R_i^* = \max\,\{R_i, n(n + m + 1) - R_i\}. \tag{41}$$

As above, the appropriate critical point r_*^α is a function of k, n, m, and α. The quantity $n(n + m + 1) - R_i$ appearing in (41) is the rank sum for sample i if the observations were assigned ranks in reverse order (1 to the largest, 2 to the next largest, etc.).

A two-sided (simultaneous) confidence interval for θ_i consists of all values θ for which $R_i^*(\theta) = \max\{R_i(\theta), n(n + m + 1) - R_i(\theta)\} < r_*^\alpha$, where $R_i(\theta)$ is the rank sum for sample i vs. sample 0 when the observations in sample i are taken to be $Y_{i1} - \theta, \ldots, Y_{in} - \theta$. In other words, the procedure for computing R_i above is applied to $Y_{01}, \ldots, Y_{0m}$ and $Y_{i1} - \theta, \ldots, Y_{in} - \theta$ to obtain $R_i(\theta)$; and it, and its conjugate $n(n + m + 1) - R_i(\theta)$, are compared with r_*^α. A one-sided (simultaneous) confidence interval for positive θ_i consists of all θ for which $R_i(\theta) < r^\alpha$.

The confidence intervals presented above are tedious to determine numerically, but can easily be obtained by graphical means. Consider the one-sided case. Compute the number

$$\#^\alpha = nm + \frac{n(n+1)}{2} - r^\alpha. \tag{42}$$

Plot the nm points representing all the different possible pairings of Y_{0j} and Y_{il}; that is, (Y_{01},Y_{i1}), (Y_{01},Y_{i2}), $\ldots$, (Y_{01},Y_{in}), (Y_{02},Y_{i1}), $\ldots$, (Y_{0m},Y_{in}). Slide the ordinate of a line with slope $+1$ (that is, a 45° line) up (or down) the Y_i axis until $\#^\alpha$ points fall below the line and one point falls on it. The interval on the Y_i axis above the ordinate of this line is the confidence interval for θ_i.†

The procedure is illustrated in Fig. 1 with the following purely hypothetical values: $Y_{01} = 1$, $Y_{02} = 3$, $Y_{03} = 4$; $Y_{i1} = 2$, $Y_{i2} = 3.5$; $r^\alpha = 8$. For the critical point $r^\alpha = 8$, expression (42) gives $\#^\alpha = 1$. From Fig. 1 the confidence interval for θ_i is $[-1,+\infty)$. The confidence coefficient in this example is .8. The null hypothesis that $\theta_i = 0$ is not rejected at this level. (Figure 1 appears on page 146.)

In the construction of the confidence interval it is sometimes helpful to subtract a convenient constant from *all* the observations. This will not affect the parametric difference $\theta_i - \theta_0 = (\theta_i - c) - (\theta_0 - c)$, and it can bring all the observations closer to the origin.

Two-sided confidence intervals are computed similarly. For the two-sided critical point r_*^α calculate

$$\#_*^\alpha = nm + \frac{n(n+1)}{2} - r_*^\alpha. \tag{43}$$

† It is theoretically impossible for more than one point to fall on any shifted 45° line. However, when this occurs in practice, the desired line is the one with the lowest ordinate which still has at least $\#^\alpha + 1$ points falling on or below it.

For the same plot of nm points $(Y_{01}, Y_{i1}), \ldots, (Y_{0m}, Y_{in})$ construct two 45° lines. The lower one should have $\#_*^\alpha$ points below it and one lying on it; and the upper should have $\#_*^\alpha$ points above it and one on it. The confidence interval for θ_i is the interval on the Y_i axis between the ordinates of the two 45° lines.

3.2 Applications

The model structure of the Steel many-one rank test is the same as Dunnett's many-one t test (Sec. 5 of Chap. 2) except that the normality assumption is dropped. Treatment populations are to be compared with a single control to see if the treatment populations are identical to the control, or whether some treatment distributions are translations of the control distribution. Simultaneous confidence intervals are available for the amounts of translation.

The test is also a reasonable procedure for testing whether any of the treatment populations are stochastically larger (or smaller) than the control.[1]

The only assumption imposed on the underlying distributions is that they have densities. This considerably broadens the range of application over the many-one t statistics.

Note that it is assumed the observations do not occur in blocks as they must for the many-one sign test (Sec. 1).

[1] Population i is *stochastically larger* than population 0 if $F_0(x) > F_i(x)$, $\forall x$, where F_0, F_i are the distributions corresponding to populations 0 and i, respectively.

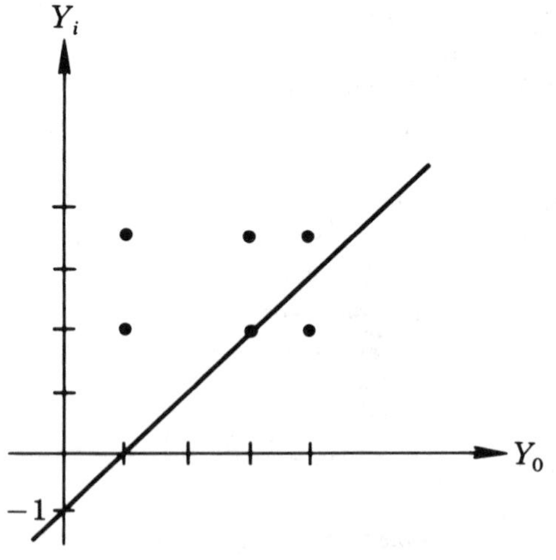

Figure 1

3.3 Comparison

The comparison between the Steel many-one rank test and Dunnett's many-one t test parallels the classic comparison between the Wilcoxon two-sample rank test and the two-sample t test. The breadth of application of the rank test is considerably greater than the t test because of no normality assumption. For normal, or near-normal, distributions the t test will perform more sharply, but the rank test will not lag far behind. [The asymptotic relative efficiency for the k-sample problem is identical with the two-sample problem; see Sherman (1965).] For nonnormal distributions the rank test is apt to be more efficient than the t test, in fact, very much more efficient [cf. Hodges and Lehmann (1956)]. In most cases the rank test will be faster to use than the t test.

The many-one sign test cannot really be considered a competitor since for it to be applied the observations must occur in blocks. If the observations do occur in blocks, the many-one rank test is not applicable. If they do not occur in blocks, it would be madness to randomly create artificial blocks when the rank test is available.

Comparison with the rank technique of Sec. 6 will be made in this later section.

3.4 Derivation

The one-sided many-one rank test (39) requires no further explanation. The form of the two-sided test (40) should be explained in somewhat more detail.

A two-sided critical region should include both large and small values of R_i (large values being indicative of $\theta_i > 0$, small values of $\theta_i < 0$). The smallest value R_i can achieve is $n(n+1)/2$, which is the sum of the n smallest ranks $1, 2, \ldots, n$. The largest value R_i can assume is $nm + n(n+1)/2$ which is the sum of the ranks $m+1, \ldots, m+n$. Depending upon whether the observations are assigned ranks in order of increasing or decreasing size, the rank sum R_i will have the value $n(n+1)/2 + \nu$ or $nm + n(n+1)/2 - \nu$, $\nu = 0, 1, \ldots, nm$. This is a symmetric pairing of the values of R_i about $n(n+m+1)/2$. If R_i has the value r, then its symmetrically paired value is $n(n+m+1) - r$.

Under the null hypothesis the random variables $Y_{01}, \ldots, Y_{0m}$, $Y_{i1}, \ldots, Y_{in}$ are mixed randomly, so any value r of R_i and its symmetric twin $n(n+m+1) - r$ have the same probability. Consequently, the critical region should be symmetrical, which is equivalent to rejecting for large values of $R_i^* = \max\{R_i, n(n+m+1) - R_i\}$.

The method of graphically constructing a confidence interval on θ_i is due to L. E. Moses [see Walker and Lev (1953, chap. 8)]. The proof that the method actually gives the confidence interval hinges on the Mann-Whitney representation of the Wilcoxon statistic. This representation is

$$R_i = \frac{n(n+1)}{2} + \sum_{j=1}^{m} \sum_{l=1}^{n} I_{Y_{0j},Y_{il}} \tag{44}$$

where

$$I_{X,Z} = \begin{cases} 1 & \text{if } X \leq Z \\ 0 & \text{if } X > Z. \end{cases} \tag{45}$$

The characterization (44) is well known, and becomes apparent from the following reasoning. If all the Y_{il}'s precede the smallest Y_{0j}, then $R_i = n(n+1)/2$. Correspondingly, $I_{Y_{0j},Y_{il}} \equiv 0$. As the Y_{0j}'s are shifted toward the left, every time a Y_{0j} hops over a Y_{il} (that is, Y_{0j} goes from $> Y_{il}$ to $< Y_{il}$) it increases the rank of that Y_{il} by 1. Correspondingly, $I_{Y_{0j},Y_{il}}$ switches from 0 to 1. Thus, the identity (44) holds for all permutations of the $n + m$ observations.

For simplicity consider the one-sided interval. Rejection of the null hypothesis for $R_i \geq r^\alpha$ is equivalent by (44) to rejection for

$$\sum_{j=1}^{m} \sum_{l=1}^{n} I_{Y_{0j},Y_{il}} \geq r^\alpha - \frac{n(n+1)}{2} \tag{46}$$

or, since

$$\sum_j \sum_l I_{Y_{0j},Y_{il}} + \sum_j \sum_l I_{Y_{il},Y_{0j}} = nm,$$

$$\sum_{j=1}^{m} \sum_{l=1}^{n} I_{Y_{il},Y_{0j}} \leq nm - \left[r^\alpha - \frac{n(n+1)}{2} \right]. \tag{47}$$

The sum on the left in (47) is simply the number of combinations (Y_{0j}, Y_{il}) in which $Y_{0j} \geq Y_{il}$. The number on the right in (47) is $\#^\alpha$ of (42).

The confidence interval for θ_i consists of all values of θ for which the rank sum $R_i(\theta)$ of the observations $Y_{i1} - \theta, \ldots, Y_{in} - \theta$ (when combined with $Y_{01}, \ldots, Y_{0m}$) is less than r^α (that is, does not reject the null hypothesis). Hence, because of (47), the confidence interval for θ_i consists of all θ for which

$$\sum_{j=1}^{m} \sum_{l=1}^{n} I_{Y_{il}-\theta,Y_{0j}} > \#^\alpha. \tag{48}$$

Consider the nm pairs $(Y_{01}, Y_{i1}), \ldots, (Y_{0m}, Y_{in})$ plotted in the (Y_0, Y_i)-plane. Any points lying on or below the 45° line through the origin have $Y_{il} \leq Y_{0j}$. Similarly, any points lying on or below the line with ordinate θ and slope $+1$ (that is, a shifted 45° line) have $Y_{il} - \theta \leq Y_{0j}$, that is, $I_{Y_{il}-\theta,Y_{0j}} = 1$. The sum $\sum_j \sum_l I_{Y_{il}-\theta,Y_{0j}}$ in (48) is just the number of points lying on or below the line with ordinate θ and slope $+1$.

The confidence interval for θ_i is thus the ordinates of all lines which have greater than $\#^\alpha$ points lying on or below them. The ordinate of the line which passes through one point and has $\#^\alpha$ points lying below it is the greatest lower bound on the ordinates θ having greater than $\#^\alpha$ points lying on or below their lines. Q.E.D.

The reason for switching from counting points above the line (46) to counting points below the line (47) is that the critical number for the latter is usually the smaller number.

By virtue of (48) this graphical method is equivalent to the following numerical method. Compute the nm differences $D_{lj} = Y_{il} - Y_{0j}$. Order them according to their size: $D_{(1)} < D_{(2)} < \cdots < D_{(nm)}$. Then the confidence interval for θ_i is $[D_{(\#^\alpha+1)}, +\infty)$. However, the graphical method should be faster than computing nm differences and ordering them (unless an electronic computer is readily available). [For additional material the reader is referred to Lehmann (1963b).]

The argument for two-sided intervals is very similar. Since it involves rejection for both large and small values of R_i (that is, large values of R_i^*), the interval will be formed from two lines, each using the critical number (43).

3.5 Distributions and tables

Under the null hypothesis the random variables $Y_{01}, \ldots, Y_{kn}$ are all independently, identically distributed. An *ordering* of the observations means a listing of which observation was smallest, which was second smallest, etc. Since there are a total of $m + kn$ observations, there are $(m + kn)!$ different orderings, and under the null hypothesis each ordering is equally likely [i.e., has probability $1/(m + kn)!$].

The ranks R_{ij} and rank sums R_i depend on the sample observations only through their ordering and not on their actual values. From each ordering the values of the rank sums $R_i = r_i$, $i = 1, \ldots, k$, can be determined and arranged in a vector $\mathbf{r} = (r_1, \ldots, r_k)$. Each ordering produces one vector $\mathbf{r}$, but there may be more than one ordering giving rise to the same vector $\mathbf{r}$. It is a many-to-one relationship. Thus, the probability structure of the random vector of rank sums

$$\mathbf{R} = (R_1, \ldots, R_k)$$

is given by

$$P\{\mathbf{R} = \mathbf{r}\} = \frac{\text{number of orderings giving } \mathbf{r}}{(m + kn)!}. \tag{49}$$

From the distribution of $\mathbf{R}$ the distribution of its maximum coordinate, $R = \max \{R_i\}$, can be determined by summing the probabilities for all vectors $\mathbf{r}$ with $\max\{r_i\} \le r$. A similar statement holds for

$$R^* = \max\{R_i^*\}$$

where R_i^* is defined by (41).

There is little that can be done to figure out the distributions of $\mathbf{R}$, R, and R^* other than to enumerate by brute force all possible orderings and to calculate $\mathbf{r}$ for each ordering. With care some of the extreme tail probabilities of R and R^* can be computed without listing all the possible orderings. Steel (1959b) gives a recursive scheme for computing the probabilities, but it appears to be only a slight aid. Steel computes some extreme tail probabilities of R for $k = 2, 3$; $n = 3, 4, 5$.

The construction of the tables of critical points rests on the large sample approximation. This approximation is most easily established through the Mann-Whitney representation (44)–(45) of the rank sums R_i. The sum $I_i = \sum_j \sum_l I_{Y_{0j},Y_{il}}$ appearing in (44) is a two-sample U statistic. Using theorems of Hoeffding (1948), Lehmann (1951) proved that a statistic of this type has a limiting normal distribution provided $n/m \to c$, $0 < c < +\infty$, as $n, m \to +\infty$ [for reference, see Fraser (1957, pp. 229–230)]. A direct generalization of these arguments will establish that the vector $\mathbf{I} = (I_1, \ldots, I_k)$ has a multivariate normal distribution provided $n/m \to c$ as $n,m \to +\infty$ [see Lehmann (1963a)].

The asymptotic moments of the multivariate normal distribution need to be determined. This, again, is accomplished most easily through the representation (44) and (45). For clarity of exposition, specialize to R_1 and R_2 since their means, variances, and covariance are the same as for any pair R_i, $R_{i'}$, $i \ne i'$.

From (44)

$$\begin{aligned} R_1 &= \frac{n(n+1)}{2} + \sum_{j=1}^{m} \sum_{l=1}^{n} I_{Y_{0j},Y_{1l}} \\ R_2 &= \frac{n(n+1)}{2} + \sum_{j=1}^{m} \sum_{l=1}^{n} I_{Y_{0j},Y_{2l}}. \end{aligned} \tag{50}$$

Under the null hypothesis all observations are identically distributed so that

$$\begin{aligned} E\{I_{Y_{0j},Y_{1l}}\} &= P\{Y_{0j} < Y_{1l}\} = \tfrac{1}{2} \\ E\{I^2_{Y_{0j},Y_{1l}}\} &= E\{I_{Y_{0j},Y_{1l}}\} = \tfrac{1}{2} \\ E\{I_{Y_{0j},Y_{1l}} I_{Y_{0j},Y_{1l'}}\} &= P\{Y_{0j} < Y_{1l},\ Y_{0j} < Y_{1l'}\} = \frac{2!}{3!} = \frac{1}{3} \qquad l \ne l' \\ E\{I_{Y_{0j},Y_{1l}} I_{Y_{0j'},Y_{1l}}\} &= P\{Y_{0j} < Y_{1l},\ Y_{0j'} < Y_{1l}\} = \frac{2!}{3!} = \frac{1}{3} \qquad j \ne j' \\ E\{I_{Y_{0j},Y_{1l}} I_{Y_{0j'},Y_{1l'}}\} &= E\{I_{Y_{0j},Y_{1l}}\}E\{I_{Y_{0j'},Y_{1l'}}\} = \tfrac{1}{4} \qquad j \ne j' \quad l \ne l'. \end{aligned} \tag{51}$$

Identical expressions hold when Y_{2l} replaces Y_{1l}.

From (50) and (51) it follows that

$$
\begin{aligned}
E\{R_1\} &= E\{R_2\} = \frac{n(n+1)}{2} + \frac{nm}{2} = \frac{n(n+m+1)}{2} \\
E\{R_1^2\} &= E\{R_2^2\} = \left[\frac{n(n+1)}{2}\right]^2 + 2\left[\frac{n(n+1)}{2}\right]\left[\frac{nm}{2}\right] \\
&\quad + \left[\frac{nm}{2} + \frac{nm(m-1)}{3} + \frac{n(n-1)m}{3} + \frac{n(n-1)m(m-1)}{4}\right] \qquad (52) \\
\mathrm{Var}\,(R_1) &= \mathrm{Var}\,(R_2) = \frac{nm(n+m+1)}{12}.
\end{aligned}
$$

These expressions are well known in nonparametric statistics.

To evaluate Cov (R_1,R_2), some expectations in addition to (51) are needed.

$$
\begin{aligned}
E\{I_{Y_{0j},Y_{1l}} I_{Y_{0j},Y_{2l'}}\} &= P\{Y_{0j} < Y_{1l},\ Y_{0j} < Y_{2l'}\} \\
&= \frac{2!}{3!} = \frac{1}{3} \qquad l = l' \text{ or } l \neq l' \qquad (53) \\
E\{I_{Y_{0j},Y_{1l}} I_{Y_{0j'},Y_{2l'}}\} &= E\{I_{Y_{0j},Y_{1l}}\}E\{I_{Y_{0j'},Y_{2l'}}\} \\
&= \tfrac{1}{4} \qquad j \neq j',\ l = l' \text{ or } l \neq l'.
\end{aligned}
$$

From (50) and (53)

$$
\begin{aligned}
E\{R_1R_2\} &= \left[\frac{n(n+1)}{2}\right]^2 + 2\left[\frac{n(n+1)}{2}\right]\left[\frac{nm}{2}\right] \\
&\qquad + \left[\frac{mn^2}{3} + \frac{m(m-1)n^2}{4}\right] \qquad (54) \\
\mathrm{Cov}\,(R_1,R_2) &= \frac{mn^2}{12}.\dagger
\end{aligned}
$$

Since under the null hypothesis **R** is asymptotically normally distributed with means, variances, and covariances given by (52) and (54), it is reasonable to approximate the critical point r^α by

$$
r^\alpha \cong \frac{n(n+m+1)}{2} + \frac{1}{2} + m^\alpha_{k(\rho)} \sqrt{\frac{nm(n+m+1)}{12}} \qquad (55)
$$

where $m^\alpha_{k(\rho)}$ is the upper α percentile point of the maximum of k equally correlated unit normal random variables with common correlation

$$
\rho = \mathrm{Cor}\,(R_i,R_{i'}) = \frac{n}{n+m+1}. \qquad (56)
$$

The constant $\frac{1}{2}$ is a standard continuity correction. Gupta (1963) has tabled $m^\alpha_{k(\rho)}$ for various values of the common correlation coefficient. In the special case $n = m$, (55) reduces to

$$
r^\alpha \cong \frac{n(2n+1)}{2} + \frac{1}{2} + m^\alpha_{k(\rho)} \sqrt{\frac{n^2(2n+1)}{12}} \qquad (57)
$$

† If populations 0, 1, and 2 have samples of sizes m, n, and p, then Cov $(R_1,R_2) = mnp/12$. This can be proved in the same way as (54).

and (56) to $\rho = n/(2n+1)$. Based on Gupta's tables the values of (57) were computed (and rounded up) for $\alpha = .05, .01$; $k = 2(1)10$; $n = 6(1)20(5)50, 100$. These appear in Table VIII in Appendix B. The agreement between (57) and the few known exact values for small k and n (the worst case) is good.

For the two-sided critical point, the appropriate approximation is

$$r_*^\alpha \cong \frac{n(n+m+1)}{2} + \frac{1}{2} + |m|_{k(\rho)}^\alpha \sqrt{\frac{nm(n+m+1)}{12}} \tag{58}$$

where $|m|_{k(\rho)}^\alpha$ is the upper α percentile of the maximum absolute value of k equally correlated unit normal random variables with common correlation (56). Unfortunately, tables of $|m|_{k(\rho)}^\alpha$ for a general common correlation coefficient do not exist. In the special case $n = m$ the correlation coefficient tends monotonically to $\frac{1}{2}$ (starting at $\frac{1}{3}$). Steel therefore suggests approximating $|m|_{k(\rho)}^\alpha$ by the critical point of Dunnett's t statistic (with d.f. $= +\infty$) which has common correlation $\rho = \frac{1}{2}$. Dunnett's (1955) values are themselves conservative approximations (see Sec. 5.5 of Chap. 2). Steel's tables based on these approximations are reproduced in Table VIII in Appendix B.

No results or tables exist to date for the power function of this test. The marginal power for one treatment is just the probability that a two-sample Wilcoxon statistic will exceed a critical number which is larger than the critical point for the ordinary two-sample test. Again, there are no general tables or results available for small samples. For large samples the marginal power is obtainable from regular normal tables.

Consider treatment 1 (with cdf F_1) and the control (with cdf F_0). Asymptotically ($n/m \to c$ as $n,m \to +\infty$) the rank sum R_1 is normally distributed under both the null and alternative hypotheses [see Lehmann (1951)]. Its asymptotic moments are

$$\begin{aligned} E\{R_1\} &= \frac{n(n+1)}{2} + nm \int_{-\infty}^{+\infty} F_0(x)\, dF_1(x) \\ \operatorname{Var}(R_1) &= nm \Big\{ \int_{-\infty}^{+\infty} F_0(x)\, dF_1(x) \\ &\qquad + (n-1) \int_{-\infty}^{+\infty} [1 - F_1(x)]^2\, dF_0(x) \\ &\qquad + (m-1) \int_{-\infty}^{+\infty} F_0^2(x)\, dF_1(x) \\ &\qquad\qquad - (m+n-1) \Big[\int_{-\infty}^{+\infty} F_0(x)\, dF_1(x) \Big]^2 \Big\} \end{aligned} \tag{59}$$

[see Mann and Whitney (1947)]. For the null case the moments in (59) reduce to (52).

For a one-sided test the critical point is r^α in (57), and the asymptotic power is approximately

$$1 - \Phi\left(\frac{r^\alpha - E\{R_1\}}{\sqrt{\text{Var}\,(R_1)}}\right) \tag{60}$$

where $E\{R_1\}$, Var (R_1) are given by (59). This approximation is likely to be fairly good even for small samples ($n = 7$, $m = 7$?). A two-tailed expression similar to (60) holds for the two-sided test.

The asymptotic moments (59) do not simplify for translation alternatives [that is, $F_1(x) = F_0(x - \theta_1)$]. However, they do simplify for Lehmann-type alternatives of stochastic shifting [that is, $F_1(x) = F_0^\gamma(x)$]. For this type of alternative, (59) reduces to

$$\begin{aligned} E\{R_1\} &= \frac{n(n+1)}{2} + nm\left(\frac{\gamma}{\gamma+1}\right) \\ \text{Var}\,(R_1) &= \frac{\gamma}{nm(\gamma+1)^2}\left[\frac{m-1}{\gamma+2} + \frac{\gamma(n-1)}{2\gamma+1} + 1\right]. \end{aligned} \tag{61}$$

Lehmann (1953) gives some power charts for alternatives of this type.

4 k-SAMPLE RANK STATISTICS

This procedure was proposed independently by Steel (1960) and Dwass (1960). It was discussed further by Steel (1961) [see also Nemenyi (1961)].

The model of this section is identical to that in the preceding section, but the problem is to make all pairwise comparisons between populations rather than comparing a group of populations with a control population.

4.1 Method

Let $\{Y_{ij};\ i = 1, \ldots, k,\ j = 1, \ldots, n\}$ be k independent samples with n independent observations in each sample. The unbalanced case of unequal sample sizes is not considered because of the impossibility of computing critical points. The observations are assumed to have the densities

$$p_{Y_{ij}}(x) = p(x - \theta_i) \tag{62}$$

where $p(x)$ is a fixed (unknown) density function. The parameters θ_i are not uniquely defined because of the arbitrariness of the density. However, the differences $\theta_i - \theta_{i'}$ are uniquely defined, and these are the parametric quantities of interest. If the reader wants θ_i to be uniquely defined, he can take $\theta_1 = 0$.

The null hypothesis is that the k populations do not differ; that is,

$$H_0: \quad \theta_i \equiv \theta_{i'} \qquad \forall\, i,\, i'. \tag{63}$$

The alternative is simply the negation of this:

$$H_1: \quad \theta_i \not\equiv \theta_{i'} \qquad \forall\, i,\, i'. \tag{64}$$

The Steel-Dwass k-sample rank statistic is the maximum Wilcoxon two-sample rank statistic over all possible pairs of populations. The observations for samples i and i' are $Y_{i1}, \ldots, Y_{in}$, and $Y_{i'1}, \ldots, Y_{i'n}$, respectively. Let $R_{i1:i'}, \ldots, R_{in:i'}$ be the ranks of the observations $Y_{i1}, \ldots, Y_{in}$ with respect to the combined sample of size $2n$. The rank sum for sample i with respect to sample i' is $R_{ii'} = \sum_{j=1}^{n} R_{ij:i'}$, and the appropriate two-sided statistic is

$$R^*_{ii'} = \max \{R_{ii'},\, n(2n + 1) - R_{ii'}\}. \tag{65}$$

For any pair i, i' only one rank sum needs to be computed, since $R_{i'i} = n(2n + 1) - R_{ii'}$ implies $R^*_{ii'} = R^*_{i'i}$. The k-sample rank statistic is the maximum of these $\binom{k}{2}$ symmetrized rank sums $R_{ii'}$; that is,

$$R^{**} = \max_{i,i'} \{R^*_{ii'}\}. \tag{66}$$

The distinction between R^* of (40) and R^{**} of (66) is that the latter is the maximum over the $\binom{k}{2}$ comparisons between k treatments and the former is the maximum over the k comparisons between k treatments and a control.

The test of the null hypothesis (63) is to reject H_0 when R^{**} equals or exceeds the appropriate critical point r^{α}_{**}. The critical point r^{α}_{**} is a function of k, n, and α.

Two-sided confidence intervals for $\theta_i - \theta_{i'}$ can be constructed as in Sec. 3.1 for the parametric difference $\theta_i - \theta_0$. Plot the n^2 points $(Y_{i'1}, Y_{i1})$, $(Y_{i'1}, Y_{i2}), \ldots, (Y_{i'n}, Y_{in})$ in the $(Y_{i'}, Y_i)$-plane, and compute the number

$$\#^{\alpha}_{**} = n^2 + \frac{n(n + 1)}{2} - r^{\alpha}_{**}. \tag{67}$$

Construct a line with slope $+1$ which passes through one point and has $\#^{\alpha}_{**}$ points lying below it. Construct another line with the same slope which passes through one point and has $\#^{\alpha}_{**}$ points lying above it. The interval on the Y_i axis between the ordinates of these two lines is the confidence interval for $\theta_i - \theta_{i'}$.

For a numerical example the reader is referred to Sec. 3.1. As in the many-one case a convenient constant can be subtracted from all the observations without affecting the confidence intervals.

4.2 Applications

The Steel-Dwass k-sample rank test can be applied to testing for differences in location of k populations when the sample sizes of all the populations are equal. It is the direct analog for ranks of the Tukey studentized range test. The rank distribution does not involve the assumption that the observations are normally distributed, as does the studentized range. However, the rank statistic is applicable only to one-way classifications. It cannot be applied in the presence of block effects as the studentized range can.

The k-sample rank test is a reasonable procedure to use in testing for general stochastic shifting as well as rigid shifts of location.

Nonparametric confidence intervals for the amount of translation between two populations are available from rank tests. Note that for both the many-one and k-sample problems the confidence intervals are applicable only to rigid translations of the distributions. There can be no distortion of the distribution except for location change.

4.3 Comparison

The comparison between the Steel-Dwass rank test and the studentized range test has already been made in Sec. 3.3 for the many-one problem. The rank test has speed, no normality assumption, and greater efficiency for nonnormal alternatives on its side. The studentized range has greater efficiency for normal or near-normal alternatives as well as the ability to be applied to two-way and higher-way classifications.

The critical point of the studentized range is approximately correct for slightly unbalanced designs. This is also true for the rank test if rank averages are used instead of rank sums.

The k-sample sign test is not really a competitor because it is designed for the two-way classification. It would be foolish to create fictitious blocks in order to apply the k-sample sign test when the k-sample rank test is available.

Sec. 6.3 will contain the comparison of the Kruskal-Wallis-type rank technique with the Dwass-Steel rank technique.

4.4 Derivation

The k-sample rank test and confidence intervals require no derivation other than what has already been given for the many-one problem. The reader is referred to Sec. 3.4.

4.5 Distributions and tables

The story for the k-sample rank statistics is very much the same as in the many-one problem.

Under the null hypothesis each of the $(kn)!$ different possible orderings of the observations is equally likely. Each ordering gives a value to the $\binom{k}{2}$-dimensional vector of rank sums $\mathbf{R}^* = (R_{12}^*, R_{13}^*, \ldots, R_{k-1,k}^*)$. The probability attached to any particular $\mathbf{r}^*$ is simply the number of orderings which give the value $\mathbf{R}^* = \mathbf{r}^*$ divided by $(kn)!$. The probability distribution of $R^{**} = \max_{i<i'} \{R_{ii'}^*\}$ can be computed from the distribution of $\mathbf{R}^*$ once this has been determined.

For small samples the calculation of the distribution of $\mathbf{R}^*$ rests essentially on the complete enumeration of the possible orderings. Some extreme tail probabilities can be computed by carefully listing the extreme orderings. Steel (1960) gives a recursive scheme for computing the probabilities, but it does not seem to be of great assistance. Steel has calculated the exact distribution of R^{**} for $k = 3$; $n = 2, 3, 4$, and the lower tail of the distribution for $k = 3$; $n = 5, 6$.

The existing tables of critical points [Table IX of Appendix B and Steel (1961)] depend on the large sample approximation. Asymptotically ($n \to +\infty$) the vector $\mathbf{R} = (R_{12}, R_{13}, \ldots, R_{k-1,k})$ has a multivariate normal distribution. This was asserted by Dwass (1961) with the details of the proof omitted, and it can be established in a variety of ways (cf. Sec. 3.5). To apply the approximation all that is required is the asymptotic moments.

The computation of the moments has already been carried out in Sec. 3.5. In particular, it was shown there (specializing to $m = n$) that

$$E\{R_{ii'}\} = \frac{n(2n+1)}{2}$$
$$\operatorname{Var}(R_{ii'}) = \frac{n^2(2n+1)}{12} \tag{68}$$

and for $i' \neq i''$

$$\operatorname{Cor}(R_{ii'}, R_{ii''}) = \frac{n}{2n+1}. \tag{69}$$

It is also true that

$$\operatorname{Cor}(R_{ii'}, R_{hh'}) = 0 \qquad i \neq h,h' \qquad i' \neq h,h' \tag{70}$$

and

$$\operatorname{Cor}(R_{ii'}, R_{i'i''}) = \frac{-n}{2n+1} \qquad i < i' < i''. \tag{71}$$

The correlation (70) is obvious because the four samples involved are all different, and (71) follows from (69).

The distribution of the maximum modulus of a multivariate normal with the correlation structure (69) to (71) has not been tabulated. However, if (69) to (71) are replaced by their limiting values $\frac{1}{2}$, 0, $-\frac{1}{2}$, respectively, then the distribution of the maximum modulus is the same as the distribution of the range of k independent normal random variables. This suggests the approximation

$$r_{**}^{\alpha} \cong \frac{n(2n+1)}{2} + \frac{1}{2} + q_{k,\infty}^{\alpha} \sqrt{\frac{n^2(2n+1)}{24}}. \tag{72}$$

The extra 2 appears in the denominator of the variance because the differences of the unit normal variables $Z_i - Z_{i'}$, whose maximum modulus is the range, have variance 2. The approximation (72) was used to compute Table IX of Appendix B.

The behavior of the overall power function of this k-sample rank test is unknown. The structure of the marginal power function is the same as for the many-one problem, except that the critical point r_{**}^{α} replaces r_{*}^{α}. The reader is referred to Sec. 3.5 for details.

5 SIGNED-RANK STATISTICS

Since a multiple comparisons rank procedure exists for observations in a one-way classification, and a sign procedure exists for observations grouped in blocks, one would naturally assume the existence of a multiple comparisons signed-rank procedure in which ranks are applied to observations occurring in blocks. This is not the case. The signed-rank procedure can be defined all right, but it is not distribution-free. The null distribution of the test statistic depends upon the distribution generating the sample observations.

There is no difference between the many-one and k-sample problems. For simplicity, consider the many-one problem with two treatment populations. Let the $3n$ observations $\{Y_{ij};\ i = 0, 1, 2, j = 1, \ldots, n\}$ be independently distributed with the densities:

$$p_{Y_{ij}}(x) = p(x - \theta_i - \beta_j) \qquad i = 0, 1, 2 \qquad j = 1, \ldots, n. \tag{73}$$

The density p is fixed and unknown. For uniqueness of the parameters define $\theta_0 = \beta_0 = 0$. The null hypothesis is

$$H_0\colon\ \theta_1 = \theta_2 = 0. \tag{74}$$

The signed-rank statistic for each treatment vs. control is defined exactly as in the usual two-sample problem. Let $U_j = Y_{1j} - Y_{0j}$,

$V_j = Y_{2j} - Y_{0j}, j = 1, \ldots, n$. For the U_i's compute $|U_1|, \ldots, |U_n|$, and rank them in order of size: $|U|_{(1)} < |U|_{(2)} < \cdots < |U|_{(n)}$. Assign rank 1 to $|U|_{(1)}$, rank 2 to $|U|_{(2)}$, etc. For the jth observation U_j let R_{1j} be the rank of its absolute value. Define

$$SR_{1j}^{+} = I_{|U_j|,U_j} R_{1j} \qquad j = 1, \ldots, n \tag{75}$$

where

$$I_{X,Z} = \begin{cases} 1 & \text{if } X \leq Z \\ 0 & \text{if } X > Z. \end{cases} \tag{76}$$

The signed-rank SR_{1j}^{+} is zero if U_j is negative, and equals the rank of $|U_j|$ if U_j is positive. Define

$$SR_1^{+} = \sum_{j=1}^{n} SR_{1j}^{+}. \tag{77}$$

The signed-rank sum SR_1^{+} is the sum of the positive signed-ranks. From the V_j's obtain the analogous statistic SR_2^{+}.

The negative signed-ranks are

$$SR_{ij}^{-} = -I_{|U_j|,-U_j} R_{ij} \qquad i = 1,2 \qquad j = 1, \ldots, n. \tag{78}$$

The negative signed-rank sums are $SR_i^{-} = \sum_{j=1}^{n} SR_{ij}^{-}$, $i = 1,2$. The positive and negative signed-rank sums are related through the identity $SR_i^{+} + |SR_i^{-}| = n(n+1)/2, i = 1,2$.

To test for positive shifts the appropriate simultaneous statistic is

$$SR^{+} = \max \{SR_1^{+}, SR_2^{+}\}. \tag{79}$$

One wants to reject H_0 for large values of SR^{+} where the critical point is a function of n and α, but not of the density $p(x)$ in (73).

The coordinate statistics SR_i^{+}, $i = 1,2$, individually are distribution-free (under the null hypothesis). Each is a standard two-sample signed-rank statistic. However, their joint distribution depends upon the underlying density $p(x)$ (except in the case $n = 1$). This was proved by Hollander (1965). Hollander studied the joint distribution of (SR_1^{+}, SR_2^{+}) in testing the null hypothesis of a k-sample problem against ordered alternatives, and he found the distribution involved the density $p(x)$.

The pair $\mathbf{SR}^{+} = (SR_1^{+}, SR_2^{+})$ is not even asymptotically $(n \rightarrow +\infty)$ distribution-free. Hollander proved that asymptotically $\mathbf{SR}^{+}$ has a bivariate normal distribution. The means and variances are free of the underlying density, but the correlation is not. The correlation varies with $p(x)$ in finite sample sizes and in the limit.

From the theory of the usual two-sample signed-rank statistic it is well known that

$$E\{SR_1^+\} = E\{SR_2^+\} = \frac{n(n+1)}{4}$$
$$\mathrm{Var}\,(SR_1^+) = \mathrm{Var}\,(SR_2^+) = \frac{n(n+1)(2n+1)}{24}. \tag{80}$$

These can be proved easily by a direct attack, or they can be obtained from the representations (81) or (82) below.

The crucial computation is the covariance, or equivalently, the cross-product moment $E\{SR_1^+SR_2^+\}$. This is obtainable from either of the following representations of SR_1^+ and SR_2^+:

$$SR_1^+ = \sum_{i=1}^{n}\sum_{h=1}^{n} I_{|U_h|,U_i} \qquad SR_2^+ = \sum_{j=1}^{n}\sum_{k=1}^{n} I_{|V_k|,V_j} \tag{81}$$

or

$$SR_1^+ = \sum_{h\le i} I_{0,U_h+U_i} \qquad SR_2^+ = \sum_{k\le j} I_{0,V_k+V_j}\,. \tag{82}$$

The indicator function $I_{X,Z}$ appearing in (81) and (82) is defined in (76). The representation (82) is due to Tukey.

The representation (81) for SR_1^+ is the sum of the coordinate representations

$$SR_{1i}^+ = \sum_{h=1}^{n} I_{|U_h|,U_i}\,. \tag{83}$$

The positive signed-rank SR_{1i}^+ is zero if U_i is negative, and accordingly all the indicator terms $I_{|U_h|,U_i}$ in (83) are zero. If U_i is positive, SR_{1i}^+ equals the rank of $|U_i|$, and the rank of $|U_i|$ is one plus the number of $|U_h|$ which are smaller than $|U_i|$. But the sum in (83) has an indicator term equal to one for each $|U_h|$ smaller than $|U_i| = U_i$, and $I_{|U_i|,U_i} = 1$. All other terms are zero. Hence, the identity (83) holds.

The representation (82) for SR_1^+ follows from (81). The diagonal terms in the double sum (81) give the diagonal terms of (82); i.e.,

$$I_{|U_i|,U_i} = I_{0,2U_i}\,. \tag{84}$$

The sum of each pair of symmetrically positioned off-diagonal terms in (81) gives the subdiagonal term in (82) in the same position; that is,

$$I_{|U_h|,U_i} + I_{|U_i|,U_h} = I_{0,U_h+U_i}\,. \tag{85}$$

The identity (85) is checked by considering the three possible cases. For $U_h, U_i < 0$, $I_{|U_h|,U_i} = I_{|U_i|,U_h} = I_{0,U_h+U_i} = 0$. If $0 < |U_h| < U_i$, then $I_{|U_h|,U_i} = 1$, $I_{|U_i|,U_h} = 0$, and $I_{0,U_h+U_i} = 1$; and if $0 < U_i < U_h$, then $I_{|U_h|,U_i} = 0$, $I_{|U_i|,U_h} = 1$, and $I_{0,U_h+U_i} = 1$.

The representations for SR_2^+ have identical proofs.

In terms of the representation (81) the cross-product moment becomes

$$E\{SR_1^+SR_2^+\} = \sum_{i=1}^{n}\sum_{h=1}^{n}\sum_{j=1}^{n}\sum_{k=1}^{n} E\{I_{|U_h|,U_i}I_{|V_k|,V_j}\}. \tag{86}$$

Since $I_{|U_h|,U_i}I_{|V_k|,V_j}$ equals one if $|U_h| \leq U_i$ and $|V_k| \leq V_j$ and equals zero otherwise,

$$E\{I_{|U_h|,U_i}I_{|V_k|,V_j}\} = P\{|U_h| \leq U_i,\ |V_k| \leq V_j\}. \tag{87}$$

The computation of (86) thus reduces to calculating the probabilities $P(h,i,k,j) = P\{|U_h| \leq U_i,\ |V_k| \leq V_j\}$ for the various possible combinations h, i, k, j.

The probabilities $P(h,i,k,j)$ are exhaustively listed below. After each combination is listed for which $P(h,i,k,j)$ is to be computed, the number of terms of this type in the quadruple sum (86) is noted in brackets. In the expressions to follow $F_{U,V}$ represents the joint cdf of (U_i,V_i) for any i. This cdf depends on $p(x)$. The marginal distributions of U_i and V_i for any i are denoted by F_U and F_V, respectively. The cdf's $F_{|U|,|V|}$, $F_{|U|,V}$, $F_{U,|V|}$, $F_{|U|}$, $F_{|V|}$ are defined analogously.

The important thing to remember in the calculations below is that U_i and V_i are not independent for the same i, but (U_i,V_i) and (U_j,V_j) are independent for $i \neq j$. The symmetry property $F_{U,V}(u,v) = F_{V,U}(u,v)$ is also used.

For h, i, k, j all different: $[n(n-1)(n-2)(n-3)]$,

$$P(h,i,k,j) = \tfrac{1}{4}\cdot\tfrac{1}{4} = \tfrac{1}{16}. \tag{88}$$

For $h = k$; $h \neq i, j$; $i \neq j$: $[n(n-1)(n-2)]$,

$$P(h,i,h,j) = \iint_0^\infty [1 - F_U(u)][1 - F_V(v)]\, dF_{|U|,|V|}\,(u,v). \tag{89}$$

For $i = j$; $i \neq h, k$; $h \neq k$: $[n(n-1)(n-2)]$,

$$P(h,i,k,i) = \iint_0^\infty F_{|U|}(u)F_{|V|}(v)\, dF_{U,V}\,(u,v). \tag{90}$$

For $h = j$; $i \neq h, k$; $k \neq h$: $[2n(n-1)(n-2)]$,

$$P(h,i,k,h) = \iint_0^\infty [1 - F_U(u)]F_{|V|}(v)\, dF_{|U|,V}\,(u,v). \tag{91}$$

For $i = j$; $h = k$; $i \neq h$: $[n(n-1)]$,

$$P(h,i,h,i) = \iint_0^\infty F_{|U|,|V|}(u,v)\, dF_{U,V}\,(u,v). \tag{92}$$

For $i = k;\ h = j;\ i \neq h$: $[n(n-1)]$,

$$P(h,i,i,h) = \iint_0^\infty [F_{|U|}(u) - F_{|U|,V}(u,v)]\, dF_{U,|V|}\,(u,v). \tag{93}$$

For $i = h;\ i \neq k,\ j;\ k \neq j$: $[2n(n-1)(n-2)]$,

$$P(i,i,k,j) = \tfrac{1}{2}\cdot\tfrac{1}{4} = \tfrac{1}{8}\cdot \tag{94}$$

For $i = h = k;\ i \neq j$: $[2n(n-1)]$,

$$P(i,i,i,j) = \iint_0^\infty [1 - F_V(v)]\, dF_{U,|V|}\,(u,v). \tag{95}$$

For $i = h = j;\ k \neq i$: $[2n(n-1)]$,

$$P(i,i,k,i) = \iint_0^\infty F_{|V|}(v)\, dF_{U,V}\,(u,v). \tag{96}$$

For $i = h;\ j = k;\ i \neq j$: $[n(n-1)]$,

$$P(i,i,j,j) = \tfrac{1}{2}\cdot\tfrac{1}{2} = \tfrac{1}{4}\cdot \tag{97}$$

For $i = h = k = j$: $[n]$,

$$P(i,i,i,i) = \tfrac{1}{3}\cdot \tag{98}$$

Multiplication of the probabilities (88) to (98) by the number of times they appear in (86), summation, and simplification yields

$$\begin{aligned} E\{SR_1^+SR_2^+\} &= n(n-1)(n-2)(n-3)[\tfrac{1}{16}] \\ &\quad + n(n-1)(n-2)\Big[\tfrac{1}{4} + \iint_0^\infty F_U(u)F_V(v)\, dF_{|U|,|V|}\,(u,v) \\ &\quad + \iint_0^\infty F_{|U|}(u)F_{|V|}(v)\, dF_{U,V}\,(u,v) \\ &\quad - 2\iint_0^\infty F_U(u)F_{|V|}(v)\, dF_{|U|,V}\,(u,v)\Big] \\ &\quad + n(n-1)\Big[\tfrac{3}{2} + \iint_0^\infty F_{|U|,|V|}(u,v)\, dF_{U,V}\,(u,v) \\ &\quad + 2\iint_0^\infty F_{|V|}(v)\, dF_{U,V}(u,v) - 2\iint_0^\infty F_V(v)\, dF_{U,|V|}\,(u,v) \\ &\quad - \iint_0^\infty F_{|U|,V}(u,v)\, dF_{U,|V|}\,(u,v)\Big] + n[\tfrac{1}{3}]. \end{aligned} \tag{99}$$

Hollander uses the representation (82) to obtain the following expression for the cross-product moment:

$$E\{SR_1^+SR_2^+\} = n(n-1)(n-2)(n-3)[\tfrac{1}{16}] + n(n-1)(n-2)[\tfrac{1}{4} + \lambda(F)] + n(n-1)[\tfrac{5}{12} + 2\mu(F)] + n[\tfrac{1}{3}] \quad (100)$$

where

$$\begin{aligned} \lambda(F) &= P\{Y_1 < Y_2 + Y_3 - Y_4,\ Y_1 < Y_5 + Y_6 - Y_7\} \\ \mu(F) &= P\{Y_1 < Y_2,\ Y_1 < Y_3 + Y_4 - Y_5\} \end{aligned} \quad (101)$$

and Y_i, $i = 1, \ldots, 7$, are independently, identically distributed according to $F(x) = \int_{-\infty}^{x} p(y)\,dy$. By juggling the probabilities the expressions (99) and (100) can be shown to be equal.

From (99) or (100) and (80) the correlation between SR_1^+ and SR_2^+ can be computed for any finite n. In the limit as $n \to +\infty$ this correlation has the limiting value

$$\rho_\infty = 12\Big[\iint_0^\infty F_U(u)F_V(v)\,dF_{|U|,|V|}(u,v) + \iint_0^\infty F_{|U|}(u)F_{|V|}(v)\,dF_{U,V}(u,v) - 2\iint_0^\infty F_U(u)F_{|V|}(v)\,dF_{|U|,V}(u,v)\Big] - 3 \quad (102a)$$

or

$$\rho_\infty = 12\lambda(F) - 3. \quad (102b)$$

The probability $\lambda(F)$ has arisen in other nonparametric problems [cf. Lehmann (1964)]. Lehmann has shown that $\lambda(F)$ must fall between $\frac{1}{4}$ and $\frac{7}{24}$ which means that

$$0 < \rho_\infty \le \tfrac{1}{2}.\dagger \quad (103)$$

He has also calculated $\lambda(F)$ for some particular F: for F normal, $\lambda(F) = .2902$; for F rectangular, $\lambda(F) = .2909$; and for F Cauchy, $\lambda(F) = .2879$. The upper bound $\frac{7}{24}$ (to four decimals) is .2917.

The signed-rank test could be made asymptotically distribution-free by estimating $\lambda(F)$. Consistent estimates of $\lambda(F)$ can be readily constructed. For instance, divide the number of blocks into groups containing three blocks each. In the triplet of blocks $n = 3m + 1, 3m + 2, 3m + 3$ let

$$I_m = \begin{cases} 1 & \text{if } Y_{0,3m+1} < Y_{1,3m+1} + Y_{1,3m+2} - Y_{0,3m+2} \text{ and} \\ & \quad Y_{0,3m+1} < Y_{2,3m+1} + Y_{2,3m+3} - Y_{0,3m+3} \\ 0 & \text{otherwise.} \end{cases} \quad (104)$$

The sequence I_m, $m = 0, 1, 2, \ldots$, is a sequence of independent Bernoulli random variables with $p = E\{I_m\} = \lambda(F)$, so the mean $\bar{I}_.$ is a

† For the strict inequality on the left see Hollander (1965).

consistent estimate of $\lambda(F)$. If for each n the critical point of the signed-rank test is computed on the basis of a bivariate normal distribution with correlation $12\bar{I} - 3$, then the test is asymptotically distribution-free.

Research on asymptotically distribution-free procedures of this type is still in progress. Until more information is available on the behavior of these procedures for small n as well as large, the reader uses these procedures at his own risk. More theory and experience with their use will eventually tell the whole story.

Application of this technique is not limited to just two treatments. For the many-one problem with k treatments or the k-sample problem where $k > 2$, the multivariate normal distribution replaces the bivariate normal with the pertinent correlation coefficient still being given by (102a,b). After (102a,b) is replaced by an estimate, the construction of the test critical point proceeds as in Secs. 3.5 and 4.5, with the relevant means and variances given by (80).

Nemenyi (1963) proposed the signed-rank test for multiple comparisons, but he failed to discover that the statistic was not distribution-free. He estimated the correlation coefficient between SR_1^+ and SR_2^+ for a uniform distribution by Monte Carlo means. He concluded that the correlation coefficient was nearly $\frac{1}{2}$ (for $n \geq 7$), and that a multivariate normal approximation with $\rho = \frac{1}{2}$ should be reasonably accurate. Although this conclusion is warranted only for the uniform distribution, Nemenyi's approximation may prove to be a reasonable one for a wide class of distributions.

The test critical point can be used to construct (simultaneous) confidence intervals for the θ_i in the usual way. There is an easy graphical method for doing this which is due to Tukey. Although the question of what precisely to use for a critical point still demands further investigation, the graphical method is described here because it is difficult to locate a description of the method even in the nonsimultaneous case. The method is also described in Walker and Lev (1953, chap. 8).

Let sr_+^α be the one-sided level α critical point for positive shifts. (The level α refers to the family probability error rate in the simultaneous case, and the usual significance level in the nonsimultaneous case.) Let $U_j(\theta) = (Y_{1j} - \theta) - Y_{0j}$, and let $SR_1^+(\theta)$ be the positive signed-rank statistic computed from the $U_j(\theta)$ as in (75) to (77). Any value of θ for which $SR_1^+(\theta) < sr_+^\alpha$ belongs to the confidence interval.

On a piece of graph paper plot the n points $U_1, \ldots, U_n$ on the ordinate axis. Through each point U_j draw two lines in the right half-plane. One should have slope $+1$, the other slope -1. These lines will intersect at $\binom{n}{2}$ points in the right half-plane. These intersections and the original n points on the ordinate axis constitute a total of $n(n+1)/2$ points.

From the bottom of the triangular array of points start upward, counting points and ordering them according to the values of their ordinates. Draw a horizontal line through the point with counting number $(n(n+1)/2) - sr_+^\alpha + 1$. The interval on the ordinate axis above the horizontal line is the confidence interval for θ_1.

This method is illustrated in Fig. 2 with the hypothetical values $n = 5$, $sr_+^\alpha = 12$, and $U_1 = 1$, $U_2 = 3.5$, $U_3 = 6$, $U_4 = 2$, $U_5 = 5$. The confidence interval for θ_1 is $(2.25, +\infty)$.

The reason this method works is simple enough. For a given θ subtraction of it from $Y_{1j} - Y_{0j}$ to get $U_j(\theta)$ is equivalent to shifting the abscissa axis to the horizontal line with ordinate θ. The new origin is at $(0,\theta)$. Any U_j above the horizontal line with ordinate θ has $U_j(\theta) > 0$; any U_j below the horizontal line has $U_j(\theta) < 0$. An intersection of a line from U_j with slope -1 and a line from $U_{j'}(U_{j'} < U_j)$ with slope $+1$ compares the relative sizes of $|U_j(\theta)|$ and $|U_{j'}(\theta)|$. If the intersection of the two sloping lines falls above the horizontal line with ordinate θ, then $|U_{j'}(\theta)| < |U_j(\theta)|$. If the intersection falls below the horizontal line, then $|U_{j'}(\theta)| > |U_j(\theta)|$.

For $U_j > \theta$ the number of points with ordinate greater than θ which lie on the negatively sloping line through U_j is the signed-rank $SR_{1j}^+(\theta)$ of $U_j(\theta)$. This follows from the preceding paragraph because each $U_{j'} > \theta$, $U_{j'} < U_j$, gives a point, and each $U_{j'} < \theta$, $|U_{j'}(\theta)| < U_j(\theta)$, gives a point. The sum of these numbers for all $U_j > \theta$ gives $SR_1^+(\theta)$. Thus, the number of points falling above the horizontal line with ordinate θ is $SR_1^+(\theta)$.

The confidence interval for θ_1 consists of all θ for which the horizontal

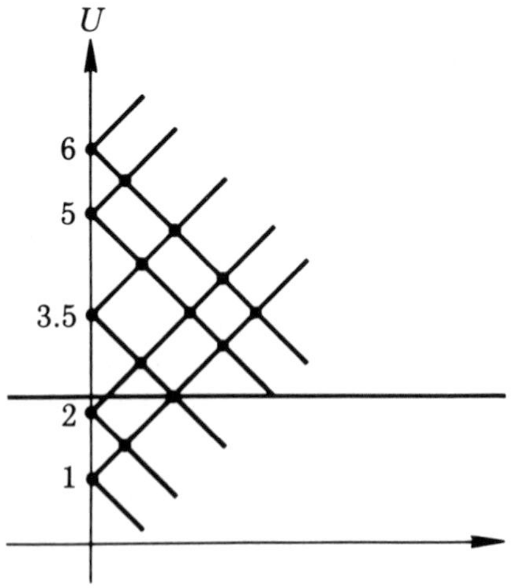

Figure 2

line through $(0,\theta)$ has $SR_1^+(\theta) < sr_+^\alpha$. Hence, it consists of all θ above the horizontal line passing through the point with counting number sr_+^α from the *top*. Since $[n(n+1)/2] - sr_+^\alpha + 1$ is usually smaller than sr_+^α, it is easier to count from the *bottom*.

For n large it is not necessary to draw all the sloping lines. Only those near the bottom are necessary and the statistician can tell visually which these will be.

By drawing two horizontal lines, one at the top and one at the bottom, the statistician can construct a two-sided interval.

6 KRUSKAL-WALLIS RANK STATISTICS (NEMENYI)

This technique, based on the Kruskal-Wallis method of ranking in a one-way classification, was proposed and analyzed by Nemenyi (1963). A similar procedure using the Bonferroni inequality was given by Dunn (1964).

6.1 Method

Consider first the k-sample problem. Let $\{Y_{ij};\ i = 1, \ldots, k,\ j = 1, \ldots, n_i\}$ be k independent samples of sizes $n_1, \ldots, n_k$ of independent observations. The total number of observations is $N = \sum_{i=1}^{k} n_i$. The observation Y_{ij} is assumed to be distributed according to the density

$$p_{Y_{ij}}(x) = p(x - \theta_i) \qquad i = 1, \ldots, k \qquad j = 1, \ldots, n_i. \tag{105}$$

The density $p(x)$ is fixed but unknown. Since p is arbitrary, the θ_i are not uniquely defined, but their differences $\theta_i - \theta_{i'}$ are uniquely defined, which is all that matters.

The null hypothesis

$$H_0\colon\quad \theta_i \equiv \theta_{i'} \qquad \forall\, i, i' \tag{106}$$

is to be tested against the general alternative

$$H_1\colon\quad \theta_i \not\equiv \theta_{i'} \qquad \forall\, i, i'. \tag{107}$$

The Kruskal-Wallis analysis-of-variance test of (106) vs. (107) is performed as follows. Rank the N observations Y_{ij} in order of size: $Y_{(1)} < Y_{(2)} < \cdots < Y_{(N)}$. Assign rank 1 to $Y_{(1)}$, rank 2 to $Y_{(2)}$, etc.

Let R_{ij} be the rank of Y_{ij}. The population and overall rank means are

$$\begin{aligned} \bar{R}_{i\cdot} &= \frac{1}{n_i} \sum_{j=1}^{n_i} R_{ij} \qquad i = 1, \ldots, k \\ \bar{R}_{\cdot\cdot} &= \frac{1}{N} \sum_{i,j} R_{ij} = \frac{N+1}{2}. \end{aligned} \tag{108}$$

The Kruskal-Wallis test at level α accepts H_0 only when

$$\begin{aligned} H_{k-1} &= \frac{12}{N(N+1)} \sum_{i=1}^{k} n_i(\bar{R}_{i\cdot} - \bar{R}_{\cdot\cdot})^2 \\ &= \frac{12}{N(N+1)} \sum_{i=1}^{k} n_i \bar{R}_{i\cdot}^2 - 3(N+1) \le h_{k-1}^{\alpha}. \end{aligned} \tag{109}$$

For N large h_{k-1}^{α} is approximately $\chi_{k-1}^{2\alpha}$, and for small N small sample tables exist (see Sec. 6.5).

The statistic (109) divided by $k - 1$ is the usual one-way analysis-of-variance F statistic, except that the observations have been replaced by their ranks and the denominator mean sum of squares has been replaced by the known expectation of the numerator.

The projection argument which derived the Scheffé simultaneous confidence intervals from the F statistic can be applied to the Kruskal-Wallis statistic (109). In particular, with probability greater than $1 - \alpha$, the inequalities

$$|\bar{R}_{i\cdot} - \bar{R}_{i'\cdot}| \le (h_{k-1}^{\alpha})^{\frac{1}{2}} \left[\frac{N(N+1)}{12}\right]^{\frac{1}{2}} \left(\frac{1}{n_i} + \frac{1}{n_{i'}}\right)^{\frac{1}{2}} \tag{110}$$

hold simultaneously for the $\binom{k}{2}$ pairs of populations (i,i'). The inference to be drawn is that any pair (i,i') for which $|\bar{R}_{i\cdot} - \bar{R}_{i'\cdot}|$ exceeds the critical constant in (110) has $\theta_i - \theta_{i'} \neq 0$. It can happen that H_{k-1} rejects the null hypothesis, but all the inequalities in (110) are satisfied.

In the special case that the n_i are all equal (i.e., $n_i \equiv n$) and large, the critical constant in (110) can be reduced in size. For large n the $\binom{k}{2}$ inequalities

$$|\bar{R}_{i\cdot} - \bar{R}_{i'\cdot}| \le q_{k,\infty}^{\alpha} \left[\frac{k(kn+1)}{12}\right]^{\frac{1}{2}} \qquad i \neq i' \tag{111}$$

hold simultaneously with probability (approximately) $1 - \alpha$. The constant $q_{k,\infty}^{\alpha}$ is the upper α percentile point of the range of k independent unit normal random variables (see Sec. 1 of Chap. 2).

The distinction between (111) and (110) exactly duplicates for ranks the distinction between the Tukey studentized range (Sec. 1 of Chap. 2) and the Scheffé F projections (Sec. 2 of Chap. 2).

Analogous procedures exist for the many-one problem. Consider the model of Sec. 3.1. Let the sample size of the control population ($i = 0$) be m, and the sample size for each of the treatment populations ($i = 1, \ldots, k$) be n. When n and m are large, the two tests to be described below have approximately the indicated significance levels.

For a one-sided test of positive shifts at level α, accept the null hypothesis when

$$\bar{R}_{i\cdot} - \bar{R}_{0\cdot} \leq m^{\alpha}_{k(\rho)} \left[\frac{N(N+1)}{12}\right]^{\frac{1}{2}} \left(\frac{1}{n} + \frac{1}{m}\right)^{\frac{1}{2}} \tag{112}$$

for $i = 1, \ldots, k$. The total sample size is $N = m + kn$. The constant $m^{\alpha}_{k(\rho)}$ is the upper α percentile point of the maximum of k unit normal random variables with common correlation

$$\rho = \frac{n}{n+m}. \tag{113}$$

If any $\bar{R}_{i\cdot} - \bar{R}_{0\cdot}$ exceeds the critical constant in (112), then the null hypothesis is rejected, and the conclusion $\theta_i > 0$ is drawn.

For a two-sided test at level α, accept the null hypothesis when

$$|\bar{R}_{i\cdot} - \bar{R}_{0\cdot}| \leq |m|^{\alpha}_{k(\rho)} \left[\frac{N(N+1)}{12}\right]^{\frac{1}{2}} \left(\frac{1}{n} + \frac{1}{m}\right)^{\frac{1}{2}} \tag{114}$$

for $i = 1, \ldots, k$. The constant $|m|^{\alpha}_{k(\rho)}$ is the upper α percentile point of the maximum absolute value of k unit normal random variables with common correlation (113). If any $|\bar{R}_{i\cdot} - \bar{R}_{0\cdot}|$ exceeds the critical constant in (114), the null hypothesis is rejected, and the conclusion $\theta_i > 0$ or $\theta_i < 0$ (depending on the sign of $\bar{R}_{i\cdot} - \bar{R}_{0\cdot}$) is drawn.

Confidence intervals for $\theta_i - \theta_{i'}$ (k-sample problem) or θ_i (many-one problem) could be constructed by determining through multivariate trial and error the parametric values for which the test statistic has a nonsignificant value. However, except for very small sample sizes (where the approximations may be poor), the labor involved is prohibitive.

Nemenyi (1963) also proposed using Fisher-Yates normal scores in place of ranks with the variances and covariances changed appropriately. Dunn (1964) proposed inequalities of the form (110) with $(h^{\alpha}_{k-1})^{\frac{1}{2}}$ replaced by $g^{\alpha/2K}$ where $K = \binom{k}{2}$ and g^{α} is the upper α percentile point of a unit normal distribution.

6.2 Applications

The Kruskal-Wallis-type simultaneous rank tests are applicable to the k-sample and many-one problems when no block effects are present.

They are nonparametric rank analogs to Scheffé's F projections for a one-way analysis of variance, Tukey's studentized range, and Dunnett's many-one t statistics. They are competitors to the rank procedures of Steel described in Secs. 3 and 4.

No assumptions are imposed on the underlying density as they are in the Scheffé, Tukey, and Dunnett procedures. This greatly expands their range of application over the normal techniques.

The Scheffé-type test (110) is more versatile than (111), (112), or (114) because it is valid for unequal sample sizes. This also makes it more versatile than the Steel rank procedures. The test (110) can be applied to the many-one problem as well as the k-sample problem. However, when (112) and (114) are applicable, they are sharper procedures than (110) because the critical constants are smaller.

6.3 Comparison

To this author the Kruskal-Wallis rank procedures finish second best to the Steel procedures of Secs. 3 and 4. The model structures for the two types are identical. The amount of time required to perform the tests should be about the same for both types. However, the rank procedures of this section have two distinctly unpleasant qualities to them.

The outcome of the comparison between population i and population i' depends upon the observations in the other populations. The test statistic for i vs. i' is $\bar{R}_{i.} - \bar{R}_{i'.}$, and the rank averages depend upon *all* the observations. For the same set of values $y_{i1}, \ldots, y_{in_i}; y_{i'1}, \ldots, y_{i'n_{i'}}$ in populations i and i' the comparison i vs. i' can be significant in one experiment and not significant in another. The observations from the other populations interspersed between the y_{ij} and $y_{i'j}$ directly affect the value of $\bar{R}_{i.} - \bar{R}_{i'.}$.

This seems unreasonable and does not occur for the normal theory procedures or the Steel rank procedures. For the normal procedures (Scheffé, Tukey, etc.) the observations in the other populations affect the value of the estimated variance s^2, but their location does not affect $\bar{Y}_{i.} - \bar{Y}_{i'.}$. For the Steel procedures the ranks for any comparison i vs. i' are computed from just the observations in the populations i and i'.

This dependence of a comparison on the extraneous observations means that the marginal power function for the comparison cannot be simply calculated as for the Steel and normal theory techniques. The marginal power will be a function of the locations of all the populations rather than just the two involved in the comparison.

The other disadvantage of the Kruskal-Wallis-type simultaneous rank tests is they do not provide confidence intervals for the differences in location of population pairs without extreme labor. There are no easy numerical or graphical shortcuts for obtaining the confidence intervals, as there are for the Steel rank procedures. Since each comparison $\bar{R}_{i.} - \bar{R}_{i'.}$ depends on all the observations, the confidence interval for $\theta_i - \theta_{i'}$ has to come from projection of a multivariate region for all the parametric differences. The calculation of the multivariate region is virtually impossible.

The one good thing that can be said for the Kruskal-Wallis-type simultaneous rank tests is the great versatility of (110). It does not require equal sample sizes for the populations, and of all the rank procedures it is the only one that does not. This means that it can be applied where others cannot.

6.4 Derivation

The only assertion (other than the distribution theory) which requires derivation is the step from (109) to (110). This follows from Lemma 2.2 which is quoted here for ease of reference.

Lemma 2.2 For $c > 0$,

$$\left| \sum_1^d a_i y_i \right| \le c \left(\sum_1^d a_i^2 \right)^{\frac{1}{2}} \qquad \text{for all } (a_1, \ldots, a_d)$$

if and only if

$$\sum_1^d y_i^2 \le c^2.$$

To obtain (110) with $i = 1$, $i' = 2$ from (109) let

$$\begin{aligned} y_i &= \sqrt{n_i}\,(\bar{R}_{i.} - \bar{R}_{..}) \qquad i = 1, \ldots, k \\ a_1 &= \frac{1}{\sqrt{n_1}} \qquad a_2 = \frac{-1}{\sqrt{n_2}} \qquad a_i = 0 \qquad i = 3, \ldots, k \\ c^2 &= h_{k-1}^{\alpha} \frac{N(N+1)}{12} \\ d &= k \end{aligned} \tag{115}$$

and apply the If part of Lemma 2.2. The proof for any other combination i, i' changes the a_i appropriately.

Since there are more linear combinations $(a_1, \ldots, a_k)$ than the $\binom{k}{2}$ combinations used to obtain (110) from (109), the probability of (110) exceeds the probability of (109), which is $1 - \alpha$. Simultaneous limits could be constructed on linear combinations $\sum_1^k a_i \sqrt{n_i}\,(\bar{R}_{i.} - \bar{R}_{..})$ other than the simple ones $\bar{R}_{i.} - \bar{R}_{i'.}$ while still maintaining the family error rate below α, but what practical use or interpretation these combinations would have is not at all clear.

The step from (109) to (110) exactly parallels for ranks the derivation of the Scheffé intervals from the F statistic.

6.5 Distributions and tables

Kruskal (1952), using the Wald-Wolfowitz (1944) limit theorem on permutations, proved that the rank sum vector $\bar{\mathbf{R}} = (\bar{R}_{1\cdot}, \bar{R}_{2\cdot}, \ldots, \bar{R}_{k\cdot})$ has a limiting (singular) multivariate normal distribution provided $n_i/N \to c_i > 0$ as $N \to +\infty$ [for reference see Fraser (1957, pp. 240–242)]. The limiting covariance matrix of $\bar{\mathbf{R}}$ has the correct form for the quadratic form

$$H_{k-1} = \frac{12}{N(N+1)} \sum_{i=1}^{k} n_i(\bar{R}_{i\cdot} - \bar{R}_{\cdot\cdot})^2 \tag{116}$$

to have a χ^2 distribution with $k - 1$ d.f. [see Kruskal (1952)]. This establishes the validity of (109) and (110) with $h_{k-1}^{\alpha} = \chi_{k-1}^{2\alpha}$ in the large sample case.

Kruskal and Wallis (1952, 1953) tabulated the upper tail of H_{k-1} for $k = 3$ and $n_i = 1(1)5$, $i = 1, 2, 3$. Values of h_{k-1}^{α} can be obtained from these tables for $k = 3$ and $n_i \leq 5$, $i = 1, 2, 3$. For $n_1 = n_2 = n_3 = 5$ the χ^2 approximation was found to be reasonably good, so the rule of thumb is that if all $n_i \geq 5$, the χ^2 approximation is satisfactory. The small sample tables for $k = 3$ can also be found in Siegel (1956), "Nonparametric Statistics for the Behavioral Sciences," and Owen (1962), "Handbook of Statistical Tables."

For approximations to the distribution of H_{k-1} which are finer than the χ^2 approximation in the range of moderate size n the reader is referred to Kruskal and Wallis (1952) and Wallace (1959).

The large sample validity of (111), (112), and (114) follows from the asymptotic normality of $\bar{\mathbf{R}}$ and its covariance structure. Kruskal (1952) gives the covariance structure, and it is also computed here.

Let Y_j and $Y_{j'}$ be two different observations from the total sample of size N, and let R_j and $R_{j'}$ be their ranks (relative to all N observations). Then

$$\begin{aligned} E\{R_j\} &= E\{R_{j'}\} = \frac{1}{N} \sum_{i=1}^{N} i = \frac{N+1}{2} \\ \text{Var}(R_j) &= \text{Var}(R_{j'}) = \frac{1}{N} \sum_{i=1}^{N} i^2 - \left(\frac{N+1}{2}\right)^2 = \frac{(N+1)(N-1)}{12} \\ \text{Cov}(R_j, R_{j'}) &= \frac{1}{N(N-1)} \sum_{i \neq j} ij - \left(\frac{N+1}{2}\right)^2 = -\frac{N+1}{12} \\ \text{Cor}(R_j, R_{j'}) &= -\frac{1}{N-1}. \end{aligned} \tag{117}$$

From (117) the moment structure of $\bar{\mathbf{R}}$ follows immediately.

$$\begin{aligned} E\{\bar{R}_{i\cdot}\} &= \frac{N+1}{2} \\ \operatorname{Var}(\bar{R}_{i\cdot}) &= \frac{N(N+1)}{12n_i} - \frac{N+1}{12} \\ \operatorname{Cov}(\bar{R}_{i\cdot},\bar{R}_{i'\cdot}) &= -\frac{N+1}{12}. \end{aligned} \tag{118}$$

Since $\bar{\mathbf{R}}$ has an asymptotic normal distribution so does the $\binom{k}{2}$-dimensional vector of pair differences $(\bar{R}_{1\cdot} - \bar{R}_{2\cdot}, \ldots, \bar{R}_{k-1\cdot} - \bar{R}_{k\cdot})$. The moments are

$$\begin{aligned} E\{\bar{R}_{i\cdot} - \bar{R}_{i'\cdot}\} &= 0 \\ \operatorname{Var}(\bar{R}_{i\cdot} - \bar{R}_{i'\cdot}) &= \frac{N(N+1)}{12}\left(\frac{1}{n_i} + \frac{1}{n_{i'}}\right) \\ \operatorname{Cov}(\bar{R}_{i\cdot} - \bar{R}_{i'\cdot}, \bar{R}_{i\cdot} - \bar{R}_{i''\cdot}) &= \frac{N(N+1)}{12n_i} \qquad i \neq i', i'' \quad i' \neq i'' \\ \operatorname{Cor}(\bar{R}_{i\cdot} - \bar{R}_{i'\cdot}, \bar{R}_{i\cdot} - \bar{R}_{i''\cdot}) &= \left(1 + \frac{n_i}{n_{i'}}\right)^{-\frac{1}{2}}\left(1 + \frac{n_i}{n_{i''}}\right)^{-\frac{1}{2}} \\ \operatorname{Cov}(\bar{R}_{i\cdot} - \bar{R}_{i'\cdot}, \bar{R}_{i''\cdot} - \bar{R}_{i'''\cdot}) &= 0 \qquad i, i', i'', i''' \text{ all different} \\ \operatorname{Cor}(\bar{R}_{i\cdot} - \bar{R}_{i'\cdot}, \bar{R}_{i''\cdot} - \bar{R}_{i'''\cdot}) &= 0. \end{aligned} \tag{119}$$

In the special case $n_i \equiv n$, (119) reduces to

$$\begin{aligned} E\{\bar{R}_{i\cdot} - \bar{R}_{i'\cdot}\} &= 0 \\ \operatorname{Var}(\bar{R}_{i\cdot} - \bar{R}_{i'\cdot}) &= \frac{k(kn+1)}{6} \\ \operatorname{Cor}(\bar{R}_{i\cdot} - \bar{R}_{i'\cdot}, \bar{R}_{i\cdot} - \bar{R}_{i''\cdot}) &= \tfrac{1}{2} \qquad i \neq i', i'' \qquad i' \neq i'' \\ \operatorname{Cor}(\bar{R}_{i\cdot} - \bar{R}_{i'\cdot}, \bar{R}_{i''\cdot} - \bar{R}_{i'''\cdot}) &= 0 \qquad i, i', i'', i''' \text{ all different.} \end{aligned} \tag{120}$$

The correlation structure of (120) is the same as the correlation structure of the $\binom{k}{2}$-dimensional vector $(Z_1 - Z_2, \ldots, Z_{k-1} - Z_k)$ where Z_i, $i = 1, \ldots, k$, are independent unit normal random variables. Since the range of $Z_1, \ldots, Z_k$ is $Q_{k,\infty} = \max_{i<i'} \{|Z_i - Z_{i'}|\}$, the distribution of

$$\frac{\max_{i<i'} \{|\bar{R}_{i\cdot} - \bar{R}_{i'\cdot}|\}}{\sqrt{\dfrac{k(kn+1)}{12}}} \tag{121}$$

coincides with the distribution of the range. This proves (111). The critical points $q^{\alpha}_{k,\infty}$ are contained in Table I of Appendix B.

From (119) it follows that the moment structure for the many-one problem is

$$E\{\bar{R}_{i.} - \bar{R}_{0.}\} = 0$$
$$\text{Var}\,(\bar{R}_{i.} - \bar{R}_{0.}) = \frac{N(N+1)}{12}\left(\frac{1}{n} + \frac{1}{m}\right) \tag{122}$$
$$\text{Cor}\,(\bar{R}_{i.} - \bar{R}_{0.}, \bar{R}_{i'.} - \bar{R}_{0.}) = \frac{n}{n+m}.$$

Since asymptotic normality continues to hold for the rank sum differences, this proves (112) to (114).

For assorted values of ρ, Gupta (1963) has tabled the probability integral of $M_{k(\rho)}$, the maximum of k unit normal random variables with common correlation ρ. Through interpolation in ρ and inverse interpolation in α, the reader can compute a good approximate value of $m^{\alpha}_{k(\rho)}$, which is required for the implementation of (112).

No tables of $|m|^{\alpha}_{k(\rho)}$ exist except for the cases $\rho = 0, \frac{1}{2}, 1$. For $\rho = 0$, $|m|^{\alpha}_{k(\rho)} = N((1-\alpha)^{1/k})$ where

$$P = \frac{1}{\sqrt{2\pi}} \int_{-N(P)}^{+N(P)} e^{-y^2/2}\, dy. \tag{123}$$

Dunnett (1955, 1964) has tabulated values of $|m|^{\alpha}_{k(\rho)}$ for $\rho = \frac{1}{2}$ since $|m|^{\alpha}_{k(\frac{1}{2})} = |d|^{\alpha}_{k,\infty}$ (see Sec. 5.5 of Chap. 2). The (1964) values of $|d|^{\alpha}_{k,\infty}$ are contained in Table IV of Appendix B. For $\rho = 1$,

$$|m|^{\alpha}_{k(\rho)} = N(1-\alpha).$$

If n and m are roughly the same size, the Dunnett tables with $\rho = \frac{1}{2}$ will provide an adequate approximation. For disproportionate values of n and m the reader may want to interpolate in the variable ρ [see also Dunnett (1964)].

How large is *large* n for the approximations (111), (112), and (114) to be reasonably good? Probably for $n \geq 5$ life approaches normalcy. Also, the larger k is (and hence N) the smaller n can be and still have life look normal.

The asymptotic relative efficiency of the Kruskal-Wallis H test relative to the F test for normal alternatives is $3/\pi$. This is essentially all that is known about the power function of the H statistic. There is no information on the power functions of (111), (112), or (114), or on any of their marginal power functions.

7 FRIEDMAN RANK STATISTICS (NEMENYI)

This procedure, which is derived from the Friedman rank analysis of a two-way classification, was proposed and analyzed by Nemenyi (1963).

7.1 Method

The k-sample problem is treated first. Let $\{Y_{ij};\ i = 1, \ldots, k, j = 1, \ldots, n\}$ be k independent samples of n independent observations each. The observation Y_{ij} is assumed to be distributed according to the density

$$p_{Y_{ij}}(x) = p(x - \theta_i - \beta_j) \qquad i = 1, \ldots, k \qquad j = 1, \ldots, n. \tag{124}$$

The density p is fixed but unknown. The population differences $\theta_i - \theta_{i'}$ and the block differences $\beta_j - \beta_{j'}$ are uniquely defined even though the individual parameters are not. To make the individual parameters unique, take $\theta_1 = \beta_1 = 0$.

The null hypothesis

$$H_0\colon\ \theta_i \equiv \theta_{i'} \qquad \forall\, i, i' \tag{125}$$

is to be tested against the general alternative

$$H_1\colon\ \theta_i \not\equiv \theta_{i'} \qquad \forall\, i, i'. \tag{126}$$

The difference between this model and the one in the preceding section is the presence of block effects. This model is a two-way classification with one observation per cell, but there is no assumption of normality on the cell densities.

The Friedman analysis ranks the observations within each block. Let R_{ij} be the rank of Y_{ij} relative to the ordered observations $Y_{(1)j} < Y_{(2)j} < \cdots < Y_{(k)j}$ in block j. The average rank over blocks for population i is

$$\bar{R}_{i\cdot} = \frac{1}{n} \sum_{j=1}^{n} R_{ij}. \tag{127}$$

The overall average is

$$\bar{R}_{\cdot\cdot} = \frac{1}{N} \sum_{i=1}^{k} \sum_{j=1}^{n} R_{ij} = \frac{k+1}{2} \tag{128}$$

where $N = nk$. The Friedman test statistic is

$$X_r^2 = \frac{12n}{k(k+1)} \sum_{i=1}^{k} (\bar{R}_{i\cdot} - \bar{R}_{\cdot\cdot})^2 = \frac{12n}{k(k+1)} \sum_{i=1}^{k} \bar{R}_{i\cdot}^2 - 3n(k+1). \tag{129}$$

The Friedman test is to reject H_0 when $X_r^2 > x_r^{2\alpha}$. The critical point $x_r^{2\alpha}$ is a function of α, k, and n.

The X_r^2 statistic divided by $k - 1$ is the usual F statistic for populations in a two-way classification except that within blocks the observations are replaced by their ranks and the denominator is the known expectation of the numerator.

The Friedman statistic gives rise to a simultaneous test through the probability statement:

$$|\bar{R}_{i\cdot} - \bar{R}_{i'\cdot}| \leq (x_r^{2\alpha})^{\frac{1}{2}} \left[\frac{k(k+1)}{6n}\right]^{\frac{1}{2}} \qquad i, i' = 1, 2, \ldots, k \qquad (130)$$

with probability greater than $1 - \alpha$. The null hypothesis is accepted when all the rank average differences fail to exceed the critical constant. Any difference which exceeds the critical value is taken to be indicative of a population difference $\theta_i - \theta_{i'} \neq 0$.

When n is large the test can be sharpened by reducing the critical constant in (130). Precisely,

$$|\bar{R}_{i\cdot} - \bar{R}_{i'\cdot}| \leq q_{k,\infty}^{\alpha} \left[\frac{k(k+1)}{12n}\right]^{\frac{1}{2}} \qquad i, i' = 1, \ldots, k \qquad (131)$$

with probability (approximately) $1 - \alpha$. The constant $q_{k,\infty}^{\alpha}$ is the upper α percentile point of the range of k independent unit normal random variables (see Sec. 1 of Chap. 2).

In the many-one problem the sample is $\{Y_{ij}; i = 0,1, \ldots, k, j = 1, \ldots, n\}$, and the population and block parameters are differences from $\theta_0 = \beta_1 = 0$. When n is large the one-sided and two-sided tests, (132) and (133), have approximately the indicated significance levels. The one-sided level α test accepts H_0 when

$$\bar{R}_{i\cdot} - \bar{R}_{0\cdot} \leq m_{k(\frac{1}{2})}^{\alpha} \left[\frac{(k+1)(k+2)}{6n}\right]^{\frac{1}{2}} \qquad i = 1, \ldots, k. \qquad (132)$$

The constant $m_{k(\frac{1}{2})}^{\alpha}$ is the upper α percentile point of the maximum of k unit normal random variables with common correlation $\rho = \frac{1}{2}$. The two-sided level α test accepts H_0 when

$$|\bar{R}_{i\cdot} - \bar{R}_{0\cdot}| \leq |m|_{k(\frac{1}{2})}^{\alpha} \left[\frac{(k+1)(k+2)}{6n}\right]^{\frac{1}{2}} \qquad i = 1, \ldots, k \qquad (133)$$

where $|m|_{k(\frac{1}{2})}^{\alpha}$ is the upper α percentile point of the maximum absolute value of k unit normal random variables with common correlation $\rho = \frac{1}{2}$.

7.2 Applications

The Friedman-type simultaneous rank tests can be applied to any two-way classification with one observation per cell in which the statistician is interested in comparing all treatment populations (rows) pairwise or in comparing all treatment populations with a control population. No internal analysis of block effects or interactions is possible. The block effects, of course, can be analyzed by applying a separate Friedman analysis in which block observations are ranked within treatments.

No assumption (viz., normality) is imposed on the underlying probability density. The model does restrict the treatment and block effects to being additive. Interactions between treatments and blocks are excluded. Further, the analysis is restricted to two-way classifications with a single observation in each cell.

7.3 Comparison

The normal theory tests which can be applied to the two-way classification with a single observation per cell are the studentized range (Sec. 1 of Chap. 2), F projections (Sec. 2 of Chap. 2), and the many-one t test (Sec. 5 of Chap. 2). When the distribution is normal, these are more powerful tests than the simultaneous Friedman-type rank tests. This is also true for near-normal distributions, but no one knows just how near *near* is.

Since the signed-rank tests are not truly nonparametric and could stand additional study in the small sample case, the primary nonparametric competitors to the simultaneous rank tests of this section are the Steel sign tests of Secs. 1 and 2. But it is difficult to compare these two methods because of their dissimilar character. It is not a simple question of signs vs. ranks, because the sign statistics are computed from paired observations and the ranks are not.

Within a block the ranks are assigned with respect to the observations from all the populations. This makes the disparity between the ranks for two populations dependent upon the values of the observations from the other populations. The significance or nonsignificance of a comparison between two populations depends upon the other populations not involved in this comparison. This is most distasteful (cf. Sec. 6.3).

Another disadvantage of the Friedman-type rank statistics is their inability (practically) to produce confidence intervals for the amounts of shift between populations. The enormity of the task involved in a multivariate trial-and-error construction of the confidence set is nightmarish (cf. Sec. 6.3).

For some underlying distributions and some configurations of the parameters (for example, $0 < \theta_1 < \theta_2 < \cdots < \theta_k$) the simultaneous rank test is likely to be more powerful than the simultaneous sign test. But no one knows for certain when this is the case, and the rank procedure has the disadvantages noted above. The rank procedure is also more time-consuming to apply. Until more information is available on the power functions (overall and marginal) of both tests, one is somewhat at a loss to make a strong recommendation for either one over the other. One feels comfortable in recommending that when examination of the marginal power function of the Steel sign test indicates it is powerful enough to do the job requested, why fool around with the rank test. In the unlucky situation that the sign test does not have the desired power, the reader is left to fend for himself.

7.4 Derivation

The probability statement (130) follows from (129) by virtue of Lemma 2.2. Since the proof exactly parallels the derivation of (110) from (109), which is given in Sec. 6.4, the details are omitted.

No other assertions in this section require derivation.

7.5 Distributions and tables

Let $\mathbf{R}_j = (R_{1j}, \ldots, R_{kj})$ be the ranks assigned to the observations $Y_{1j}, \ldots, Y_{kj}$ in block j. The rank sums for the treatments 1 to k are the coordinates of the vector $\mathbf{R} = \sum_{j=1}^{n} \mathbf{R}_j$. The average ranks appearing in expressions (129) to (133) are the coordinates of $\bar{\mathbf{R}} = \left(\sum_{j=1}^{n} \mathbf{R}_j\right)/n$.

Since $\mathbf{R}_j$, $j = 1, \ldots, n$, are independently, identically distributed random vectors, the vector $\mathbf{R}$ or $\bar{\mathbf{R}}$ has an asymptotic normal distribution. From (117) the identification $N = k$ gives (under the null hypothesis)

$$\begin{aligned} E\{R_{ij}\} &= \frac{k+1}{2} \\ \operatorname{Var}(R_{ij}) &= \frac{(k+1)(k-1)}{12} \\ \operatorname{Cov}(R_{ij},R_{i'j}) &= -\frac{k+1}{12} \qquad i \neq i' \\ \operatorname{Cor}(R_{ij},R_{i'j}) &= -\frac{1}{k-1} \qquad i \neq i'. \end{aligned} \tag{134}$$

This shows that the moments of $\bar{\mathbf{R}}$ are

$$\begin{aligned} E\{\bar{R}_{i\cdot}\} &= \frac{k+1}{2} \\ \operatorname{Var}(\bar{R}_{i\cdot}) &= \frac{(k+1)(k-1)}{12n} \\ \operatorname{Cov}(\bar{R}_{i\cdot},\bar{R}_{i'\cdot}) &= -\frac{k+1}{12n} \qquad i \neq i' \\ \operatorname{Cor}(\bar{R}_{i\cdot},\bar{R}_{i'\cdot}) &= -\frac{1}{k-1} \qquad i \neq i'. \end{aligned} \tag{135}$$

From (135)

$$
\begin{aligned}
E\{\bar{R}_{i.} - \bar{R}_{i'.}\} &= 0 \\
\text{Var}\,(\bar{R}_{i.} - \bar{R}_{i'.}) &= \frac{k(k+1)}{6n} \\
\text{Cor}\,(\bar{R}_{i.} - \bar{R}_{i'.}, \bar{R}_{i.} - \bar{R}_{i''.}) &= \tfrac{1}{2} \qquad i \neq i', i'' \qquad i' \neq i'' \\
\text{Cor}\,(\bar{R}_{i.} - \bar{R}_{i'.}, \bar{R}_{i''.} - \bar{R}_{i'''.}) &= 0 \qquad i, i', i'', i''' \text{ all different.}
\end{aligned}
\tag{136}
$$

This means that the covariance matrix of the differences $\bar{R}_{i.} - \bar{R}_{i'.}$ is the same as the covariance matrix of the differences $Z_i - Z_{i'}$ where $Z_1, \ldots, Z_k$ are independent random variables with zero means and variances $k(k+1)/12n$. Thus, the asymptotic distribution of

$$
\frac{\max_{i,i'} \{|\bar{R}_{i.} - \bar{R}_{i'.}|\}}{\sqrt{\dfrac{k(k+1)}{12n}}} \tag{137}
$$

coincides with the distribution of the range ($Q_{k,\infty}$) of k independent, unit normal random variables. This proves (131). The critical points $q^{\alpha}_{k,\infty}$ appear in tables of the studentized range (Table I of Appendix B).

Similarly, the covariance structure of $\bar{R}_{i.} - \bar{R}_{0.}$, $i = 1, \ldots, k$, coincides with the covariance structure of $Z_i - Z_0$, $i = 1, \ldots, k$, where $Z_0, Z_1, \ldots, Z_k$ are independent random variables with zero means and variances $(k+1)(k+2)/12n$. This establishes (132) and (133).

Gupta (1963) has tabulated the distribution of $M_{k(\frac{1}{2})}$, the maximum of k equally correlated ($\rho = \frac{1}{2}$) unit normal random variables. From Gupta's tables $m^{\alpha}_{k(\frac{1}{2})}$ can be easily obtained by inverse interpolation. Since $m^{\alpha}_{k(\frac{1}{2})} = d^{\alpha}_{k,\infty}$ (the critical point of Dunnett's one-sided many-one statistics; see Sec. 5 of Chap. 2), the critical points $m^{\alpha}_{k(\frac{1}{2})}$ can be found in Table IV of Appendix B.

Since $|m|^{\alpha}_{k(\frac{1}{2})} = |d|^{\alpha}_{k,\infty}$ (see Sec. 5 of Chap. 2), values of $|m|^{\alpha}_{k(\frac{1}{2})}$ appear in Table IV of Appendix B.

That X^2_r of (129) has asymptotically ($n \to \infty$) a χ^2 distribution with $k - 1$ d.f. is a well-known result due to Friedman (1937). This result follows from the moment calculations (135) and some matrix manipulation (given below because of its lack of availability elsewhere).

The quadratic form X^2_r can be written as

$$
X^2_r = \sum_{i=1}^{k} Z^2_i \tag{138}
$$

where

$$
Z_i = \frac{\bar{R}_{i.} - \dfrac{k+1}{2}}{\sqrt{\dfrac{k(k+1)}{12n}}} \qquad i = 1, \ldots, k. \tag{139}
$$

Asymptotically ($n \to +\infty$) $\mathbf{Z}$ has a (singular) multivariate normal distribution; the moments are

$$\begin{aligned} E\{Z_i\} &= 0 \\ \operatorname{Var}(Z_i) &= 1 - \frac{1}{k} \\ \operatorname{Cov}(Z_i, Z_{i'}) &= -\frac{1}{k} \qquad i \neq i'. \end{aligned} \tag{140}$$

Let $\mathbf{\Sigma}$ be the covariance matrix of $\mathbf{Z}$ given by (140). There exists an orthogonal matrix $\mathbf{O}$ such that $\mathbf{O\Sigma O}^T = \mathbf{D}$ where $\mathbf{D}$ is a diagonal matrix. Define the transformation $\mathbf{W} = \mathbf{OZ}$. Then

$$X_r^2 = \mathbf{Z}^T\mathbf{Z} = \mathbf{W}^T\mathbf{O}\mathbf{O}^T\mathbf{W} = \mathbf{W}^T\mathbf{W} \qquad \text{and} \qquad \mathbf{W} \sim N(\mathbf{0},\mathbf{D}). \tag{141}$$

If $\mathbf{\Sigma}$ is idempotent ($\mathbf{\Sigma}^2 = \mathbf{\Sigma}$) and has rank $k - 1$, then

$$\mathbf{D} = \begin{pmatrix} & & & 0 \\ & & & \cdot \\ & \mathbf{I}_{k-1} & & \cdot \\ & & & \cdot \\ & & & 0 \\ 0 & \cdots & 0 & 0 \end{pmatrix} \tag{142}$$

where $\mathbf{I}_{k-1}$ is a $(k - 1) \times (k - 1)$ identity matrix because

$$\mathbf{D}^2 = \mathbf{O\Sigma O}^T\mathbf{O\Sigma O}^T = \mathbf{O\Sigma}^2\mathbf{O}^T = \mathbf{O\Sigma O}^T = \mathbf{D} \tag{143}$$

and $\mathbf{D}$ has the rank of $\mathbf{\Sigma}$. (The zero on the diagonal is taken in the last position for convenience.) And if $\mathbf{D}$ has the form (142), then

$$X_r^2 = \sum_{i=1}^{k-1} W_i^2 \sim \chi_{k-1}^2 \tag{144}$$

since $W_1, \ldots, W_{k-1}$ are independent, unit normal random variables.

But it is well known (and easily checked) that $\mathbf{\Sigma}$ is idempotent. It is also well known (and easily checked) that $\mathbf{\Sigma}$ has rank $k - 1$. {The rank of $\mathbf{\Sigma}$ is, for instance, the sum of its diagonal elements [see Graybill (1961, p. 7)]}. Q.E.D.

By the above $x_r^{2\alpha} \cong \chi_{k-1}^{2\alpha}$ for n large. For small n the exact distribution of X_r^2 has been tabulated by Friedman (1937). He gives the exact distribution for $k = 3$, $n = 2(1)9$; $k = 4$, $n = 2, 3, 4$. These tables are reproduced in Siegel (1956), "Nonparametric Statistics for the Behavioral Sciences." Extended tables [$k = 3$, $n = 2(1)15$; $k = 4$, $n = 2(1)8$] appear in Owen (1962), "Handbook of Statistical Tables."

8 OTHER TECHNIQUES

8.1 Permutation tests

Permutation tests form a large vertebra in the backbone of nonparametric statistics.[1] The primary permutation tests are Fisher's (1935, sec. 21) test for the single-sample problem, Pitman's (1937a) spread test for the two-sample problem, and the Welch (1937) and Pitman (1937b) tests for analysis-of-variance problems. These tests are adequately described in their original sources, and also in textbooks on nonparametric statistics such as Siegel (1956) and Fraser (1957). Discussions also appear in Kempthorne (1952) and Scheffé (1959).

The philosophy behind these tests can be applied to simultaneous inference. The fundamental assumption is that the assignment of treatments to the basic experimental units has been random. A basic experimental unit is a plot of ground, a technician, a position in time, etc., or combinations of several of these. The assumption requires that random number tables or some random (truly random, not haphazard) device has been used to determine which treatment is applied to which plot, which technician analyzes which test tube, which treatment is tested first, etc.

Under the randomness assumption any differences between experimental observations are due solely to chance if there are no treatment effects (assuming block effects have been eliminated). Given the values of the order statistics (within a block if block effects are present) the conditional distribution of any arrangement of the ordered observations is the uniform distribution, i.e., for n observations the probability of $(Y_1, \ldots, Y_n)$ given $Y_{(1)} < \cdots < Y_{(n)}$ is $1/n!$.

A permutation test is performed by choosing an appropriate test statistic, calculating its value for each possible arrangement of the ordered observations, and rejecting the null hypothesis of no treatment effects for the $\alpha \cdot n!$ arrangements whose test statistic values are most discordant with the null hypothesis. This test is a conditional test. The critical region depends upon the values of the order statistics, and is not characterized by one critical point for the test statistic as in all previous cases. The experimenter cannot determine his critical region until he knows what his observations are.

Under the null hypothesis of no treatment effects the conditional probability of falsely rejecting the null hypothesis, given the order

[1] The term *randomization test* is sometimes used instead of permutation test.

statistics, is $\alpha \cdot n!/n! = \alpha$. Since this is true for all values of the order statistics, the unconditional significance level is also α.

The permutation test for all mean comparisons within a one-way classification is described below. It is the simultaneous k-sample analog to the Pitman spread test. A simultaneous permutation test analogous to Fisher's test can be defined for a two-way classification with block effects. Permutation tests for many-one problems and permutation tests based on F type statistics exist as well.

Let $\{Y_{ij}; i = 1, \ldots, k, j = 1, \ldots, n\}$ be k samples for the populations $i = 1, \ldots, k$ with n independent observations in each sample. The assignment of populations to experimental units is assumed to be random, and block effects are assumed to be absent. The null hypothesis is that the k populations are identical.

Let $\mathbf{Y} = (Y_{11}, \ldots, Y_{kn})$, and let $\mathbf{Y}_{(\ )} = (Y_{(1)}, \ldots, Y_{(kn)})$ be the vector of ordered observations $Y_{(1)} < \cdots < Y_{(kn)}$. Under the null hypothesis

$$P\{\mathbf{Y} = \mathbf{y} | \mathbf{Y}_{(\ )} = \mathbf{y}_{(\ )}\} = \frac{1}{(kn)!} \tag{145}$$

where $\mathbf{y} = (y_{11}, \ldots, y_{kn})$ is any permutation of the coordinates of $\mathbf{y}_{(\ )} = (y_{(1)}, \ldots, y_{(kn)})$.

If the statistician is interested in shift alternatives, a statistic appropriate for testing the null hypothesis is the range of the mean values $\bar{Y}_{1.}, \ldots, \bar{Y}_{k.}$. Define the variable

$$Q(\mathbf{Y}|\mathbf{Y}_{(\ )}) = \max_{i,i'} \{|\bar{Y}_{i.} - \bar{Y}_{i'.}|\} = \frac{\max_{i,i'} \{|Y_{i.} - Y_{i'.}|\}}{n}. \tag{146}$$

For a fixed order statistic vector $\mathbf{Y}_{(\ )} = \mathbf{y}_{(\ )}$ there are $(kn)!$ different arrangements $\mathbf{y}$ for the random vector $\mathbf{Y}$. Each arrangement gives a value $Q(\mathbf{y}|\mathbf{y}_{(\ )})$. Thus, for fixed $\mathbf{Y}_{(\ )} = \mathbf{y}_{(\ )}$ there are $(kn)!$ or less different values $Q(\mathbf{y}|\mathbf{y}_{(\ )})$. Less than $(kn)!$ different values occur when two or more arrangements give the same value.

For the order statistics $\mathbf{Y}_{(\ )} = \mathbf{y}_{(\ )}$ compute the value $Q(\mathbf{y}|\mathbf{y}_{(\ )})$ for each arrangement $\mathbf{y}$ of $\mathbf{y}_{(\ )}$. Arrange the values in order of size: $Q_{(1)}(\mathbf{y}_{(\ )}) \le Q_{(2)}(\mathbf{y}_{(\ )}) \le \cdots \le Q_{((kn)!)}(\mathbf{y}_{(\ )})$. Select the $\alpha \cdot (kn)!$ largest values: $Q_{(j^\alpha)}(\mathbf{y}_{(\ )}) \le \cdots \le Q_{((kn)!)}(\mathbf{y}_{(\ )})$ where $j^\alpha = (kn)! - \alpha \cdot (kn)! + 1$.† The critical region for a fixed $\mathbf{y}_{(\ )}$ consists of those arrangements $\mathbf{y}$ which yield the $\alpha \cdot (kn)!$ highest values $Q_{(j^\alpha)}(\mathbf{y}_{(\ )}) \le \cdots \le Q_{((kn)!)}(\mathbf{y}_{(\ )})$; that is, reject the null hypothesis if

$$\max_{i,i'} \{|\bar{y}_{i.} - \bar{y}_{i'.}|\} \ge Q_{(j^\alpha)}(\mathbf{y}_{(\ )}). \tag{147}$$

† Rarely will $\alpha \cdot (kn)!$ be an integer. But $\alpha \cdot (kn)!$ is understood to be the nearest integer or the largest integer not exceeding $\alpha \cdot (kn)!$.

Any mean difference $\bar{y}_{i.} - \bar{y}_{i'.}$ for which

$$|\bar{y}_{i.} - \bar{y}_{i'.}| \geq Q_{(j^\alpha)}(\mathbf{y}_{(\ )}) \tag{148}$$

is inferred to be indicative of a difference between populations i and i'.

In the unhappy circumstance that more arrangements $\mathbf{y}$ give the same value $Q_{(j^\alpha)}(\mathbf{y}_{(\ )})$ than can be included in the critical region, the statistician must either increase α to include all these arrangements, decrease α to exclude all of them, or establish some additional criterion for including or excluding arrangements from the critical region.

Because of (145)

$$P\{\text{Rejecting } H_0|\mathbf{Y}_{(\ )} = \mathbf{y}_{(\ )}\} = \frac{\alpha \cdot (kn)!}{(kn)!} = \alpha \tag{149}$$

under the null hypothesis. Since this is true for all values of $\mathbf{y}$, the unconditional probability of rejecting H_0 under the null hypothesis is α.

The numbers k and n do not have to be very sizable before $(kn)!$ becomes astronomically large (from a computational point of view). For example, $k = 3$, $n = 3$ gives 362,880 permutations. A listing by hand of all these permutations is impossible. However, not all permutations need be listed. Only those giving the $\alpha \cdot (kn)!$ highest values to Q are required. This reduces the labor somewhat if one can guess the extreme permutations, and in some instances, the observed value of Q is one of the most extreme. Nevertheless, except in rare circumstances, the amount of labor required to actually carry out the permutation test renders it impractical.

The great merit of permutation tests is the courage they infuse into the statistician to use normal theory tests in the face of nonnormality. This comes from the asymptotic equivalence of a permutation test and its corresponding normal theory test under mild restrictions on the moments of the underlying distribution. When the statistician uses a normal theory test without knowing precisely whether the observations are normally distributed, he can rationalize his analysis by reasoning that he was really using an approximation to the exact permutation test.

This can be illustrated for the permutation test based on (146). The definition (146) could be changed to

$$Q(\mathbf{Y}|\mathbf{Y}_{(\ )}) = \frac{\max\limits_{i,i'} \{|\bar{Y}_{i.} - \bar{Y}_{i'.}|\}}{\sqrt{\dfrac{1}{n(kn-1)} \sum\limits_{i,j} (Y_{ij} - \bar{Y}_{..})^2}} \tag{150}$$

without altering the permutation test because the denominator is the same constant for every permutation of the order statistic $\mathbf{Y}_{(\ )}$. This looks almost like the Tukey studentized range statistic, which is

$$Q_{k,k(n-1)} = \frac{\max_{i,i'} \{|\bar{Y}_{i.} - \bar{Y}_{i'.}|\}}{\sqrt{\frac{1}{kn(n-1)} \sum_{i,j} (Y_{ij} - \bar{Y}_{i.})^2}}. \tag{151}$$

The denominators in (150) and (151) differ by the factor

$$\left\{\frac{k(n-1)}{kn-1}\left[1 + \frac{n \sum_i (\bar{Y}_{i.} - \bar{Y}_{..})^2}{\sum_{i,j} (Y_{ij} - \bar{Y}_{i.})^2}\right]\right\}^{\frac{1}{2}}. \tag{152}$$

But under the null hypothesis this factor converges to one in probability as $n \to +\infty$ when $E\{Y_{ij}^2\} < +\infty$.

As $n \to \infty$ the conditional distribution of $Q(\mathbf{Y}|\mathbf{Y}_{(\ )})$, given $\mathbf{Y}_{(\ )}$, converges in probability to the distribution of the range $Q_{k,\infty}$ under the null hypothesis provided $E\{|Y_{ij}|^3\} < +\infty$. This follows from Noether's (1949) extension of the Wald and Wolfowitz (1944) limit theorem or from a theorem of Hoeffding (1952) [for reference, see Fraser (1957, chap. 6, sec. 6)]. The α-level critical point of $Q(\mathbf{Y}|\mathbf{Y}_{(\ )})$ converges in probability to $q_{k,\infty}^{\alpha}$ [see Fraser (1957, chap. 7, sec. 4)]. Thus, asymptotically, the Tukey studentized range test is equivalent under the null hypothesis to the permutation test based on (150).

Similar arguments establish the asymptotic equivalence of other permutation tests and their normal theory counterparts.

Theoretically, a multivariate confidence region for $(\theta_1 - \theta_2, \ldots, \theta_{k-1} - \theta_k)$ could be constructed from the permutation test. However, this is impossible to carry out in practice.

Nemenyi (1963) gives a limited discussion of simultaneous permutation tests.

8.2 Median tests (Nemenyi)

Mood (1950) proposed a test based on a median statistic to detect a difference in location of two populations. [The identical discussion appears in Mood and Graybill (1963).] Let $Y_{11}, \ldots, Y_{1n_1}$ and $Y_{21}, \ldots, Y_{2n_2}$ be two samples of independent observations from populations 1 and 2, respectively. Let $Y_{(1)} < Y_{(2)} < \cdots < Y_{(n_1+n_2)}$ be the ordered combined sample. The median of the combined sample is

$$\text{med}\,(Y_{ij}) = \begin{cases} Y_{((n_1+n_2+1)/2)} & \text{if } n_1 + n_2 \text{ is odd} \\ \frac{1}{2}[Y_{((n_1+n_2)/2} + Y_{((n_1+n_2)/2+1))}] & \text{if } n_1 + n_2 \text{ is even.} \end{cases} \tag{153}$$

The Mood median test lists the frequencies of how many observations are from populations 1 and 2 and how many exceed or fail to exceed med (Y_{ij}). The frequencies are displayed in a 2×2 contingency table (Table 1).

Table 1

	Population		
	1	2	
$> \text{med}(Y_{ij})$	M_1	M_2	$M_1 + M_2$
$\leq \text{med}(Y_{ij})$	$n_1 - M_1$	$n_2 - M_2$	$n_1 + n_2 - M_1 - M_2$
	n_1	n_2	$n_1 + n_2$

The random variable M_i is the number of observations from population i which exceed med (Y_{ij}). Because of the way the row classification is defined

$$M_1 + M_2 = \begin{cases} \dfrac{n_1 + n_2 - 1}{2} & \text{if } n_1 + n_2 \text{ is odd} \\ \dfrac{n_1 + n_2}{2} & \text{if } n_1 + n_2 \text{ is even.} \end{cases} \tag{154}$$

Under the null hypothesis that populations 1 and 2 are identical the probability attached to a pair of values $(M_1, M_2) = (m_1, m_2)$ is a hypergeometric probability:

$$P\{M_1 = m_1,\, M_2 = m_2\} = \frac{\dbinom{n_1}{m_1}\dbinom{n_2}{m_2}}{\dbinom{n_1 + n_2}{m_1 + m_2}}. \tag{155}$$

The null hypothesis can be tested by Fisher's exact test for 2×2 contingency tables which is based on the hypergeometric probabilities (155). For a description of the exact test the reader is referred to Fisher (1934, sec. 21.02), Siegel (1956, pp. 96–101), Finney, et al. (1963), or innumerable other sources. For large samples (n_1, $n_2 \to +\infty$) the exact test can be replaced by the χ^2 test with 1 d.f. for 2×2 tables which is to reject if

$$\frac{(n_1 + n_2)[M_1(n_2 - M_2) - M_2(n_1 - M_1)]^2}{n_1 n_2 (M_1 + M_2)\,(n_1 + n_2 - M_1 - M_2)} \geq \chi_1^{2\alpha}. \tag{156}$$

Except for slight disagreement in the denominator when $n_1 + n_2$ is odd, the test statistic in (156) is equal to the square of

$$\frac{\left|\dfrac{M_1}{n_1} - \dfrac{M_2}{n_2}\right|}{\sqrt{\dfrac{1}{4}\left(\dfrac{1}{n_1} + \dfrac{1}{n_2}\right)}} = \frac{|\bar{Y}_{1.}^* - \bar{Y}_{2.}^*|}{\sqrt{\dfrac{1}{4}\left(\dfrac{1}{n_1} + \dfrac{1}{n_2}\right)}} \tag{157}$$

where
$$Y_{ij}^* = \begin{cases} 1 & \text{if } Y_{ij} > \text{med}\,(Y_{ij}) \\ 0 & \text{if } Y_{ij} \leq \text{med}\,(Y_{ij}). \end{cases} \tag{158}$$

The observations Y_{ij} have been replaced by 1's or 0's depending on whether they do or do not exceed the common median. The statistic (157) is the normalized difference of the averages of the transformed observations for the two samples. For large samples the approximate critical point for (157) is $\sqrt{\chi_1^{2\alpha}}$, the upper $(\alpha/2)$ percentile point of a unit normal random variable. In the case of small samples the exact distribution of (157) could be tabulated from (155).

Nemenyi (1963) proposed an extension of (157) to multiple comparisons problems. Let $\{Y_{ij};\ i = 1, \ldots, k,\ j = 1, \ldots, n\}$ be k independent samples of n independent observations each from populations $1, 2, \ldots, k$. Let $Y_{(1)} < Y_{(2)} < \cdots < Y_{(kn)}$ be the ordered combined sample, and let med (Y_{ij}) be the median of the combined sample.

$$\text{med}\,(Y_{ij}) = \begin{cases} Y_{((kn+1)/2)} & \text{if } kn \text{ is odd} \\ \frac{1}{2}[Y_{(kn/2)} + Y_{((kn/2)+1)}] & \text{if } kn \text{ is even.} \end{cases} \tag{159}$$

Replace the observations Y_{ij} by the counting variables Y_{ij}^* which score Y_{ij} as 1 or 0 depending on whether it does or does not exceed med (Y_{ij}):

$$Y_{ij}^* = \begin{cases} 1 & \text{if } Y_{ij} > \text{med}\,(Y_{ij}) \\ 0 & \text{if } Y_{ij} \leq \text{med}\,(Y_{ij}). \end{cases} \tag{160}$$

The simultaneous test statistic for testing the hypothesis of homogeneity among the populations is

$$\frac{\max\limits_{i,i'} \{|\bar{Y}_{i.}^* - \bar{Y}_{i'.}^*|\}}{\sqrt{\dfrac{1}{2n}}}. \tag{161}$$

Asymptotically, as $n \to +\infty$, the statistic (161) is distributed as $Q_{k,\infty}/\sqrt{2}$ where $Q_{k,\infty}$ is a studentized range statistic with d.f. $= +\infty$ for the denominator. For small samples the distribution of (161) can be calculated from

$$P\{M_1 = m_1, \ldots, M_k = m_k\} = \frac{\dbinom{n}{m_1}\dbinom{n}{m_2}\cdots\dbinom{n}{m_k}}{\dbinom{kn}{m_1 + \cdots + m_k}} \tag{162}$$

where M_i is the number of observations in population i which exceed med (Y_{ij}). Nemenyi has tabled the exact 5 percent and 1 percent critical points of (161) for $k = 3, 4, 5$; $n = 4(1)15$. The agreement between the asymptotic and exact critical points is good for $n \geq 6$.

Appropriate continuity corrections can be applied to (156), (157), and (161) to improve the fit of the asymptotic distributions.

Nemenyi (1963) also treats the corresponding many-one problem by defining a statistic analogous to (161). He gives exact critical points for $\alpha = .05, .01$; $k = 2, 3, 4$; $n = 4(1)15$ for both the one-sided and two-sided problems.

Nemenyi (1963) includes an extension of the median test in the presence of block effects. He gives exact critical points for small samples.

8.3 Kolmogorov-Smirnov statistics

Let $Y_1, Y_2, \ldots, Y_n$ be independent observations distributed according to the cdf $F(x)$. A classical problem in statistics is to test whether the unknown cdf $F(x)$ is equal to a specified cdf $G(x)$; that is,

$$H_0\colon \quad F \equiv G. \tag{163}$$

This problem is interpretable as a problem in simultaneous inference. For a given x, $P_F(x) = F(x)$ is the probability associated with the Bernoulli trial of whether an observation has value $\leq x$ or $> x$, when the observation is distributed according to $F(x)$. The probability

$$P_G(x) = G(x)$$

has a similar interpretation. The hypothesis (163) is that these Bernoulli probabilities agree for all values of x; that is,

$$H_0\colon \quad P_F(x) = P_G(x) \qquad \forall\, x. \tag{164}$$

Kolmogorov proposed to test (163) on the basis of the statistic

$$K_n = \sqrt{n} \sup_x |S_n(x) - G(x)| \tag{165}$$

where $S_n(x)$ is the sample cdf of $Y_1, \ldots, Y_n$, that is,

$$S_n(x) = \frac{\text{number of } Y_1, \ldots, Y_n \leq x}{n}. \tag{166}$$

When the null hypothesis is true, the statistic K_n is distribution-free, and the test rejects H_0 for large values of K_n (that is, $K_n \geq k_n^\alpha$). The statistic $S_n(x)$ is just the usual binomial estimate $\widehat{P_F}(x)$ of $P_F(x)$ based on n independent Bernoulli trials. The Kolmogorov statistic can thus be interpreted as the maximum discrepancy between the hypothesized Bernoulli probabilities $P_G(x)$ and the observed proportions $\widehat{P_F}(x)$; that is,

$$K_n = \sqrt{n} \sup_x |\widehat{P_F}(x) - P_G(x)|. \tag{167}$$

If $K_n \geq k_n^\alpha$, then those values of x for which $\sqrt{n}\, |\widehat{P_F}(x) - P_G(x)| \geq k_n^\alpha$ are inferred to have probabilities $P_F(x)$ which differ from $P_G(x)$.

Kolmogorov obtained the limiting distribution of K_n:

$$\lim_{n\to\infty} P\{K_n < k\} = \sum_{j=-\infty}^{+\infty} (-1)^j e^{-2j^2k^2} \qquad 0 < k < +\infty. \tag{168}$$

Tables of critical points $k_n^\alpha/\sqrt{n}$ based on (168) have been computed by L. H. Miller (1956) for $\alpha = .10, .05, .02, .01$; $n = 1(1)100$. Exact small sample tables of the distribution of K_n are given by Z. W. Birnbaum (1952) for $n = 1(1)100$, and exact tables of $k_n^\alpha/\sqrt{n}$ are given by Massey (1951) for $\alpha = .20, .15, .10, .05, .01$; $n = 1(1)20(5)35$, and $n > 35$. These tables are reproduced in Siegel (1956) and Owen (1962).

The critical constants k_n^α can be used to construct a confidence band $S_n(x) \pm (k_n^\alpha/\sqrt{n})$ for the unknown cdf $F(x)$. This band constitutes a family of simultaneous confidence intervals $\widehat{P_F}(x) \pm (k_n^\alpha/\sqrt{n})$ for the unknown Bernoulli probabilities $\widehat{P_F}(x) = F(x)$.

For a one-sided test of (163) or (164) the appropriate statistic is

$$K_n^+ = \sqrt{n} \sup_x [S_n(x) - G(x)] = \sqrt{n} \sup_x [\widehat{P_F}(x) - P_G(x)]. \tag{169}$$

The limiting distribution of K_n^+ is

$$\lim_{n\to\infty} P\{K_n^+ < k\} = 1 - e^{-2k^2} \qquad 0 < k < +\infty. \tag{170}$$

Critical points based on (170) have been tabled by L. H. Miller (1956) for $\alpha = .10, .05, .025, .01$, and $.005$; $n = 1(1)100$. Birnbaum and Tingey (1951) derived explicitly the exact distribution of K_n^+ and gave tables of critical points for $\alpha = .10, .05, .01, .001$; $n = 5, 8, 10, 20, 40, 50$.

The two-sample comparison is also a classical problem. Let $Y_{11}, \ldots, Y_{1n_1}$ and $Y_{21}, \ldots, Y_{2n_2}$ be independent samples distributed according to the cdf's $F_1(x)$ and $F_2(x)$, respectively. The null hypothesis is

$$H_0'\colon \quad F_1 \equiv F_2. \tag{171}$$

If $P_i(x) = F_i(x)$, then $P_i(x)$ is the probability an observation from population i falls less than or equal to x. The hypothesis (171) is identical to the simultaneous hypothesis

$$H_0'\colon \quad P_1(x) = P_2(x) \quad \forall\, x. \tag{172}$$

Smirnov proposed the statistic

$$D_{n_1n_2} = \sqrt{\frac{n_1n_2}{n_1 + n_2}} \sup_x |S_{1n_1}(x) - S_{2n_2}(x)| \tag{173}$$

for testing (171) against two-sided alternatives. The functions $S_{in_i}(x)$, $i = 1, 2$, are the sample cdf's for populations 1 and 2, respectively. For fixed x the ratio $S_{in_i}(x)$ is just the usual binomial estimate $\widehat{P_i}(x)$ of $P_i(x)$, so $D_{n_1n_2}$ represents the maximum disparity between the families of binomial estimates; that is,

$$D_{n_1n_2} = \sqrt{\frac{n_1n_2}{n_1 + n_2}} \sup_x |\widehat{P_1}(x) - \widehat{P_2}(x)|. \tag{174}$$

For those x's with

$$\sqrt{\frac{n_1n_2}{n_1 + n_2}} |\widehat{P_1}(x) - \widehat{P_2}(x)| \geq d^\alpha_{n_1n_2} \tag{175}$$

the probabilities $P_1(x)$ and $P_2(x)$ are inferred to be different.

Smirov proved the limiting distribution of $D_{n_1n_2}$ as $n_1, n_2 \to \infty$ $(n_1/n_2 \to c, 0 < c < +\infty)$ coincides with the limiting distribution of K_n given in (168). This limiting distribution was tabled by Smirnov (1948) from which a large sample approximation to $d^\alpha_{n_1n_2}$ can be obtained. The exact distribution of D_{nn} was computed by Birnbaum and Hall (1960) for $n = 1(1)40$. Both sets of tables appear in Owen (1962).

The corresponding one-sided statistic is

$$D^+_{n_1n_2} = \sqrt{\frac{n_1n_2}{n_1 + n_2}} \sup_x [S_{1n_1}(x) - S_{2n_2}(x)]. \tag{176}$$

The limiting distribution of $D^+_{n_1n_2}$ coincides with (170). Small sample tables for D^+_{nn} can be found in Birnbaum and Hall (1960) and Owen (1962).

From the point of view of simultaneous inference a statistic which is more natural than K_n for testing (164) is

$$K^*_n = \sqrt{n} \sup_x \frac{|S_n(x) - G(x)|}{\sqrt{G(x)[1 - G(x)]}}. \tag{177}$$

In this statistic each binominal estimate $S_n(x)$ is normalized by its theoretical standard deviation. However, the distributional problems associated with this statistic are far more complicated than for K_n, both for limiting and finite sample sizes.

For reference to results related to Kolmogorov-Smirnov statistics a good survey article is Darling (1957).

The hypothesis (163) can be generalized to several populations. Let $\{Y_{ij}; i = 1, \ldots, k, j = 1, \ldots, n_i\}$ be k independent samples from populations $1, \ldots, k$ with cdf's $F_1, \ldots, F_k$, respectively. The question could be posed as to whether all the population cdf's agree with a specified cdf G; that is,

$$H_0\colon \quad F_1 \equiv F_2 \equiv \cdots \equiv F_k \equiv G. \tag{178}$$

This hypothesis is simultaneous in both the populations and the Bernoulli probabilities; that is,

$$H_0\colon \quad P_i(x) = G(x) \qquad \forall\, x \qquad \forall\, i. \tag{179}$$

Kiefer (1959) proposed the statistic

$$T' = \sup_x \sum_{i=1}^{k} n_i[S_{in_i}(x) - G(x)]^2 \tag{180}$$

for testing H_0 and obtained its limiting distribution. Tables are given from which large sample critical points can be determined.

The generalization of (171) is

$$H_0'\colon\ F_1 \equiv \cdots \equiv F_k. \tag{181}$$

For this hypothesis Kiefer (1959) proposed the statistic

$$T = \sup_x \sum_{i=1}^{k} n_i[S_{in_i}(x) - \bar{S}_{\cdot}(x)]^2 \tag{182}$$

where $\bar{S}_{\cdot}(x)$ is the sample cdf for the combined observations. Kiefer derived the limiting distribution of T and tabled it. David (1958) and Birnbaum and Hall (1960) proposed the statistic

$$D(n_1, n_2, \ldots, n_k) = \sup_{x,i,i'} |S_{in_i}(x) - S_{i'n_{i'}}(x)|. \tag{183}$$

Birnbaum and Hall tabled the upper tail of the distribution of $D(n,n,n)$ for $n = 1(1)20(2)40$. David studied the corresponding one-sided statistics.

There are other statistics used in testing the null hypotheses (163), (171), (178), and (181). In particular, those of the Cramér-von Mises-type are important. However, these statistics do not admit of a simultaneous interpretation so they are not discussed here.

The author of this book has not laid greater stress on the statistics in this section because he has often found the critical points to be so large as to be of no practical use. The alternative hypotheses are too general to give good power to the tests. Better results are usually obtained by formulating alternatives more specific to the problem at hand and constructing tests specifically for these alternatives.

CHAPTER 5

Multivariate Techniques

A preponderance of the work on multivariate simultaneous confidence intervals and tests is contained in two articles: Roy and Bose (1953) and Dunn (1958). The remaining material appears in Roy (1954,1956), Scheffé (1956,1959), Healy (1956), and Anderson (1965). A summary of the work of Roy and Bose is given in Roy (1957).

The basic model of this chapter is that a p-dimensional sample vector $\mathbf{Y}$ has a multivariate normal distribution with mean vector $\boldsymbol{\mu}$ and covariance matrix $\boldsymbol{\Sigma}$. This is abbreviated by $\mathbf{Y} \sim N(\boldsymbol{\mu},\boldsymbol{\Sigma})$. Different assumptions are made about $\boldsymbol{\Sigma}$. Usually $\boldsymbol{\Sigma}$ is either completely unknown, or it is known except for a scalar constant (that is, $\boldsymbol{\Sigma} = \sigma^2\boldsymbol{\Sigma}$, $\boldsymbol{\Sigma}$ known, σ^2 unknown).

Interest is focused on confidence intervals and tests on $\boldsymbol{\mu}$ for a single population, and on $\boldsymbol{\mu}_1, \ldots, \boldsymbol{\mu}_k$ for k populations. For a single population simultaneous confidence intervals are sought for the linear combinations $\boldsymbol{l}^T\boldsymbol{\mu}$ for all $\boldsymbol{l} \in \mathfrak{L}_{lc}$. The space $\mathfrak{L}_{lc}$ is the p-dimensional space of arbitrary linear combinations; that is, $\mathfrak{L}_{lc} = \{\boldsymbol{l} = (l_1, \ldots, l_p): l_1, \ldots, l_p \text{ arbitrary}\}$. Included within $\mathfrak{L}_{lc}$ is the subspace of all contrasts $\mathfrak{L}_c = \left\{\mathbf{c} = (c_1, \ldots, c_p): \sum_{i=1}^{p} c_i = 0\right\}$ (cf. Secs. 1 and 2 of Chap. 2). Analogous linear combinations are investigated in the k populations problem.

The best general reference on multivariate (nonsimultaneous) techniques is Anderson (1958).

1 SINGLE POPULATION; COVARIANCE SCALAR UNKNOWN

1.1 Method

Let $\mathbf{Y}_1, \ldots, \mathbf{Y}_n$ be n independent observations, identically distributed according to $N(\boldsymbol{\mu},\sigma^2\boldsymbol{\Sigma})$. The matrix $\boldsymbol{\Sigma}$ is assumed known, and positive-definite. The scalar σ^2 and the mean vector $\boldsymbol{\mu}$ are unknown.

Simultaneous confidence intervals are sought for the linear combinations $\boldsymbol{l}^T\boldsymbol{\mu}$, $\forall\, \boldsymbol{l} \in \mathcal{L}_{lc}$. These can be used to test the hypothesis

$$H_0: \quad \boldsymbol{\mu} = \boldsymbol{\mu}^0 \tag{1}$$

since (1) is equivalent to the hypothesis

$$H_0: \quad \boldsymbol{l}^T\boldsymbol{\mu} = \boldsymbol{l}^T\boldsymbol{\mu}^0 \qquad \forall\, \boldsymbol{l} \in \mathcal{L}_{lc}. \tag{2}$$

The sample mean vector $\hat{\boldsymbol{\mu}} = \bar{\mathbf{Y}}_.$ is the maximum likelihood estimator of $\boldsymbol{\mu}$ and the estimator customarily used in practice. The variance σ^2 is estimated by

$$s^2 = \frac{1}{p(n-1)} \sum_{j=1}^{n} (\mathbf{Y}_j - \bar{\mathbf{Y}}_.)^T\boldsymbol{\Sigma}^{-1}(\mathbf{Y}_j - \bar{\mathbf{Y}}_.) \tag{3}$$

which is the maximum likelihood estimator adjusted to be unbiased. The mean $\bar{\mathbf{Y}}_.$ is distributed as $N(\boldsymbol{\mu},(\sigma^2/n)\boldsymbol{\Sigma})$, s^2 is distributed as $\sigma^2\chi^2_{p(n-1)}/p(n-1)$, and $\bar{\mathbf{Y}}_.$ and s^2 are independent.

From these distributional properties it follows that the quadratic form

$$n(\bar{\mathbf{Y}}_. - \boldsymbol{\mu})^T\boldsymbol{\Sigma}^{-1}(\bar{\mathbf{Y}}_. - \boldsymbol{\mu}) \tag{4}$$

is distributed as $\sigma^2\chi^2_p$ (when $\boldsymbol{\mu}$ is the mean of $\mathbf{Y}$). Furthermore, s^2 and the quadratic form (4) are independent, so the ratio

$$\frac{n}{p}\,\frac{(\bar{\mathbf{Y}}_. - \boldsymbol{\mu})^T\boldsymbol{\Sigma}^{-1}(\bar{\mathbf{Y}}_. - \boldsymbol{\mu})}{s^2} \tag{5}$$

has an F distribution with d.f. p, $p(n-1)$.

When the confidence ellipsoid for $\boldsymbol{\mu}$ based on (5) is defined in terms of its planes of support, the desired simultaneous confidence intervals on $\boldsymbol{\mu}$ appear. With probability $1-\alpha$

$$\boldsymbol{l}^T\boldsymbol{\mu} \in \boldsymbol{l}^T\bar{\mathbf{Y}}_. \pm (pF^{\alpha}_{p,p(n-1)})^{\frac{1}{2}} \left(\frac{s^2}{n}\,\boldsymbol{l}^T\boldsymbol{\Sigma}\boldsymbol{l}\right)^{\frac{1}{2}} \qquad \forall\, \boldsymbol{l} \in \mathcal{L}_{lc}. \tag{6}$$

The critical constant $F^{\alpha}_{p,p(n-1)}$ is the upper α percentile point of the F distribution with $p,p(n-1)$ d.f.

The ratio (5) with $\boldsymbol{\mu}^0$ substituted for $\boldsymbol{\mu}$ does not exceed $F^{\alpha}_{p,p(n-1)}$ if and only if $\boldsymbol{l}^T\boldsymbol{\mu}^0$ falls in its interval (6) for all $\boldsymbol{l} \in \mathfrak{L}_{lc}$. The null hypothesis (1) is actually tested by computing (5) with $\boldsymbol{\mu} = \boldsymbol{\mu}^0$ and comparing it with $F^{\alpha}_{p,p(n-1)}$, rather than searching for linear combinations $\boldsymbol{l}$ which give values $\boldsymbol{l}^T\boldsymbol{\mu}^0$ outside their respective intervals. If the null hypothesis is rejected by the test, any linear combination with $\boldsymbol{l}^T\boldsymbol{\mu}^0$ outside its confidence interval gives some indication of how or where the observed population seems to differ from the hypothesized one.

The expressions for the intervals (6) may look formidable, but they are not. The variance of the linear combination $\boldsymbol{l}^T\bar{\mathbf{Y}}_.$ is $(\sigma^2/n)\boldsymbol{l}^T\boldsymbol{\Sigma}\boldsymbol{l}$, which for some linear combinations may be so well known that the matrix multiplication is unnecessary. The last term in (6) is just the square root of an estimate of this variance obtained by replacing σ^2 with s^2. The intervals (6) could thus be expressed in the form

$$\boldsymbol{l}^T\boldsymbol{\mu} \in \boldsymbol{l}^T\bar{\mathbf{Y}}_. \pm [pF^{\alpha}_{p,p(n-1)}]^{\frac{1}{2}}[\widehat{\text{Var}}\,(\boldsymbol{l}^T\bar{\mathbf{Y}}_.)]^{\frac{1}{2}} \qquad \forall\, \boldsymbol{l} \in \mathfrak{L}_{lc}. \tag{7}$$

If the statistician is only interested in a coordinate by coordinate comparison of $\bar{\mathbf{Y}}_.$ with $\boldsymbol{\mu}^0$, or in confidence intervals on the coordinate means μ_i, $i = 1, \ldots, p$, the appropriate intervals from (6) are

$$\mu_i \in \bar{Y}_{i.} \pm (pF^{\alpha}_{p,p(n-1)})^{\frac{1}{2}} \frac{s}{\sqrt{n}} \sqrt{\phi_{ii}} \qquad i = 1, \ldots, p \tag{8}$$

where ϕ_{ii} is the ith diagonal element of $\boldsymbol{\Sigma}$. Since not all possible $\boldsymbol{l} \in \mathfrak{L}_{lc}$ are used in (8), the probability that the p statements in (8) are simultaneously true is greater than $1 - \alpha$.

Instead of the full p-dimensional hypothesis (1) the statistician might be solely interested in a lower dimensional hypothesis such as contrasts, that is,

$$H_0\colon\quad \mu_i \equiv \mu_{i'} \qquad \forall\, i, i'. \tag{9}$$

This hypothesis is equivalent to

$$H_0\colon\quad \mathbf{c}^T\boldsymbol{\mu} = 0 \quad \forall\, \mathbf{c} \in \mathfrak{L}_c. \tag{10}$$

Confidence intervals for all $\mathbf{c}^T\boldsymbol{\mu}$ can be obtained from (6) since contrasts are special linear combinations. However, when only contrasts are of interest, the critical constant can be reduced in size, thereby producing tighter intervals. The appropriate intervals are

$$\mathbf{c}^T\boldsymbol{\mu} \in \mathbf{c}^T\bar{\mathbf{Y}}_. \pm [(p-1)F^{\alpha}_{p-1,p(n-1)}]^{\frac{1}{2}} \left(\frac{s^2}{n}\mathbf{c}^T\boldsymbol{\Sigma}\mathbf{c}\right)^{\frac{1}{2}} \qquad \forall\, \mathbf{c} \in \mathfrak{L}_c. \tag{11}$$

The probability that all statements in (11) hold simultaneously is $1 - \alpha$.

The null hypothesis (9) can be tested without examining the infinity of intervals in (11). Let the $(p-1) \times p$ matrix $\mathbf{L}$ be any basis for the constrasts (see Sec. 2 of Chap. 2). The vector $\mathbf{L}\bar{\mathbf{Y}}_.$ is distributed as $N(\mathbf{0},(\sigma^2/n)\mathbf{L}\boldsymbol{\Sigma}\mathbf{L}^T)$ under the null hypothesis. Rejection of the null hypothesis when the quadratic form

$$\frac{n}{p-1}\frac{(\mathbf{L}\bar{\mathbf{Y}}_.)^T(\mathbf{L}\boldsymbol{\Sigma}\mathbf{L}^T)^{-1}(\mathbf{L}\bar{\mathbf{Y}}_.)}{s^2} \tag{12}$$

exceeds $F^{\alpha}_{p-1,p(n-1)}$ is equivalent to examining all the intervals in (11) for inclusion of zero. The value of the quadratic form (12) is invariant under the choice of $\mathbf{L}$ so the statistician is free to choose a convenient one.

In the event the covariance matrix is known completely, the F-critical constants should be replaced by the appropriate χ^2 constants. Let $\mathbf{Y}_1, \ldots, \mathbf{Y}_n$ be independently, identically distributed $N(\boldsymbol{\mu},\boldsymbol{\Sigma})$ random variables where $\boldsymbol{\Sigma}$ is known. Then the intervals (6) become

$$l^T\boldsymbol{\mu} \in l^T\bar{\mathbf{Y}}_. \pm (\chi^{2\alpha}_p)^{\frac{1}{2}}\left(\frac{1}{n}\,l^T\boldsymbol{\Sigma}l\right)^{\frac{1}{2}} \qquad \forall\, l \in \mathcal{L}_{lc} \tag{13}$$

and the intervals (11) become

$$\mathbf{c}^T\boldsymbol{\mu} \in \mathbf{c}^T\bar{\mathbf{Y}}_. \pm (\chi^{2\alpha}_{p-1})^{\frac{1}{2}}\left(\frac{1}{n}\,\mathbf{c}^T\boldsymbol{\Sigma}\mathbf{c}\right)^{\frac{1}{2}} \qquad \forall\, \mathbf{c} \in \mathcal{L}_c. \tag{14}$$

1.2 Applications

When this technique can be applied should be very clear since the model is quite simple. The only question is whether or not all the assumptions are satisfied. The heaviest restriction is the assumption that the covariance matrix is known except for a multiplicative constant and is the same for all vectors.

An example in which the covariance matrix is known except for a multiplicative constant is where the vectors $\mathbf{Y}_j$ are themselves sets of regression estimates from different experiments. For experiment j, $j = 1, \ldots, n$, the regression surface might be $\beta_{0j} + \beta_1 x + \beta_2 x^2$, where experimental effect may alter the location β_{0j} of the regression surface in each experiment but not its shape. If the independent variable is identically spaced in each experiment, then the n estimates of (β_1,β_2) could be combined and tested by the technique of this section because the covariance matrices are identical and involve an unknown scalar.

In this example there is an internal estimate of σ^2 within each experiment from deviations about the regression curve. These n estimates of σ^2 could be combined with the estimate (3) between experiments to increase the denominator degrees of freedom. Or the n estimates of σ^2 could be used to test the homogeneity of the regression estimates between experiments by comparing them with (3).

1.3 Comparison

The technique of this section is reminiscent of the extension of the Scheffé F projections to correlated variables discussed at the end of Sec. 2.1 of Chap. 2 [see (2.56) to (2.58)]. In fact, with the proper identification, this section is just a special case of the results stated there. The $\boldsymbol{\Sigma}$ of Sec. 2.1 of Chap. 2 would be $np \times np$ and consist of n diagonal $p \times p$ matrices $\boldsymbol{\Sigma}$ of this section.

If the only combinations of interest are the p coordinate intervals (8), then a Bonferroni technique is likely to give shorter intervals. With probability $1 - (\alpha/p)$

$$\mu_i \in \bar{Y}_{i\cdot} \pm [t_{p(n-1)}^{\alpha/2p}] \frac{s}{\sqrt{n}} \sqrt{\sigma_{ii}} \tag{15}$$

where $t_{p(n-1)}^{\alpha/2p}$ is the upper $(\alpha/2p)$ percentile point of a t distribution with $p(n-1)$ d.f. By the Bonferroni inequality (1.13) the probability is at least $1 - \alpha$ that the p statements (15) for $i = 1, \ldots, p$ hold simultaneously. In many instances the critical constant $t_\nu^{\alpha/2p}$ is smaller than $(pF_{p,\nu}^{\alpha})^{\frac{1}{2}}$ so the intervals (15) are shorter than (8). (See Sec. 3 of Chap. 2.)

When combinations other than just the coordinate means are of interest, the technique of this section is likely to be preferable. For a total of K comparisons the Bonferroni constant is $t_{p(n-1)}^{\alpha/2k}$ whereas $(pF_{p,p(n-1)}^{\alpha})^{\frac{1}{2}}$ remains fixed for any number of comparisons. For $K > p$, $(pF_{p,p(n-1)}^{\alpha})^{\frac{1}{2}}$ is usually the smaller critical constant.

Similar comments apply for a limited number of contrasts.

1.4 Derivation

The proof that (6) follows from

$$P\left\{\frac{n}{p}\frac{(\bar{\mathbf{Y}}_{\cdot} - \boldsymbol{\mu})^T\boldsymbol{\Sigma}^{-1}(\bar{\mathbf{Y}}_{\cdot} - \boldsymbol{\mu})}{s^2} \leq F_{p,p(n-1)}^{\alpha}\right\} = 1 - \alpha \tag{16}$$

depends on Lemma 2.2 which is reproduced here for ease of reference.

Lemma 2.2 For $c > 0$,

$$\Big|\sum_1^d a_i y_i\Big| \leq c\left(\sum_1^d a_i^2\right)^{\frac{1}{2}} \qquad \text{for all } (a_1, \ldots, a_d)$$

if and only if

$$\sum_1^d y_i^2 \leq c^2.$$

Since $\boldsymbol{\Sigma}$ is positive-definite there exists a nonsingular matrix $\mathbf{P}$ such that

$$\mathbf{P}\boldsymbol{\Sigma}\mathbf{P}^T = \mathbf{I}. \tag{17}$$

Let $\mathbf{Z} = \mathbf{P}\bar{\mathbf{Y}}_.$ Then, $\mathbf{Z} \sim N(\mathbf{P}\boldsymbol{\mu},(\sigma^2/n)\mathbf{I})$, and

$$(\bar{\mathbf{Y}}_. - \boldsymbol{\mu})^T\boldsymbol{\Sigma}^{-1}(\bar{\mathbf{Y}}_. - \boldsymbol{\mu}) = (\mathbf{Z} - \mathbf{P}\boldsymbol{\mu})^T(\mathbf{Z} - \mathbf{P}\boldsymbol{\mu}). \tag{18}$$

By Lemma 2. 2

$$|\boldsymbol{\lambda}^T(\mathbf{Z} - \mathbf{P}\boldsymbol{\mu})| \leq (pF^{\alpha}_{p,p(n-1)})^{\frac{1}{2}}\left(\frac{s^2}{n}\right)^{\frac{1}{2}}(\boldsymbol{\lambda}^T\boldsymbol{\lambda})^{\frac{1}{2}} \qquad \forall\, \boldsymbol{\lambda} \in \mathfrak{L}_{lc} \tag{19}$$

if and only if
$$(\mathbf{Z} - \mathbf{P}\boldsymbol{\mu})^T(\mathbf{Z} - \mathbf{P}\boldsymbol{\mu}) \leq pF^{\alpha}_{p,p(n-1)}\,\frac{s^2}{n}. \tag{20}$$

But (20) occurs with probability $1 - \alpha$, so (19) holds with probability $1 - \alpha$.

Let $\boldsymbol{l} = \mathbf{P}^T\boldsymbol{\lambda}$. Since $\mathbf{P}$ is nonsingular there is a $1 - 1$ correspondence between $\boldsymbol{\lambda}$ and $\boldsymbol{l}$. In terms of $\boldsymbol{l}$ (19) becomes

$$|\boldsymbol{l}^T(\bar{\mathbf{Y}}_. - \boldsymbol{\mu})| \leq (pF^{\alpha}_{p,p(n-1)})^{\frac{1}{2}}\left(\frac{s^2}{n}\right)^{\frac{1}{2}}[\boldsymbol{l}^T(\mathbf{P}^{-1})(\mathbf{P}^{-1})^T\boldsymbol{l}]^{\frac{1}{2}} \qquad \forall\, \boldsymbol{l} \in \mathfrak{L}_{lc} \tag{21}$$

or, since $\boldsymbol{\Sigma} = (\mathbf{P}^{-1})(\mathbf{P}^{-1})^T$,

$$|\boldsymbol{l}^T(\bar{\mathbf{Y}}_. - \boldsymbol{\mu})| \leq (pF^{\alpha}_{p,p(n-1)})^{\frac{1}{2}}\left(\frac{s^2}{n}\right)^{\frac{1}{2}}(\boldsymbol{l}^T\boldsymbol{\Sigma}\boldsymbol{l})^{\frac{1}{2}} \qquad \forall\, \boldsymbol{l} \in \mathfrak{L}_{lc}.\ \text{Q.E.D.} \tag{22}$$

The proof of (11) follows immediately from (22). Let $\mathbf{L}$ be a basis for $\mathfrak{L}_c$. Then, for any $\mathbf{c} \in \mathfrak{L}_c$ there exists a $(p - 1)$-dimensional linear combination $\boldsymbol{\lambda}$ such that $\mathbf{c}^T = \boldsymbol{\lambda}^T\mathbf{L}$, and any $(p - 1)$-dimensional linear combination $\boldsymbol{\lambda}$ gives a contrast $\mathbf{c}^T = \boldsymbol{\lambda}^T\mathbf{L}$. The vector $\mathbf{L}\bar{\mathbf{Y}}_.$ is distributed as $N(\mathbf{L}\boldsymbol{\mu},(\sigma^2/n)\mathbf{L}\boldsymbol{\Sigma}\mathbf{L}^T)$ so (22), applied to $\mathbf{L}\bar{\mathbf{Y}}_.$, gives

$$|\boldsymbol{\lambda}^T(\mathbf{L}\bar{\mathbf{Y}}_. - \mathbf{L}\boldsymbol{\mu})| \leq [(p - 1)F^{\alpha}_{p-1,p(n-1)}]^{\frac{1}{2}}\left(\frac{s^2}{n}\right)^{\frac{1}{2}}[\boldsymbol{\lambda}^T(\mathbf{L}\boldsymbol{\Sigma}\mathbf{L}^T)\boldsymbol{\lambda}]^{\frac{1}{2}} \qquad \forall\, \boldsymbol{\lambda} \in \mathfrak{L}_{lc}. \tag{23}$$

When (23) is written in terms of $\mathbf{c}$ through $\mathbf{c}^T = \boldsymbol{\lambda}^T\mathbf{L}$, it becomes (11).

A geometric interpretation and proof can be given to (6) since the intervals are just Scheffé F projections. The set of $\boldsymbol{\mu}$ satisfying

$$\frac{n}{p}\,\frac{(\bar{\mathbf{Y}}_. - \boldsymbol{\mu})^T\boldsymbol{\Sigma}^{-1}(\bar{\mathbf{Y}}_. - \boldsymbol{\mu})}{s^2} \leq F^{\alpha}_{p,p(n-1)} \tag{24}$$

is just an ellipsoid centered at $\bar{\mathbf{Y}}_.$ in p-dimensional Euclidean space. The two $(p - 1)$-dimensional hyperplanes which are tangent to the ellipsoid and perpendicular to the line generated by $\boldsymbol{l}$ intersect this line at the values

$$\boldsymbol{l}^T\bar{\mathbf{Y}}_. \pm (pF^{\alpha}_{p,p(n-1)})^{\frac{1}{2}}\left(\frac{s^2}{n}\right)^{\frac{1}{2}}(\boldsymbol{l}^T\boldsymbol{\Sigma}\boldsymbol{l})^{\frac{1}{2}} \tag{25}$$

(divided by the length of $\boldsymbol{l}$). A point $\boldsymbol{\mu}$ is contained in the ellipsoid if and only if it lies between all pairs of tangent planes perpendicular to $\boldsymbol{l}$ as $\boldsymbol{l}$

varies over $\mathcal{L}_{lc}$, that is, if and only if $l^T\mathbf{\mu}$ is between the values (25), $\forall\, l \in \mathcal{L}_{lc}$. For greater detail the reader is referred to Sec. 2 of Chap. 2.

1.5 Distributions and tables

The distribution theory for $\bar{\mathbf{Y}}_{.}$, s^2, and the quadratic form (4) is well known. The reader is referred to Anderson (1958).

The critical constant $F^{\alpha}_{p,p(n-1)}$ requires no discussion.

Under the alternative hypothesis $\mathbf{\mu} \neq \mathbf{\mu}^0$ the ratio (5) has a noncentral F distribution with d.f. $p,p(n-1)$ and noncentrality parameter $\delta^2 = n(\mathbf{\mu} - \mathbf{\mu}^0)^T\mathbf{\Sigma}^{-1}(\mathbf{\mu} - \mathbf{\mu}^0)/\sigma^2$. The power of the test of (1) can be calculated from tables of the noncentral F distribution. The marginal power for any single interval depends on the noncentral t distribution. For greater detail the reader is referred to Sec. 9 of Chap. 2.

2 SINGLE POPULATION; COVARIANCE MATRIX UNKNOWN

This technique was introduced by Roy and Bose (1953). An important application to a mixed model analysis-of-variance problem is due to Scheffé (1956). Scheffé (1959) also discusses this application in his book.

2.1 Method

Let $\mathbf{Y}_1, \ldots, \mathbf{Y}_n$ be n independent observations, identically distributed according to $N(\mathbf{\mu},\mathbf{\Sigma})$. Both the mean vector $\mathbf{\mu}$ and covariance matrix $\mathbf{\Sigma}$ are unknown. Assume $n > p$.

Simultaneous confidence intervals can be obtained for the linear combinations $l^T\mathbf{\mu}$, $\forall\, l \in \mathcal{L}_{lc}$. They are based on the T^2 statistic which tests the null hypothesis

$$H_0\colon \quad \mathbf{\mu} = \mathbf{\mu}^0 \tag{26}$$

against a general alternative. The hypothesis (26) is equivalent to the simultaneous hypothesis

$$H_0\colon \quad l^T\mathbf{\mu} = l^T\mathbf{\mu}^0 \qquad \forall\, l \in \mathcal{L}_{lc}. \tag{27}$$

The maximum likelihood estimator of $\mathbf{\mu}$ is $\hat{\mathbf{\mu}} = \bar{\mathbf{Y}}_{.}$, and the maximum likelihood estimator of $\mathbf{\Sigma}$ (adjusted to be unbiased) is

$$\mathbf{S} = \frac{1}{n-1}\sum_{i=1}^{n}(\mathbf{Y}_i - \bar{\mathbf{Y}}_{.})(\mathbf{Y}_i - \bar{\mathbf{Y}}_{.})^T. \tag{28}$$

The sample mean $\bar{\mathbf{Y}}_.$ is distributed as $N\left(\boldsymbol{\mu}, \frac{1}{n}\boldsymbol{\Sigma}\right)$. The sample covariance matrix $\mathbf{S}$ has a Wishart distribution $W\left(\frac{1}{n-1}\boldsymbol{\Sigma}, n-1\right)$. The vector $\bar{\mathbf{Y}}_.$ and the matrix $\mathbf{S}$ are independently distributed.

The likelihood ratio test of (26) against a general alternative is a monotone function of Hotelling's T^2 statistic:

$$T^2 = n(\bar{\mathbf{Y}}_. - \boldsymbol{\mu}^0)^T\mathbf{S}^{-1}(\bar{\mathbf{Y}}_. - \boldsymbol{\mu}^0). \tag{29}$$

Under the null hypothesis

$$\frac{n-p}{p}\frac{T^2}{n-1} \sim F_{p,n-p} \tag{30}$$

and under the alternative hypothesis the ratio in (30) has a noncentral F distribution with noncentrality parameter $\delta^2 = n(\boldsymbol{\mu} - \boldsymbol{\mu}^0)^T\boldsymbol{\Sigma}^{-1}(\boldsymbol{\mu} - \boldsymbol{\mu}^0)$. The null hypothesis (26) is rejected when

$$T^2 > F^{\alpha}_{p,n-p}\frac{p(n-1)}{n-p} \tag{31}$$

where $F^{\alpha}_{p,n-p}$ is the upper α percentile point of the F distribution with d.f. p, $n-p$.

The T^2 statistic (29) with $\boldsymbol{\mu}$ replacing $\boldsymbol{\mu}^0$ can be used to construct simultaneous confidence intervals on $\boldsymbol{l}^T\boldsymbol{\mu}$, $\forall\, \boldsymbol{l} \in \mathfrak{L}_{lc}$. With probability $1 - \alpha$

$$\boldsymbol{l}^T\boldsymbol{\mu} \in \boldsymbol{l}^T\bar{\mathbf{Y}}_. \pm (pF^{\alpha}_{p,n-p})^{\frac{1}{2}}\left(\frac{n-1}{n-p}\right)^{\frac{1}{2}}\left(\frac{1}{n}\boldsymbol{l}^T\mathbf{S}\boldsymbol{l}\right)^{\frac{1}{2}} \qquad \forall\, \boldsymbol{l} \in \mathfrak{L}_{lc}. \tag{32}$$

The null hypothesis (26) is accepted if and only if $\boldsymbol{l}^T\boldsymbol{\mu}^0$ falls inside its interval (32) for all linear combinations $\boldsymbol{l}$.

The intervals (32) closely resemble the Scheffé intervals of Sec. 2 of Chap. 2 and the multivariate intervals of the previous section. The constant $pF^{\alpha}_{p,\nu}$ is the usual constant for Scheffé-type intervals. The middle term converts the constant multiplier of the sums of squares and cross-products to the proper denominator degrees of freedom. The last term is an estimate of the variance of $\boldsymbol{l}^T\bar{\mathbf{Y}}_.$, which is $\boldsymbol{l}^T\boldsymbol{\Sigma}\boldsymbol{l}/n$.

The intervals (32) have the usual interpretation in terms of projections. The $100(1-\alpha)$ percent confidence ellipsoid for $\boldsymbol{\mu}$ is

$$E_{\boldsymbol{\mu}} = \left\{\boldsymbol{\mu}\colon\; n(\bar{\mathbf{Y}}_. - \boldsymbol{\mu})^T\mathbf{S}^{-1}(\bar{\mathbf{Y}}_. - \boldsymbol{\mu}) \le pF^{\alpha}_{p,n-p}\left(\frac{n-1}{n-p}\right)\right\}. \tag{33}$$

In contrast to the confidence ellipsoids dealt with previously in this book, this ellipsoid has random shape and orientation since the matrix $\mathbf{S}$ is random. Previously, the matrix of the quadratic form has been fixed except for a random scale factor s^2. Nevertheless, the interval (32) for an $\mathbf{l} \in \mathfrak{L}_{lc}$ is the projection of the ellipsoid $E_{\boldsymbol{\mu}}$ onto the line generated by $\mathbf{l}$.

For the p linear combinations giving the p coordinate means $\mu_1, \ldots,$ μ_p the intervals (32) reduce to

$$\mu_i \in \bar{Y}_{i\,.} \pm (pF^{\alpha}_{p,n-p})^{\frac{1}{2}} \left(\frac{n-1}{n-p}\right)^{\frac{1}{2}} \left(\frac{s_i^2}{n}\right)^{\frac{1}{2}} \qquad i = 1, \ldots, p \tag{34}$$

where s_i^2 is the usual variance estimate for the ith coordinate variable, that is,

$$s_i^2 = \frac{1}{n-1} \sum_{j=1}^{n} (Y_{ij} - \bar{Y}_{i\,.})^2. \tag{35}$$

If the statistician is interested solely in contrasts, the length of the confidence intervals can be reduced. The appropriate intervals are

$$\mathbf{c}^T\boldsymbol{\mu} \in \mathbf{c}^T\bar{\mathbf{Y}}_{.} \pm [(p-1)F^{\alpha}_{p-1,n-p+1}]^{\frac{1}{2}} \left(\frac{n-1}{n-p+1}\right)^{\frac{1}{2}} \left(\frac{1}{n}\,\mathbf{c}^T\mathbf{S}\mathbf{c}\right)^{\frac{1}{2}} \qquad \forall\, \mathbf{c} \in \mathfrak{L}_c. \tag{36}$$

2.2 Applications

An important application of this method is to a two-way analysis-of-variance problem under the assumption of a mixed model. The model is

$$Y_{ijk} = \mu + \alpha_i + b_j + c_{ij} + e_{ijk} \qquad i = 1, \ldots, r \quad j = 1, \ldots, c \quad k = 1, \ldots, n. \tag{37}$$

The general mean μ and the row effects α_i are assumed to be constants with the row effects satisfying the constraint $\sum_{i=1}^{r} \alpha_i = 0$. The column effects b_j, interactions c_{ij}, and observational errors e_{ijk} are random. The observational errors are assumed to be independent of the column effects and interactions, and to be independent among themselves. The column effects and interactions are independent between blocks, and within each block $(j = 1, \ldots, c)$ the vector

$$(b_j, c_{1j}, \ldots, c_{rj}) \tag{38}$$

is distributed according to a multivariate normal distribution which is singular due to the constraint $\sum_{i=1}^{r} c_{ij} = 0$. This multivariate distribution is the same for each block.

An exact test for the presence of row effects,

$$H_0\colon \quad \alpha_i = 0 \qquad \forall\, i \tag{39}$$

can be constructed through Hotelling's T^2 statistic. Define

$$D_{ij} = \bar{Y}_{ij.} - \bar{Y}_{rj.} \qquad i = 1, \ldots, r-1 \qquad j = 1, \ldots, c. \tag{40}$$

Then the c vectors $\mathbf{D}_j = (D_{1j}, \ldots, D_{r-1,j})$ are independently, identically, normally distributed with mean vector

$$\boldsymbol{\Delta} = \begin{pmatrix} \Delta_1 \\ \cdot \\ \cdot \\ \cdot \\ \Delta_{r-1} \end{pmatrix} = \begin{pmatrix} \alpha_1 - \alpha_r \\ \cdot \\ \cdot \\ \cdot \\ \alpha_{r-1} - \alpha_r \end{pmatrix} \tag{41}$$

and covariance matrix $\boldsymbol{\Sigma}_D$. The matrix $\boldsymbol{\Sigma}_D$ could be expressed in terms of the variances and covariances of the observational errors and (38), but this is unimportant since it does not aid the analysis. Under the assumptions on the distributional structure of the column effects, interactions, and observational errors, the covariance matrix $\boldsymbol{\Sigma}_D$ is positive-definite.

The mean vector $\boldsymbol{\Delta}$ is estimated by

$$\hat{\boldsymbol{\Delta}} = \begin{pmatrix} \bar{Y}_{1..} - \bar{Y}_{r..} \\ \cdot \\ \cdot \\ \cdot \\ \bar{Y}_{r-1..} - \bar{Y}_{r..} \end{pmatrix} \tag{42}$$

and $\boldsymbol{\Sigma}_D$ is estimated by

$$\mathbf{S}_D = \frac{1}{c-1} \sum_{j=1}^{c} (\mathbf{D}_j - \bar{\mathbf{D}}_.)(\mathbf{D}_j - \bar{\mathbf{D}}_.)^T. \tag{43}$$

Under the null hypothesis (39) $\boldsymbol{\Delta} = \mathbf{0}$, so rejection of H_0 when

$$c\hat{\boldsymbol{\Delta}}^T \mathbf{S}_D^{-1} \hat{\boldsymbol{\Delta}} > F^{\alpha}_{r-1,c-r+1} \frac{(r-1)(c-1)}{c-r+1} \tag{44}$$

gives an exact level α test. Note the requirement that the number of columns c must be at least as large as the number of rows.

Simultaneous confidence intervals for all linear combinations $\boldsymbol{l}^T\boldsymbol{\Delta}$ follow directly from (44) or from (32) with the proper identifications. With probability $1 - \alpha$

$$\boldsymbol{l}^T\boldsymbol{\Delta} \in \boldsymbol{l}^T\hat{\boldsymbol{\Delta}} \pm [(r-1)F^{\alpha}_{r-1,c-r+1}]^{\frac{1}{2}} \left(\frac{c-1}{c-r+1}\right)^{\frac{1}{2}} \left(\frac{1}{c}\,\boldsymbol{l}^T\mathbf{S}_D\boldsymbol{l}\right)^{\frac{1}{2}} \qquad \forall\, \boldsymbol{l} \in \mathfrak{L}_{lc}. \tag{45}$$

Any linear combination $\boldsymbol{l}^T\boldsymbol{\Delta}$ is, in fact, a contrast among the original coordinate means, and vice versa. For each $\boldsymbol{l}$ there is a $\mathbf{c} \in \mathfrak{L}_c$ such that

$\boldsymbol{l}^T\boldsymbol{\Delta} = \mathbf{c}^T\boldsymbol{\alpha}$ where $\boldsymbol{\alpha} = (\alpha_1, \ldots, \alpha_r)$, and vice versa. The confidence intervals (45) are thus really confidence intervals on row differences or more general contrasts.

Although at first glance it might seem that the selection of the rth row for subtraction in the definition of D_{ij} might affect the test (44) and the intervals (45), this is not so. No matter which row is selected for subtraction (or, more generally, which basis is selected for the space of contrasts) the confidence interval will be the same for any contrast $\mathbf{c}^T\boldsymbol{\alpha}$ although the $\boldsymbol{l}$ giving this interval will change. In fact, if the statistician is not interested in actually testing the null hypothesis (39), but only interested in confidence intervals for a variety of contrasts, the intervals can be obtained directly from (36) without ever defining the D_{ij}. The intervals expressed in this form are:

$$\mathbf{c}^T\boldsymbol{\alpha} \in \mathbf{c}^T\bar{\mathbf{Y}}_{..} \pm [(r-1)F^{\alpha}_{r-1,c-r+1}]^{\frac{1}{2}} \left(\frac{c-1}{c-r+1}\right)^{\frac{1}{2}} \left(\frac{1}{c}\,\mathbf{c}^T\mathbf{S}\mathbf{c}\right)^{\frac{1}{2}} \qquad \forall\, \mathbf{c} \in \mathcal{L}_c \tag{46}$$

where

$$\bar{\mathbf{Y}}_{j.} = \begin{pmatrix} \bar{Y}_{1j.} \\ \cdot \\ \cdot \\ \cdot \\ \bar{Y}_{rj.} \end{pmatrix} \qquad \bar{\mathbf{Y}}_{..} = \begin{pmatrix} \bar{Y}_{1..} \\ \cdot \\ \cdot \\ \cdot \\ \bar{Y}_{r..} \end{pmatrix} \tag{47}$$

$$\mathbf{S} = \frac{1}{c-1} \sum_{j=1}^{c} (\bar{\mathbf{Y}}_{j.} - \bar{\mathbf{Y}}_{..})(\bar{\mathbf{Y}}_{j.} - \bar{\mathbf{Y}}_{..})^T.$$

The difference between (45) and (46) is that the estimated variance in (46) is based on the $r \times r$ sample covariance matrix of $(\bar{Y}_{1j.}, \ldots, \bar{Y}_{rj.})$, $j = 1, \ldots, c$, whereas (45) uses the $(r-1) \times (r-1)$ sample covariance matrix of the differences $(\bar{Y}_{1j.} - \bar{Y}_{rj.}, \ldots, \bar{Y}_{r-1,j.} - \bar{Y}_{rj.})$, $j = 1, \ldots, c$. The differencing operation in (40) is replaced by having to compute a larger covariance matrix with $2r - 1$ more elements.

The test (44) is equivalent to rejecting H_0 if any interval in (45) does not contain zero. To convert the intervals (46) into a single test statistic it is necessary to switch to a basis for the contrast space. One such basis is the set of differences (40). Whatever basis is chosen the value of the test statistic will be the same as in (44).

Scheffé (1959) questions whether in practice this analysis is worth all the effort required to carry it out. The computation of the covariance matrix $\mathbf{S}$ or $\mathbf{S}_D$ and the quadratic forms $\mathbf{c}^T\mathbf{S}\mathbf{c}$ or $\boldsymbol{l}^T\mathbf{S}_D\boldsymbol{l}$ is lengthy unless an electronic computer is readily available. Scheffé suggests the use of an approximate F test based on the ratio

$$\frac{\text{Mean Squares (Rows)}}{\text{Mean Squares (Interactions)}} \tag{48}$$

where the numerator and denominator are the usual entries in an analysis-of-variance table. Approximate simultaneous confidence intervals are

$$\mathbf{c}^T\boldsymbol{\alpha} \in \mathbf{c}^T\bar{\mathbf{Y}}_{..} \pm [(r-1)F^{\alpha}_{r-1,(r-1)(c-1)}]^{\frac{1}{2}} \left(\sum_1^r c_i^2\right)^{\frac{1}{2}} \left(\frac{1}{cn} MSI\right)^{\frac{1}{2}} \qquad \forall\, \mathbf{c} \in \mathfrak{L}_c \tag{49}$$

where MSI = Mean Squares (Interactions).

2.3 Comparison

There are essentially no techniques competitive to this one other than Bonferroni intervals. If the number of confidence intervals required is small, the Bonferroni intervals may be shorter. For a family of K linear combinations the appropriate Bonferroni intervals are

$$\boldsymbol{l}^T\boldsymbol{\mu} \in \boldsymbol{l}^T\bar{\mathbf{Y}}_{.} \pm t_{n-1}^{\alpha/2K} \left(\frac{1}{n}\, \boldsymbol{l}^T\mathbf{S}\boldsymbol{l}\right)^{\frac{1}{2}}. \tag{50}$$

These intervals should be contrasted with (32). In any experimental situation the statistician can easily determine whether the constant $t_{n-1}^{\alpha/2K}$ in (50) or the constant

$$(pF^{\alpha}_{p,n-p})^{\frac{1}{2}} \left(\frac{n-1}{n-p}\right)^{\frac{1}{2}} \tag{51}$$

in (32) is smaller.

A similar comparison applies to the contrast intervals (36).

2.4 Derivation

The derivation of (32) from (31) depends on Lemma 2.2, which was reproduced earlier in this chapter in Sec. 1.4.

For fixed $\bar{\mathbf{Y}}_{.}$ and $\mathbf{S}$

$$|\boldsymbol{l}^T(\bar{\mathbf{Y}}_{.} - \boldsymbol{\mu})| \le (pF^{\alpha}_{p,n-p})^{\frac{1}{2}} \left(\frac{n-1}{n-p}\right)^{\frac{1}{2}} \left(\frac{1}{n}\, \boldsymbol{l}^T\mathbf{S}\boldsymbol{l}\right)^{\frac{1}{2}} \tag{52}$$

for all linear combinations $\boldsymbol{l}$ if and only if

$$n(\bar{\mathbf{Y}}_{.} - \boldsymbol{\mu})^T\mathbf{S}^{-1}(\bar{\mathbf{Y}}_{.} - \boldsymbol{\mu}) \le F^{\alpha}_{p,n-p}\, \frac{p(n-1)}{n-p}. \tag{53}$$

This follows from Lemma 2.2 by diagonalizing $\mathbf{S}$; the proof is the following. Since $\mathbf{S}$ is positive-definite (with probability 1 since $n > p$), there exists a nonsingular matrix $\mathbf{P}$ (which depends on $\mathbf{S}$) such that

$$\mathbf{PSP}^T = \mathbf{I}. \tag{54}$$

Let $\boldsymbol{l} = \mathbf{P}^T\boldsymbol{\lambda}$. Then (52) holds for all $\boldsymbol{l}$ if and only if

$$|\boldsymbol{\lambda}^T(\mathbf{P}\bar{\mathbf{Y}}_. - \mathbf{P}\boldsymbol{\mu})| \leq (pF^{\alpha}_{p,n-p})^{\frac{1}{2}} \left(\frac{n-1}{n-p}\right)^{\frac{1}{2}} \left(\frac{1}{n}\boldsymbol{\lambda}^T\boldsymbol{\lambda}\right)^{\frac{1}{2}} \tag{55}$$

holds for all $\boldsymbol{\lambda}$. But by Lemma 2.2, (55) holds for all $\boldsymbol{\lambda}$ if and only if

$$(\mathbf{P}\bar{\mathbf{Y}}_. - \mathbf{P}\boldsymbol{\mu})^T(\mathbf{P}\bar{\mathbf{Y}}_. - \mathbf{P}\boldsymbol{\mu}) \leq F^{\alpha}_{p,n-p} \frac{p(n-1)}{n(n-p)} \tag{56}$$

which is (53) since $\mathbf{S}^{-1} = \mathbf{P}^T\mathbf{P}$. Thus, the event (52), $\forall\, \boldsymbol{l} \in \mathfrak{L}_{lc}$, and the event (53) coincide.

For $\bar{\mathbf{Y}}_.$ and $\mathbf{S}$ random the probability of (53) occurring is $1 - \alpha$. Hence, the probability that (52) holds for all linear combinations $\boldsymbol{l}$ is $1 - \alpha$. This establishes (32).

The intervals for contrasts (36) follow from (32) by transforming to a basis for $\mathfrak{L}_c$. The proof is analogous to the one for contrasts in Sec. 1.4.

2.5 Distributions and tables

The distribution theory for $\bar{\mathbf{Y}}_.$, $\mathbf{S}$, and T^2 is classical and is not reviewed here. For reference the reader is referred to Anderson (1958).

Under the alternative hypothesis the statistic $T^2(n-p)/p(n-1)$ has a noncentral F distribution with noncentrality parameter

$$\delta^2 = n(\boldsymbol{\mu} - \boldsymbol{\mu}^0)^T\boldsymbol{\Sigma}^{-1}(\boldsymbol{\mu} - \boldsymbol{\mu}^0)$$

and d.f. p, $n - p$. The overall power of the test of (26) could be calculated from tables or charts of the noncentral F distribution (for greater detail see Sec. 9 of Chap. 2). A single linear combination will be declared significant by (32) if

$$\frac{\sqrt{n}\,|\boldsymbol{l}^T\bar{\mathbf{Y}}_. - \boldsymbol{l}^T\boldsymbol{\mu}^0|}{\sqrt{\boldsymbol{l}^T\mathbf{S}\boldsymbol{l}}} > (pF^{\alpha}_{p,n-p})^{\frac{1}{2}} \left(\frac{n-1}{n-p}\right)^{\frac{1}{2}}. \tag{57}$$

Considered by itself (not in conjunction with other $\boldsymbol{l}$) the ratio on the left in (57) has a noncentral t distribution with d.f. $n - 1$ and noncentrality parameter

$$\delta = \frac{\sqrt{n}\,|\boldsymbol{l}^T\boldsymbol{\mu} - \boldsymbol{l}^T\boldsymbol{\mu}^0|}{\sqrt{\boldsymbol{l}^T\boldsymbol{\Sigma}\boldsymbol{l}}} \tag{58}$$

when $\boldsymbol{\mu}$ is the true mean. The marginal power for any single linear combination can thus be calculated from tables of the noncentral t distribution (see Sec. 9 of Chap. 2).

3 k POPULATIONS; COVARIANCE MATRIX UNKNOWN

This technique is due to Roy and Bose (1953).

3.1 Method

Let $\{\mathbf{Y}_{hj}; h = 1, \ldots, k, j = 1, \ldots, n\}$ be k independent samples of n independent random vectors each from the k populations $N(\boldsymbol{\mu}_h, \boldsymbol{\Sigma})$, $h = 1, \ldots, k$. The mean vectors $\boldsymbol{\mu}_h$, $h = 1, \ldots, k$, and the covariance matrix $\boldsymbol{\Sigma}$ are unknown. The matrix $\boldsymbol{\Sigma}$ is assumed to be positive-definite. Assume $k \geq p$ and $n > 1$.

Interest is centered on double linear combinations of the population coordinate means: $\sum_{h=1}^{k} b_h \mathbf{a}^T \boldsymbol{\mu}_h$. The vectors $\mathbf{a} = (a_1, \ldots, a_p)$ and $\mathbf{b} = (b_1, \ldots, b_k)$ are arbitrary p-dimensional and k-dimensional linear combinations, respectively; i.e., they belong to the p-dimensional and k-dimensional $\mathcal{L}_{lc}$ spaces. These double combinations could be interpreted as comparisons *between* populations of linear combinations *within* populations. For a linear combination $\mathbf{a}$ the parametric value in population h is $\mathbf{a}^T\boldsymbol{\mu}_h$. If $\mathbf{b}$ is a contrast, then $\sum_{h=1}^{k} b_h(\mathbf{a}^T\boldsymbol{\mu}_h)$ is the contrast between populations of the parametric combination $\mathbf{a}^T\boldsymbol{\mu}$. This includes, in particular, the pairwise comparisons $\mathbf{a}^T\boldsymbol{\mu}_h - \mathbf{a}^T\boldsymbol{\mu}_{h'}$ for all pairs of populations h and h'.

Within population h the mean $\boldsymbol{\mu}_h$ is estimated by $\bar{\mathbf{Y}}_{h\cdot}$, and the covariance matrix $\boldsymbol{\Sigma}$ is estimated by

$$\mathbf{S}_h = \frac{1}{n-1} \sum_{j=1}^{n} (\mathbf{Y}_{hj} - \bar{\mathbf{Y}}_{h\cdot})(\mathbf{Y}_{hj} - \bar{\mathbf{Y}}_{h\cdot})^T. \tag{59}$$

Since $\boldsymbol{\Sigma}$ is assumed to be the same for all populations, the covariance estimates can be pooled over populations to give the common estimate

$$\mathbf{S} = \frac{1}{k} \sum_{h=1}^{k} \mathbf{S}_h = \frac{1}{k(n-1)} \sum_{h,j} (\mathbf{Y}_{hj} - \bar{\mathbf{Y}}_{h\cdot})(\mathbf{Y}_{hj} - \bar{\mathbf{Y}}_{h\cdot})^T. \tag{60}$$

Let

$$\mathbf{S}^* = \frac{1}{k} \sum_{h=1}^{k} (\bar{\mathbf{Y}}_{h\cdot} - \boldsymbol{\mu}_h)(\bar{\mathbf{Y}}_{h\cdot} - \boldsymbol{\mu}_h)^T. \tag{61}$$

The random matrix $\mathbf{S}^*$ is not a statistic [as is (60)] because it involves the values of the unknown mean parameters. However, the matrix $\mathbf{S}^*$ will be useful in describing the confidence intervals.

The simultaneous confidence interval for the double linear combination $\sum_{h=1}^{k} b_h \mathbf{a}^T \boldsymbol{\mu}_h$ is

$$\sum_{h=1}^{k} b_h \mathbf{a}^T \boldsymbol{\mu}_h \in \sum_{h=1}^{k} b_h \mathbf{a}^T \bar{\mathbf{Y}}_{h.} \pm (k\lambda_{\max}^{\alpha})^{\frac{1}{2}} \left(\sum_{h=1}^{k} b_h^2 \right)^{\frac{1}{2}} \left(\frac{1}{n} \mathbf{a}^T \mathbf{S} \mathbf{a} \right)^{\frac{1}{2}}. \tag{62}$$

The random variable $\Lambda_{\max}$ is the maximum characteristic root of the determinantal equation

$$\det \left(\mathbf{S}^* - \lambda \left(\frac{1}{n} \mathbf{S} \right) \right) = 0 \tag{63}$$

and $\lambda_{\max}^{\alpha}$ is the upper α percentile point of the distribution of $\Lambda_{\max}$.† With probability $1 - \alpha$ the intervals (62) contain the true parametric values for all $\mathbf{a}$ and $\mathbf{b}$.

The critical constant $\lambda_{\max}^{\alpha}$ is a function of α, p, k, and n, and it has been tabulated for $p = 2(1)7$. For details see Sec. 3.5.

The likelihood ratio test of the homogeneity of k multivariate populations with common covariance matrix produces a statistic involving a product of the roots of (63). Since (62) depends on the maximum root, it differs from the likelihood ratio approach. However, the maximum root is frequently used to test the homogeneity hypothesis.

For the case of unequal sample sizes the intervals (62) can be generalized. Let population h have n_h independent observations. Then, the matrices (59) to (61) are replaced by

$$\mathbf{U}_h = \frac{1}{n_h - 1} \sum_{j=1}^{n_h} (\mathbf{Y}_{hj} - \bar{\mathbf{Y}}_{h.})(\mathbf{Y}_{hj} - \bar{\mathbf{Y}}_{h.})^T \tag{64}$$

$$\mathbf{U} = \frac{1}{\sum_{h=1}^{k} (n_h - 1)} \sum_{h,j} (\mathbf{Y}_{hj} - \bar{\mathbf{Y}}_{h.})(\mathbf{Y}_{hj} - \bar{\mathbf{Y}}_{h.})^T \tag{65}$$

$$\mathbf{U}^* = \frac{1}{k} \sum_{h=1}^{k} n_h (\bar{\mathbf{Y}}_{h.} - \boldsymbol{\mu}_h)(\bar{\mathbf{Y}}_{h.} - \boldsymbol{\mu}_h)^T. \tag{66}$$

The intervals (62) become

$$\sum_{h=1}^{k} b_h \sqrt{n_h}\, \mathbf{a}^T \boldsymbol{\mu}_h \in \sum_{h=1}^{k} b_h \sqrt{n_h}\, \mathbf{a}^T \bar{\mathbf{Y}}_{h.} \pm (k u_{\max}^{\alpha})^{\frac{1}{2}} \left(\sum_{h=1}^{k} b_h^2 \right)^{\frac{1}{2}} (\mathbf{a}^T \mathbf{U} \mathbf{a})^{\frac{1}{2}} \qquad \forall\, \mathbf{a} \in \mathcal{L}_{lc} \quad \forall\, \mathbf{b} \in \mathcal{L}_{lc} \tag{67}$$

† The symbol det A denotes the determinant of A.

where $u_{\max}^{\alpha}$ is the upper α percentile point of the distribution of the largest characteristic root of the determinantal equation

$$\det(\mathbf{U}^* - u\mathbf{U}) = 0. \tag{68}$$

The intervals (67) are not as nice as (62) because of the $\sqrt{n_h}$, but they are exact intervals.

Included within the intervals (62) and (67) are contrasts between populations. Since the space of arbitrary linear combinations is larger than the space of contrasts, one can reduce the confidence interval lengths somewhat if attention is restricted to contrasts. The reader is referred to Roy and Bose (1953) for the appropriate intervals. These authors also discuss the appropriate region for pairwise differences alone. However, the distribution theory has not been worked out in this case.

When the covariance matrix $\mathbf{\Sigma}$ is known except for a multiplicative constant (that is, $\mathbf{\Sigma} = \sigma^2\mathbf{\Sigma}$; σ^2 unknown, $\mathbf{\Sigma}$ known), intervals similar to (62) or (67) arise. For equal sample sizes the intervals are

$$\sum_{h=1}^{k} b_h \mathbf{a}^T \boldsymbol{\mu}_h \in \sum_{h=1}^{k} b_h \mathbf{a}^T \bar{\mathbf{Y}}_{h.} \pm (k\lambda_{k,kp(n-1)}^{\alpha})^{\frac{1}{2}} \left(\sum_{h=1}^{k} b_h^2\right)^{\frac{1}{2}} \left(\frac{s^2}{n} \mathbf{a}^T \mathbf{\Sigma} \mathbf{a}\right)^{\frac{1}{2}} \tag{69}$$

where
$$s^2 = \frac{1}{kp(n-1)} \sum_{h=1}^{k} \sum_{j=1}^{n} (\mathbf{Y}_{hj} - \bar{\mathbf{Y}}_{h.})^T \mathbf{\Sigma}^{-1} (\mathbf{Y}_{hj} - \bar{\mathbf{Y}}_{h.}) \tag{70}$$

and $\lambda_{k,kp(n-1)}^{\alpha}$ is the α percentile of the studentized maximum characteristic root of

$$\det\left(\mathbf{S}^* - \lambda\left(\frac{1}{n}\mathbf{\Sigma}\right)\right) = 0. \tag{71}$$

Let Λ_k be the maximum root of (71). Then, $\lambda_{k,\nu}^{\alpha}$ is the upper α percentile point of the random variable Λ_k/s^2 where $\nu s^2/\sigma^2$ has a χ^2 distribution with ν d.f. and Λ_k and s^2 are independent.

The case of unknown covariance scalar is discussed here rather than in a separate section because of the lack of availability of the critical points. Tables of $\lambda_{k,\nu}^{\alpha}$ do not exist, which renders the technique useless from a practical point of view. Even if $\mathbf{\Sigma}$ were known completely, the situation is not much better. In this case the appropriate critical constant is just the upper α percentile of the largest root of (71) with $\mathbf{\Sigma} = \mathbf{\Sigma}$, or for $\mathbf{\Sigma} = \mathbf{I}$ the largest root of $\det(\mathbf{S}^* - \lambda(1/n)\mathbf{I}) = 0$. However, even this distribution has not been tabulated.

3.2 Applications

When this technique can be applied should be clear from the assumptions set forth at the beginning of the method section. For a coordinate by coordinate comparison of k multivariate populations the intervals (62) reduce to

$$\mu_{ih} - \mu_{ih'} \in \bar{Y}_{ih.} - \bar{Y}_{ih'.} \pm (k\lambda_{\max}^{\alpha})^{\frac{1}{2}}(2)^{\frac{1}{2}}\left(\frac{s_i^2}{n}\right)^{\frac{1}{2}} \qquad (72)$$
$$i = 1, \ldots, p \qquad h, h' = 1, \ldots, k \qquad h \neq h'.$$

In the last term of (72) s_i^2 is the pooled estimate of the variance of coordinate i (that is, the ith diagonal element of $\mathbf{S}$). For a pairwise comparison between populations of a linear combination $\mathbf{a}$ of the coordinate means the intervals (62) are

$$\mathbf{a}^T\boldsymbol{\mu}_h - \mathbf{a}^T\boldsymbol{\mu}_{h'} \in \mathbf{a}^T\bar{\mathbf{Y}}_{h.} - \mathbf{a}^T\bar{\mathbf{Y}}_{h'.} \pm (k\lambda_{\max}^{\alpha})^{\frac{1}{2}}(2)^{\frac{1}{2}}\left(\frac{1}{n}\,\mathbf{a}^T\mathbf{S}\mathbf{a}\right)^{\frac{1}{2}} \qquad (73)$$
$$h, h' = 1, \ldots, k \qquad h \neq h'.$$

3.3 Comparison

The only alternative to this technique is to piece together individual t intervals by the Bonferroni inequality (1.13). For a total of K combinations of $\mathbf{a}$ and $\mathbf{b}$ the Bonferroni intervals comparable to (62) are

$$\sum_{h=1}^{k} b_h\mathbf{a}^T\boldsymbol{\mu}_h \in \sum_{h=1}^{k} b_h\mathbf{a}^T\bar{\mathbf{Y}}_{h.} \pm t_{k(n-1)}^{\alpha/2K}\left(\sum_{h=1}^{k} b_h^2\right)^{\frac{1}{2}}\left(\frac{1}{n}\,\mathbf{a}^T\mathbf{S}\mathbf{a}\right)^{\frac{1}{2}} \qquad (74)$$

where $t_{k(n-1)}^{\alpha/2K}$ is the upper $(\alpha/2K)$ percentile point of the t distribution with $k(n-1)$ d.f. In any practical problem the statistician can determine which of the critical constants $[(k\lambda_{\max}^{\alpha})^{\frac{1}{2}}$ or $t_{k(n-1)}^{\alpha/2K}]$ is smaller. Similar comments apply to unequal sample sizes.

Bonferroni intervals are applicable in the case of unknown covariance scalar, whereas the technique discussed at the end of Sec. 3.1 is not because of the lack of appropriate tables. For a family of K combinations of $\mathbf{a}$ and $\mathbf{b}$ the Bonferroni intervals are

$$\sum_{h=1}^{k} b_h\mathbf{a}^T\boldsymbol{\mu}_h \in \sum_{h=1}^{k} b_h\mathbf{a}^T\bar{\mathbf{Y}}_{h.} \pm t_{kp(n-1)}^{\alpha/2K}\left(\sum_{h=1}^{k} b_h^2\right)^{\frac{1}{2}}\left(\frac{s^2}{n}\,\mathbf{a}^T\boldsymbol{\Sigma}\mathbf{a}\right)^{\frac{1}{2}} \qquad (75)$$

where s^2 is given by (70).

3.4 Derivation

The proof of (62) proceeds as follows. For fixed $\bar{\mathbf{Y}}_{h.}(h = 1, \ldots, k)$, $\mathbf{S}$, and $\mathbf{a}$, the event

$$\left|\sum_{h=1}^{k} b_h\mathbf{a}^T(\bar{\mathbf{Y}}_{h.} - \boldsymbol{\mu}_h)\right| \leq (k\lambda_{\max}^{\alpha})^{\frac{1}{2}}\left(\sum_{h=1}^{k} b_h^2\right)^{\frac{1}{2}}\left(\frac{1}{n}\,\mathbf{a}^T\mathbf{S}\mathbf{a}\right)^{\frac{1}{2}} \qquad \forall\, \mathbf{b} \in \mathcal{L}_{lc} \quad (76)$$

is identical to the event

$$\sum_{h=1}^{k} [\mathbf{a}^T(\bar{\mathbf{Y}}_{h\cdot} - \boldsymbol{\mu}_h)][(\bar{\mathbf{Y}}_{h\cdot} - \boldsymbol{\mu}_h)^T\mathbf{a}] \leq \frac{k}{n} \lambda^{\alpha}_{\max} \mathbf{a}^T\mathbf{S}\mathbf{a}. \tag{77}$$

This follows from Lemma 2.2 (see Sec. 1.4 or Sec. 2.4 of Chap. 2) with the identifications

$$\begin{aligned} c^2 &= \frac{k}{n} \lambda^{\alpha}_{\max} \mathbf{a}^T\mathbf{S}\mathbf{a} \\ d &= k \qquad \mathbf{a} = \mathbf{b} \\ y_i &= \mathbf{a}^T(\bar{\mathbf{Y}}_{i\cdot} - \boldsymbol{\mu}_i). \end{aligned} \tag{78}$$

Expression (77) can be rewritten as

$$\mathbf{a}^T\mathbf{S}^*\mathbf{a} \leq \frac{1}{n} \lambda^{\alpha}_{\max} \mathbf{a}^T\mathbf{S}\mathbf{a}. \tag{79}$$

With $\bar{\mathbf{Y}}_{h\cdot}\,(h = 1, \ldots, k)$ and $\mathbf{S}$ still fixed, the event

$$\frac{\mathbf{a}^T\mathbf{S}^*\mathbf{a}}{\mathbf{a}^T\left(\frac{1}{n}\mathbf{S}\right)\mathbf{a}} \leq \lambda^{\alpha}_{\max} \qquad \forall\, \mathbf{a} \in \mathcal{L}_{lc} \tag{80}$$

is identical to the event

$$\max_{\mathbf{a}} \frac{\mathbf{a}^T\mathbf{S}^*\mathbf{a}}{\mathbf{a}^T\left(\frac{1}{n}\mathbf{S}\right)\mathbf{a}} \leq \lambda^{\alpha}_{\max}. \tag{81}$$

The maximum in (81) is attained and is equal to $\Lambda_{\max}$, the maximum characteristic root of (63). This is a standard result in matrix theory [see, for example, Gantmacher (1959, p. 322)] and is easily proved (proof omitted here) by diagonalizing both $\mathbf{S}^*$ and $\frac{1}{n}\mathbf{S}$. Since

$$P\{\Lambda_{\max} \leq \lambda^{\alpha}_{\max}\} = 1 - \alpha \tag{82}$$

the result (62) is established.

The proof of (67) is identical to the previous paragraph with $\sqrt{n_h}$ affixed to $\bar{\mathbf{Y}}_{h\cdot} - \boldsymbol{\mu}_h$ and $1/n$ deleted everywhere.

The proof of (69) is also identical to the above with $\boldsymbol{\Sigma}$ replacing $\mathbf{S}$ and $s^2\lambda^{\alpha}_{k,\nu}$ replacing $\lambda^{\alpha}_{\max}$.

3.5 Distributions and tables

Let $\mathbf{W}_1$ and $\mathbf{W}_2$ be two independent random matrices with the Wishart distributions $W\left(\frac{1}{\nu_1}\boldsymbol{\Sigma},\ \nu_1\right)$ and $W\left(\frac{1}{\nu_2}\boldsymbol{\Sigma},\ \nu_2\right)$, respectively. $\mathbf{W}_1$ and $\mathbf{W}_2$

are independent estimators of the same covariance matrix $\mathbf{\Sigma}$ based on ν_1 and ν_2 d.f., respectively. The characteristic roots of the determinantal equation

$$\det(\mathbf{W}_1 - \lambda \mathbf{W}_2) = 0 \tag{83}$$

are related to the roots of the determinantal equation

$$\det(\nu_1 \mathbf{W}_1 - \theta(\nu_1 \mathbf{W}_1 + \nu_2 \mathbf{W}_2)) = 0 \tag{84}$$

by

$$\lambda = \frac{\nu_2}{\nu_1} \frac{\theta}{1 - \theta}. \tag{85}$$

The distribution of the characteristic roots of (84) is a classic result in multivariate analysis [see Anderson (1958) or Wilks (1962)]. The density function of the ordered roots $0 < \theta_1 < \theta_2 < \cdots < \theta_p < 1$ for $\nu_1, \nu_2 \geq p$ is

$$C \left(\prod_{i=1}^{p} \theta_i \right)^{\frac{1}{2}(\nu_1 - p - 1)} \left[\prod_{i=1}^{p} (1 - \theta_i) \right]^{\frac{1}{2}(\nu_2 - p - 1)} \prod_{i>j} (\theta_i - \theta_j) \tag{86}$$

where

$$C = \pi^{\frac{p}{2}} \prod_{i=1}^{p} \frac{\Gamma\left(\dfrac{\nu_1 + \nu_2 + 1 - i}{2}\right)}{\Gamma\left(\dfrac{\nu_1 + 1 - i}{2}\right) \Gamma\left(\dfrac{\nu_2 + 1 - i}{2}\right) \Gamma\left(\dfrac{p + 1 - i}{2}\right)}. \tag{87}$$

The density of the maximum root θ_p is the integral of (86), integrated with respect to $\theta_1, \ldots, \theta_{p-1}$. Simplification of this integral produces functions of incomplete beta functions.

With the aid of approximations and recursion relations, tables of $\theta_p^\alpha(\nu_1,\nu_2)$, the upper α percentile point of the maximum root, have been computed by various investigators. Pillai (1956) tabled $\theta_p^\alpha(\nu_1,\nu_2)$ for $p = 2$; $\alpha = .05, .01$; $\nu_1 = 3(2)11$; $\nu_2 = 13(10)63, 83(40)203, 263, 323, 403, 603, 1003, 2003$. Foster and Rees (1957) extended this table to $p = 2$; $\alpha = .20, .15, .10, .05, .01$; $\nu_1 = 2, 3(2)21$; $\nu_2 = 5(2)41(10)101, 121, 161$. Foster (1957) tabulated $\theta_p^\alpha(\nu_1,\nu_2)$ for $p = 3$; $\alpha = .20, .15, .10, .05, .01$; $\nu_1 = 3(1)10$; $\nu_2 = 4(2)194$. Foster (1958) gave $\theta_p^\alpha(\nu_1,\nu_2)$ for $p = 4$; $\alpha = .20, .15, .10, .05, .01$; $\nu_1 = 4(1)11$; $\nu_2 = 5(2)195$. For $p = 6$ Pillai and Bantegui (1959) computed the $\alpha = .05$ and $.01$ points for $\nu_1 = 7(2)15$; $\nu_2 = 17(10)67, 87(40)207, 267, 327, 407, 607, 1007, 2007$. For $p = 7$ Pillai (1964) gave the $\alpha = .05$ and $.01$ points for $\nu_1 = 8(2)18, 22, 28$; $\nu_2 = 18(10)68, 88(40)208, 268, 328, 408, 608, 1008, 2008$.

Pillai (1957) has compiled tables for $p = 2, 3, 4, 5$; $\alpha = .05, .01$; and various ν_1, ν_2. Heck (1960) gives charts for $p = 2(1)5$; $\alpha = .05, .025, .01$; various ν_1; and a continuous scale for ν_2 from which $\theta_p^\alpha(\nu_1,\nu_2)$ can be read.

Other tables of percentage points have been published, but they are not as readily available as the above.

The reader is advised to check very carefully the meaning of the notation used in any tables. For the Pillai tables $\nu_1 = 2m + p + 1$, $\nu_2 = 2n + p + 1$. In the Foster tables ν_1 and ν_2 are interchanged from what they are here. The Heck notation is the same as Pillai.

To apply these tables to (62) the correspondence [using (85)] is

$$\lambda^{\alpha}_{\max} = \frac{(n-1)\theta^{\alpha}_{p}(k,k(n-1))}{1-\theta^{\alpha}_{p}(k,k(n-1))}. \tag{88}$$

For (67)

$$u^{\alpha}_{\max} = \frac{\sum_{h=1}^{k}(n_h-1)}{k} \frac{\theta^{\alpha}_{p}\left(k, \sum_{h=1}^{k}(n_h-1)\right)}{1-\theta^{\alpha}_{p}\left(k, \sum_{h=1}^{k}(n_h-1)\right)}. \tag{89}$$

The distribution theory for the roots of (71) can be found in Anderson (1958). The percentage points of Λ_k, the maximum root, have not been tabulated, nor have the percentage points of the studentized maximum root.

No results on the power function have come forth. Since the statistic for an individual interval has a noncentral t distribution, tables of the noncentral t distribution can be applied in marginal power investigations.

4 OTHER TECHNIQUES

4.1 Variances known; covariances unknown

Dunn (1958) presented several alternative methods for treating a single multivariate normal population with known variances but unknown covariances. Let $\mathbf{Y}_1, \ldots, \mathbf{Y}_n$ be independently distributed according to $N(\boldsymbol{\mu},\boldsymbol{\Sigma})$. Assume $\sigma_{ii} = \sigma_i^2$, $i = 1, \ldots, p$, are known. The covariances $\sigma_{ii'}$, $i \neq i'$, and means $\mu_1, \ldots, \mu_p$ are unknown. Dunn derived alternative methods for placing simultaneous confidence intervals on $\mu_1, \ldots, \mu_p$.

The first method involves the construction of a matrix $\mathbf{A} = (a_{ij})$ with a specified form so that transformation to the variables $Z_i = \sum_{j=1}^{n} a_{ji}Y_{ij}$, $i = 1, \ldots, p$, gives independent $N(\mu_i,p\sigma_i^2/n)$ random variables. The confidence intervals are then

$$\mu_i \in Z_i \pm N((1-\alpha)^{1/p})\sigma_i\sqrt{\frac{p}{n}} \qquad i = 1, \ldots, p \tag{90}$$

where

$$(1-\alpha)^{1/p} = \frac{1}{\sqrt{2\pi}} \int_{-N((1-\alpha)^{1/p})}^{+N((1-\alpha)^{1/p})} e^{-y^2/2}\,dy. \tag{91}$$

The probability that the p intervals in (90) are simultaneously correct is $1 - \alpha$.

The second method hinges on the fact that the projections of the confidence ellipsoid

$$E_{\mathbf{\mu}} = \{\mathbf{\mu}: n(\bar{\mathbf{Y}}. - \mathbf{\mu})\mathbf{\Sigma}^{-1}(\bar{\mathbf{Y}}. - \mathbf{\mu}) \leq \chi_p^{2^\alpha}\} \tag{92}$$

on the coordinate axes are

$$\mu_i \in \bar{Y}_i. \pm \frac{\sigma_i}{\sqrt{n}} \sqrt{\chi_p^{2^\alpha}} \qquad i = 1, \ldots, p. \tag{93}$$

Although the quadratic form in (92) involves the unknown covariances, the projected intervals (93) do not, so they are bona fide confidence intervals. The probability of the simultaneous veracity of the intervals (93) is greater than or equal to $1 - \alpha$.

The third set of intervals, the Bonferroni intervals, is

$$\mu_i \in \bar{Y}_i. \pm g^{\alpha/2p} \frac{\sigma_i}{\sqrt{n}} \qquad i = 1, \ldots, p \tag{94}$$

where

$$1 - \frac{\alpha}{2p} = \frac{1}{\sqrt{2\pi}} \int_{-\infty}^{g^{\alpha/2p}} e^{-y^2/2}\, dy. \tag{95}$$

These follow immediately from the Bonferroni inequality (1.13).

Numerical comparison of the lengths of the three different types of intervals indicates that the Bonferroni intervals are best for $\alpha = .05$, $p = 2(2)10$.

Dunn (1958) gives the analogous techniques for unknown but equal variances. Some specialized results for $p = 2$ or 3, and for special covariances, are also derived.

4.2 Variance-covariance intervals

Roy and Bose (1953) treated the problem of putting simultaneous confidence intervals on the characteristic roots of the covariance matrix of a single multivariate normal population, and on the characteristic roots of $\mathbf{\Sigma}_1\mathbf{\Sigma}_2^{-1}$ where $\mathbf{\Sigma}_1$, $\mathbf{\Sigma}_2$ are the covariance matrices from two different multivariate normal populations. Later, Roy (1954) elaborated and expanded on these results. Anderson (1965) gives intervals which are tighter than the Roy intervals.

4.3 Two-sample confidence intervals of predetermined length

In Sec. 8.4 of Chap. 2 the two-sample simultaneous procedures of Healy (1956) for univariate problems were described. These procedures are extensions of Stein's two-sample procedure to simultaneous problems. To obtain a confidence interval of predetermined length, the size of the second sample is determined from the variance estimate in the first sample.

In the multivariate situation Healy considers a single multivariate normal population $N(\boldsymbol{\mu},\boldsymbol{\Sigma})$. He gives one technique which produces confidence intervals of length l for all linear combinations $\boldsymbol{l}^T\boldsymbol{\mu}(\boldsymbol{l}^T\boldsymbol{l} = 1)$ and which has probability exceeding $1 - \alpha$ that all the intervals contain the true value. The second technique yields an exact confidence region for $\boldsymbol{\mu}$ whose maximum diameter does not exceed l.

CHAPTER 6

Miscellaneous Techniques

This chapter groups together a variety of techniques whose discussions are too short to constitute separate chapters. For the most part the techniques are unrelated, hence the title of this chapter.

The last section outlines alternative approaches to the type of problem considered in this book. The discussion is somewhat brief because the author does not find these methods as useful as those already presented. However, the reader is invited to examine them in detail in the original articles, and use whichever technique is most appropriate for him.

1 OUTLIER DETECTION

An outlier is a single observation or single mean which does not conform with the rest of the data. (West of the Missouri the term *maverick* is more common.) Conformity is measured by how close the observation or mean lies to the other observations or means, respectively.

The detection and correction of outliers is a statistical problem with a long history. Anscombe (1960) gives an historical sketch of the attempts of nineteenth century scientists to cope with outliers. Irwin (1925) worked on outlier detection, and the earlier work of Galton (1902) and K. Pearson (1902) had bearing on this topic. Since the 1920's there has been a good deal of literature on outliers. A number of methods have been proposed for detecting them, and a number of methods have been proposed for correcting them. No attempt is made here to recount or expound all the possibilities. One detection technique, however, is based on a statistic (the studentized extreme deviate from the sample mean) which closely resembles statistics appearing in this book. This statistic is singled out for discussion.

Let $Y_1, Y_2, \ldots, Y_k$ be k independent random variables. For example, they might be individual observations, or class means in a balanced one-way classification. Presumably, they all have the same distribution, which is assumed to be a normal distribution with mean μ and variance σ^2. Let s_ν^2 be an estimate of the variance σ^2 for which $\nu s_\nu^2/\sigma^2$ has a χ^2 distribution with ν d.f. In addition, assume s_ν^2 is *independent* of $Y_1, \ldots, Y_k$. This means that for individual observations s_ν^2 has to be something other than the sample variance of $Y_1, \ldots, Y_k$. For the balanced one-way classification s_ν^2 can be the within-class sample variance (divided by the number of observations in a class).

A natural criterion for deciding whether any Y_i differs too much from the group is the distance of Y_i from the mean $\bar{Y}.$ of $Y_1, \ldots, Y_k$. This criterion leads to one of the following statistics:

$$
\begin{aligned}
O_{\pm} &= \frac{\max\limits_{1 \le i \le k} \{|Y_i - \bar{Y}.|\}}{s_\nu} = \frac{\max\{|Y_{(1)} - \bar{Y}.|, Y_{(k)} - \bar{Y}.\}}{s_\nu} \\
O_{+} &= \frac{Y_{(k)} - \bar{Y}.}{s_\nu} \qquad (1) \\
O_{-} &= \frac{|Y_{(1)} - \bar{Y}.|}{s_\nu}
\end{aligned}
$$

where $Y_{(1)} < \cdots < Y_{(k)}$ are the order statistics for $Y_1, \ldots, Y_k$.

The statistic $O_{\pm}$ seems more natural than O_+ or O_- for detecting outliers, since it inspects both tails of the sample. The statistic $O_+(O_-)$ just examines the upper (lower) tail. The two statistics O_+ and O_- can of course be combined to test both tails, but then the combined significance level is not known exactly. The precise way of combining them is to use $O_{\pm}$.

Tabulation of O_+ (and hence O_-) was attacked first. The distribution or density of O_+ can be written in integral form, but it cannot be simplified analytically to any substantial degree. The tabulation of the distribution depends on approximations, recursion relations, and numerical integration. Nair (1948) published the first tables of the upper percentile points o_+^α for $\alpha = .05, .01$; $k = 3(1)9$; $\nu = 10(1)20, 24, 30, 40, 60, 120, \infty$. Nair (1952) expanded this table to $\alpha = .10, .05, .025, .01, .005, .001$. David (1956) corrected Nair's 1952 tables and expanded them to $k = 10, 12$. To cover small d.f. Pillai and Tienzo (1959) and Pillai (1959) developed tables of o_+^α for $\alpha = .05, .01$; $k = 2(1)10, 12$; $\nu = 1(1)10$. Nair's tables also include some lower percentage points.

Halperin, et al. (1955) extensively tabulated the critical points of $O_\pm$. They give $o_\pm^\alpha$ to two decimals for $\alpha = .05, .01$; $k = 3(1)10(5)20(10)40, 60$; $\nu = 3(1)10(5)20(10)40, 60, 120, \infty$. They also give some lower tail percentage points.

In the gap-straggler-variance test (Sec. 8.1 of Chap. 2) Tukey (1949) proposed the statistic $O_\pm$ (or O_+) as a test for straggler means. Since adequate tables were not available at the time, Tukey resorted to approximate distribution theory.

For detecting maverick means in a two-way classification, Moshman (1952) studied a shortcut variant of O_+ for which residual ranges are used in the denominator instead of s. Bliss, Cochran, and Tukey (1956) proposed a rejection rule for a one-way classification which employs ranges in both the numerator and the denominator.

The aforementioned critical points of (1) are not suitable for use in a single sample problem unless there is outside information about σ^2. This is because s_ν^2 is assumed to be independent of the numerator, and consequently cannot be the sample variance. The distribution of the statistics (1) when s_ν^2 is the sample variance $s^2 = \sum_{j=1}^{k} (Y_j - \bar{Y}.)^2/(k-1)$ was treated first by Thompson (1935) and Pearson and Chandra Sekar (1936). Thompson actually considered error rates for a rejection criterion based on $|Y_i - \bar{Y}.|/s$ where Y_i is an *arbitrary* observation (not the maximum nor minimum), but his work influenced the later development. Pearson and Chandra Sekar produced the first tables of o_+^α. McKay (1935) studied the distribution of $(Y_{(k)} - \bar{Y}.)/\sigma$.

Grubbs (1950) tabulates to three decimals the upper percentile points of O_+ for $\alpha = .10, .05, .025, .01$; $k = 3(1)25$. He also gives a table (for the same α and k) from which the upper percentile points $o_\pm^\alpha$ can be obtained.

Borenius (1958) studied the same distributions from a geometrical point of view and calculated some of the frequency curves.

Quesenberry and David (1961) studied a generalization of the statistics (1). In addition to the sample variance s^2 from $Y_1, \ldots, Y_k$, there may be extra information about σ^2 available in the form of an independent estimator s_ν^2 based on a χ^2 distribution with ν d.f. In this event it makes sense to pool s^2 and s_ν^2 to give a combined estimator

$$s^{*2} = \frac{(k-1)s^2 + \nu s_\nu^2}{k + \nu - 1}. \tag{2}$$

Quesenberry and David studied the distributions of

$$O_\pm^* = \frac{\max_i \{|Y_i - \bar{Y}.|\}}{s^*} \quad \text{and} \quad O_+^* = \frac{Y_{(k)} - \bar{Y}.}{s^*}. \tag{3}$$

They give the upper α percentile points of O_+^* for $\alpha = .05, .01$; $k = 3(1)10, 12, 15, 20$; $\nu = 0(1)10, 12, 15, 20, 24, 30, 40, 50$. Bounds on $o_\pm^{*\alpha}$ are given for the same ranges of α, k, and ν.

Criteria other than (1) or (3) have been proposed for outlier detection. David, Hartley, and Pearson (1954) studied the range of the sample divided by the standard deviation of the sample [that is, $(Y_{(k)} - Y_{(1)})/s$], and they tabled its upper percentile points. These tables were revised and extended by Pearson and Stephens (1964). Thomson (1955) gave upper and lower bounds on the possible values this statistic can achieve.

Dixon (1950, 1951) proposed statistics based on ratios of distances between tail observations; he considered the following ratios in detail:

$$\begin{aligned} R_{10} &= \frac{Y_{(2)} - Y_{(1)}}{Y_{(k)} - Y_{(1)}} & R_{11} &= \frac{Y_{(2)} - Y_{(1)}}{Y_{(k-1)} - Y_{(1)}} \\ R_{12} &= \frac{Y_{(2)} - Y_{(1)}}{Y_{(k-2)} - Y_{(1)}} & R_{20} &= \frac{Y_{(3)} - Y_{(1)}}{Y_{(k)} - Y_{(1)}} \\ R_{21} &= \frac{Y_{(3)} - Y_{(1)}}{Y_{(k-1)} - Y_{(1)}} & R_{22} &= \frac{Y_{(3)} - Y_{(1)}}{Y_{(k-2)} - Y_{(1)}}. \end{aligned} \tag{4}$$

In his 1951 paper Dixon calculated the upper percentile points of the statistics (4) for samples up to size 30. In his 1950 paper he compared the performance of these ratio criteria with the extreme studentized deviates.

The question still remains of what to do with an outlier once it has been found. In some situations just the finding of the outlier is the entire problem. An experimenter might then want to pursue investigation of the treatment leading to the outlier mean. Or in maintaining quality control an inspector might want to discard the material giving rise to the outlying observation. In some instances the physical or chemical analysis producing the outlier should be repeated.

In other situations, however, the outlier is not important per se. The outlying observation is just part of the data which are to be used to estimate parameters or test hypotheses. How the outlier should be handled in this context is the perplexing problem. There are three main possibilities:

1. *Ignore:* Simply forget that the outlier is strange compared with the other observations and proceed with the analysis.
2. *Trim:* Discard the outlier from the sample and proceed with the analysis.
3. *Winsorize:* Replace the value of the outlying observation by the value of the nearest nonoutlying observation and proceed with the analysis.

The latter two techniques can be applied to one or more observations on one or both tails of the sample.

The effects of these correction devices (or lack thereof) on the statistical analysis have been explored in various articles. Of the more important ones the following have been of special interest to this author: Dixon (1953,1960), Anscombe (1960), Tukey (1962), Tukey and McLaughlin (1963), and Huber (1964). Since this book is not directed at handling outliers, the reader is left to study the results and details of these articles on his own.

The discussion of outliers in this section was presented in the format of detection followed by correction. In data analysis, however, it is often more convenient and sensible just to routinely correct for outliers without a preliminary detection analysis. When no outliers are present, some good observations will be affected, but this will not hurt the analysis very much. When outliers are present, they will be corrected whether or not they would have been detected. The virtues and dangers of routine correction are examined in the articles mentioned in the preceding paragraph.

2 MULTINOMIAL POPULATIONS

In all the previous sections the sample observations were assumed to be quantitative random variables. Except in the nonparametric chapter the observations were assumed to be normally distributed, and in the nonparametric sections the random variables were assumed to at least have a distribution with a density.

This section deals with qualitative random variables. An individual or item which is being observed falls into one of c categories. The sample observation for an individual or item is the specification of which category the individual or item belongs to.

No order is assumed among the categories. The categories may, of course, be ordered, but this is not taken into account in the analysis. For some ordered categories it may be advantageous to take account of the ordering in the analysis, which would lead to an analysis different from any to be discussed.

All the results to be presented in this section are large sample results. For each pivotal statistic the only distribution theory available is the limiting distribution as the sample size tends to infinity. No small sample distributions or critical points have been tabulated.

2.1 Single population

For the categories $1, 2, \ldots, c$ let $p_1, p_2, \ldots, p_c$ $\left(p_j \geq 0, \sum_{j=1}^{c} p_j = 1\right)$ be the probabilities an individual or item falls in each of the categories, respectively. For a sample of n individuals or items let N_j be the number which fall in category j $\left(n = \sum_{j=1}^{c} N_j\right)$. The maximum likelihood estimators of the probabilities p_j are

$$\hat{p}_j = \frac{N_j}{n} \qquad j = 1, \ldots, c. \tag{5}$$

Asymptotically $(n \to +\infty)$ the random vector $\hat{\mathbf{p}} = (\hat{p}_1, \ldots, \hat{p}_{c-1})$ has a (nonsingular) multivariate normal distribution with mean vector $\mathbf{p} = (p_1, \ldots, p_{c-1})$ and covariance matrix $(1/n)\mathbf{\Sigma}$ with elements

$$\begin{aligned} \sigma_{jj} &= p_j(1 - p_j) \\ \sigma_{jj'} &= -p_j p_{j'} \qquad j \neq j'. \end{aligned} \tag{6}$$

The covariance matrix $\mathbf{\Sigma}$ is consistently estimated by replacing each p_j in its elements by its estimate $\hat{p}_j$. Denote this estimated matrix by $\hat{\mathbf{\Sigma}}$. Asymptotically, the quadratic form $n(\hat{\mathbf{p}} - \mathbf{p})^T\hat{\mathbf{\Sigma}}^{-1}(\hat{\mathbf{p}} - \mathbf{p})$ has a χ^2 distribution with $c - 1$ d.f., so the ellipsoid

$$E_{\mathbf{p}} = \{\mathbf{p}\colon n(\hat{\mathbf{p}} - \mathbf{p})^T\hat{\mathbf{\Sigma}}^{-1}(\hat{\mathbf{p}} - \mathbf{p}) \leq \chi^{2\alpha}_{c-1}\} \tag{7}$$

is asymptotically a $100(1 - \alpha)$ percent confidence region for $\mathbf{p}$ in $(c - 1)$-dimensional space. Projection of this region (*à la* Scheffé) onto the coordinate axes yields the simultaneous confidence intervals

$$p_j \in \hat{p}_j \pm (\chi^{2\alpha}_{c-1})^{\frac{1}{2}} \left[\frac{\hat{p}_j(1 - \hat{p}_j)}{n}\right]^{\frac{1}{2}} \qquad j = 1, \ldots, c - 1. \tag{8}$$

The simultaneous confidence interval for p_c is obtained from the projection of the ellipsoid onto the vector $-\mathbf{1} = (-1, \ldots, -1)$ because $p_c = 1 - \sum_{j=1}^{c-1} p_j$. As expected, the interval is

$$p_c \in \hat{p}_c \pm (\chi^{2\alpha}_{c-1})^{\frac{1}{2}} \left[\frac{\hat{p}_c(1 - \hat{p}_c)}{n}\right]^{\frac{1}{2}}. \tag{9}$$

These intervals were proposed by Gold (1963). Goodman (1965) also considered them, but found that for the usual values of α and c the Bonferroni intervals

$$p_j \in \hat{p}_j \pm g^{\alpha/2c} \left[\frac{\hat{p}_j(1 - \hat{p}_j)}{n}\right]^{\frac{1}{2}} \qquad j = 1, \ldots, c \tag{10}$$

are shorter. The critical constant g^p is defined by

$$1 - p = \frac{1}{\sqrt{2\pi}} \int_{-\infty}^{g^p} e^{-y^2/2}\, dy. \tag{11}$$

Quesenberry and Hurst (1964) considered the same problem, but based their intervals on the customary χ^2 statistic

$$\sum_{j=1}^{c} \frac{(N_j - np_j)^2}{np_j} \tag{12}$$

which is asymptotically distributed as χ^2_{c-1}. The χ^2 statistic (12) is identical to the quadratic form $n(\hat{\mathbf{p}} - \mathbf{p})^T \mathbf{\Sigma}^{-1} (\hat{\mathbf{p}} - \mathbf{p})$ in which the covariance matrix is not replaced by its estimate. The upper and lower simultaneous confidence limits for p_j, $j = 1, \ldots, c$, based on (12) are the solutions of the quadratic equations

$$(\hat{p}_j - p_j)^2 = \chi^{2\alpha}_{c-1} \frac{p_j(1 - p_j)}{n} \qquad j = 1, \ldots, c. \tag{13}$$

Goodman (1965) has pointed out that in the usual cases the Bonferroni intervals [with $\chi_1^{2\alpha/c} = (g^{\alpha/2c})^2$ replacing $\chi^{2\alpha}_{c-1}$ in (13)] are shorter. The Bonferroni intervals are

$$p_j \in \frac{\chi_1^{2\alpha/c} + 2N_j \pm \sqrt{\chi_1^{2\alpha/c}[\chi_1^{2\alpha/c} + 4N_j(n - N_j)/n]}}{2(n + \chi_1^{2\alpha/c})} \qquad j = 1, \ldots, c. \tag{14}$$

Asymptotically the intervals (14) and (10) are equivalent.

The confidence ellipsoid (7) will also give simultaneous confidence intervals on differences $p_j - p_{j'}$, $j \neq j'$ (and other linear combinations of the p_j) by projection of the ellipsoid onto the appropriate vectors. For $p_j - p_{j'}$, $j \neq j'$, the projected interval is

$$p_j - p_{j'} \in \hat{p}_j - \hat{p}_{j'} \pm (\chi^{2\alpha}_{c-1})^{\frac{1}{2}} \left[\frac{\hat{p}_j + \hat{p}_{j'} - (\hat{p}_j - \hat{p}_{j'})^2}{n}\right]^{\frac{1}{2}}. \tag{15}$$

For a general linear combination $\boldsymbol{l} = (l_1, \ldots, l_c)$ the interval is

$$\sum_{j=1}^{c} l_j p_j \in \sum_{j=1}^{c} l_j \hat{p}_j \pm (\chi^{2\alpha}_{c-1})^{\frac{1}{2}} \left[\sum_{j=1}^{c} l_j^2 \hat{p}_j - \left(\sum_{j=1}^{c} l_j \hat{p}_j\right)^2\right]^{\frac{1}{2}} \left(\frac{1}{n}\right)^{\frac{1}{2}}. \tag{16}$$

The term $\left[\sum_j l_j^2 \hat{p}_j - \left(\sum_j l_j \hat{p}_j\right)^2\right] / n$ is the estimated variance of $\sum_j l_j \hat{p}_j$.

The intervals (15) and (16) are due to Gold (1963). Goodman (1965) has indicated that for most α and c the intervals (15) can be shortened by replacing $(\chi^{2\alpha}_{c-1})^{\frac{1}{2}}$ by $g^{\alpha/2C}$ where $C = \binom{c}{2}$. By the Bonferroni inequality the intervals still have a family probability error rate less than α. For L linear combinations of the p_j (including perhaps individual parameters and differences) the Bonferroni critical constant $g^{\alpha/2L}$ replaces $(\chi^{2\alpha}_{c-1})^{\frac{1}{2}}$. However, for large L the latter will likely be smaller.

Goodman (1965) also gives simultaneous intervals for the ratios $p_j/p_{j'}$, $j \neq j'$. However, the technique is the same as Goodman uses for cross-product ratios in contingency tables (Sec. 2.3) so no discussion is presented here.

2.2 Several populations

Let there be r populations. For population i let p_{ij}, $j = 1, \ldots, c$, be the probability an individual or item belongs to category j. $\left(\text{For each } i, p_{ij} \geq 0, \sum_j p_{ij} = 1.\right)$ The number of observations on population i is n_i, and N_{ij} is the number belonging to category j in population i $\left(n_i = \sum_j N_{ij}\right)$. The total sample size is $n = \sum_{i=1}^{r} n_i$. When taken together the observations form an $r \times c$ contingency table.

One hypothesis that is frequently of interest in $r \times c$ tables is the homogeneity or independence hypothesis, i.e.,

$$H_0\colon \quad p_{ij} = p_j \qquad i = 1, \ldots, r \qquad j = 1, \ldots, c. \tag{17}$$

The null hypothesis (17) is that the probability structure over the categories does not change from population to population, and this naturally leads to a study of contrasts between populations. A contrast in this context is defined to be any double linear combination $\sum_{i,j} c_{ij}p_{ij}$ which satisfies $\sum_i c_{ij} = 0$ for $j = 1, \ldots, c$.

The $(r - 1)(c - 1)$ differences $\Delta_{ij} = p_{ij} - p_{1j}$, $i = 2, \ldots, r$, $j = 2, \ldots, c$, are contrasts, and conversely, any contrast is a linear combination of Δ_{ij}. The converse holds because the constraints on the c_{ij} and p_{ij} produce the identity

$$\sum_{i=1}^{r} \sum_{j=1}^{c} c_{ij}p_{ij} = \sum_{i=2}^{r} \sum_{j=2}^{c} (c_{ij} - c_{i1})\Delta_{ij}. \tag{18}$$

The maximum likelihood estimators of th· p_{ij} are $\hat{p}_{ij} = N_{ij}/n_i$, $i = 1, \ldots, r$, $j = 1, \ldots, c$. A contrast $\sum_{i,j} c_{ij}p_{ij}$ is estimated by $\sum_{i,j} c_{ij}\hat{p}_{ij}$, which is an unbiased estimator and has variance

$$\sigma_c^2 = \sum_{i=1}^{r} \frac{1}{n_i} \left[\sum_{j=1}^{c} c_{ij}^2 p_{ij} - \left(\sum_{j=1}^{c} c_{ij}p_{ij} \right)^2 \right]. \tag{19}$$

This variance can be estimated by replacing p_{ij} by $\hat{p}_{ij}$ in (19). Denote this estimator by s_c^2.

Asymptotically, the $\hat{\Delta}_{ij} = \hat{p}_{ij} - \hat{p}_{1j}$, $i = 2, \ldots, r$, $j = 2, \ldots, c$, have a (nonsingular) multivariate normal distribution with means Δ_{ij} and covariance matrix $\mathbf{\Sigma}$ which involves the p_{ij} and n_i. The matrix $\mathbf{\Sigma}$ is consistently estimated by substituting $\hat{p}_{ij}$ for p_{ij} in it. Call this matrix $\hat{\mathbf{\Sigma}}$. Then, asymptotically the ellipsoid

$$E_{\mathbf{\Delta}} = \{\mathbf{\Delta}: (\hat{\mathbf{\Delta}} - \mathbf{\Delta})^T\hat{\mathbf{\Sigma}}^{-1}(\hat{\mathbf{\Delta}} - \mathbf{\Delta}) \leq \chi^{2\alpha}_{(r-1)(c-1)}\} \tag{20}$$

is a $100(1 - \alpha)$ percent confidence region for $\mathbf{\Delta}$ (where $\hat{\mathbf{\Delta}}$, $\mathbf{\Delta}$ are the vectors of the $\hat{\Delta}_{ij}$, Δ_{ij}, respectively). Projection of this ellipsoid (*à la* Scheffé) onto the various one-dimensional subspaces gives the following simultaneous confidence intervals:

$$\sum_{i,j} c_{ij}p_{ij} \in \sum_{i,j} c_{ij}\hat{p}_{ij} \pm (\chi^{2\alpha}_{(r-1)(c-1)})^{\frac{1}{2}} s_{\mathbf{c}} \tag{21}$$

for all $\mathbf{c} = (c_{11}, \ldots, c_{rc})$ satisfying

$$\sum_{i=1}^{r} c_{ij} = 0 \quad j = 1, \ldots, c.$$

The intervals (21) were derived by Goodman (1964b). He is also careful to point out that when only a small finite number of contrasts L are of interest the substitution of $g^{\alpha/2L}$ for $(\chi^{2\alpha}_{(r-1)(c-1)})^{\frac{1}{2}}$ may reduce the size of the intervals.

Simultaneous inclusion of zero in all the intervals (21) is not equivalent to

$$\sum_{i,j} \frac{(N_{ij} - n_i\hat{p}_j)^2}{n_i\hat{p}_j} \leq \chi^{2\alpha}_{(r-1)(c-1)} \tag{22}$$

where $\hat{p}_j = \sum_i n_i\hat{p}_{ij}/n$. Consequently, testing the hypothesis (17) through the intervals (21) is not equivalent to the usual χ^2 test of homogeneity. However, simultaneous inclusion of zero in the intervals (21) is equivalent to the nonsignificance of a related χ^2 statistic. Asymptotically, the two χ^2 statistics are equivalent under the null hypothesis. For details the reader is referred to Goodman (1964b).

Gold (1963) gives simultaneous confidence intervals for all (not just contrasts) linear combinations of the p_{ij}. The intervals are:

$$\sum_{i,j} l_{ij}p_{ij} \in \sum_{i,j} l_{ij}\hat{p}_{ij} \pm (\chi^{2\alpha}_{r(c-1)})^{\frac{1}{2}} s_l \qquad \forall\, \mathbf{l} = (l_{11}, \ldots, l_{rc}) \in \mathcal{L}_{lc} \tag{23}$$

where

$$s_l^2 = \sum_{i=1}^{r} \frac{1}{n_i} \Big[\sum_{j=1}^{c} l_{ij}^2\hat{p}_{ij} - \Big(\sum_{j=1}^{c} l_{ij}\hat{p}_{ij}\Big)^2\Big]. \tag{24}$$

The critical constant is increased from $\chi^{2\alpha}_{(r-1)(c-1)}$ to $\chi^{2\alpha}_{r(c-1)}$.

2.3 Cross-product ratios

In an $r \times c$ contingency table a ratio

$$\beta_{iji'j'} = p_{ij}p_{i'j'}/p_{ij'}p_{i'j} \qquad i < i' \qquad j < j' \tag{25}$$

is called a *cross-product ratio.* In a 2×2 table there is one cross-product ratio, and it is usually called the *relative risk* or *odds ratio.* The $\binom{r}{2}\binom{c}{2}$ cross-product ratios in a $r \times c$ table measure the degree of association in the table.

Goodman (1964a) constructed simultaneous confidence intervals for all the cross-product ratios $\beta_{iji'j'}$. Let $\delta_{iji'j'} = \log \beta_{iji'j'}$. A natural estimator of $\delta_{iji'j'}$ is

$$\hat{\delta}_{iji'j'} = \log N_{ij} + \log N_{i'j'} - \log N_{ij'} - \log N_{i'j} \tag{26}$$

(assuming all entries in the table are nonzero). The approximate variance of $\hat{\delta}_{iji'j'}$ is

$$\frac{1}{N_{ij}} + \frac{1}{N_{i'j'}} + \frac{1}{N_{i'j}} + \frac{1}{N_{ij'}}. \tag{27}$$

From asymptotic distribution theory and projection arguments (identical to those used in the two preceding sections) Goodman establishes that

$$\delta_{iji'j'} \in \hat{\delta}_{iji'j'} \pm (\chi^{2\alpha}_{(r-1)(c-1)})^{\frac{1}{2}} \left(\frac{1}{N_{ij}} + \frac{1}{N_{i'j'}} + \frac{1}{N_{ij'}} + \frac{1}{N_{i'j}}\right)^{\frac{1}{2}} \qquad i < i' \qquad j < j' \tag{28}$$

are approximate simultaneous confidence intervals. The intervals (28) can be converted to intervals on $\beta_{iji'j'}$ by taking antilogarithms.

Goodman extends the above technique to give simultaneous confidence intervals on all *first-order interactions,* which are specialized contrasts among the log p_{ij}.

The intervals (28) are simultaneous descendents of the technique introduced by Woolf (1955) for studying relative risks in 2×2 tables.

2.4 Logistic response curves

Reiersol (1961) applies Scheffé-type multiple comparisons procedures to the study of families of logistic response curves. In a quantal response experiment an individual or item responds in one of two ways ($c = 2$). For bioassay the responses are usually *dead* or *alive.* In general, the categories are labeled *response* and *no response.* The probability $p(= p_1)$ of the individual responding is a function of an independent variable x. In bioassay this variable is frequently dosage or log dosage. Reiersol

(1961) assumes the relationship between the probability of response and the independent variable is a logistic curve:

$$p = \frac{1}{1 + e^{-(\alpha+\beta x)}} \tag{29}$$

or

$$\alpha + \beta x = \log \frac{p}{1-p}. \tag{30}$$

Consider r populations, each with a possibly different response curve. For population i let n_{ij} individuals or items be tested at level x_{ij}, $j = 1, \ldots, m_i$, of the independent variable. If p_{ij} is the probability an individual or item in population i responds to level x_{ij}, then

$$p_{ij} = \frac{1}{1 + e^{-(\alpha_i+\beta_i x_{ij})}}. \tag{31}$$

For N_{ij} equal to the number of individuals or items responding to level x_{ij}, the proportions of responses $\hat{p}_{ij} = N_{ij}/n_{ij}$ estimate the unknown probabilities p_{ij}. The method of minimum logit χ^2 can be applied to estimate α_i, β_i, $i = 1, \ldots, r$.

By the method of χ^2 projections identical to that employed in the preceding sections, Reiersol (1961) obtains simultaneous confidence intervals for the differences

$$\beta_i - \beta_{i'} \qquad \forall\, i \neq i' \tag{32}$$

and

$$(\alpha_i + \beta_i x) - (\alpha_{i'} + \beta_{i'} x) \qquad \forall\, i \neq i' \qquad \forall\, x. \tag{33}$$

The intervals have the customary form involving the estimates of α_i and β_i, the critical point $(\chi^{2\alpha}_{2r-2})^{\frac{1}{2}}$, and the estimated standard deviations. The reader is referred to Reiersol's paper for details.

3 EQUALITY OF VARIANCES

With the exception of an extremely brief discussion in the multivariate chapter (Sec. 4.2 of Chap. 5) this entire book has been concerned with tests and confidence intervals for means, differences of means, linear combinations of means, etc. These problems all involve the *locations* of populations as opposed to the *dispersions* of populations. There are, of course, corresponding problems for dispersions of populations. The classic problem of this type is whether the variances of r different populations are the same.

There are two reasons why these dispersion problems have not been discussed before this. The first is that the vast majority of simultaneous techniques has been proposed for location problems. Location seems to be the natural setting for multiple comparisons, and in practice, the interest in means is frequently much greater than the interest in variances. The second reason is that except for Box's test, tests on variances tend to be extremely nonrobust. This makes the actual confidence or significance levels extremely sensitive to the form of the usually unknown underlying distribution.

Some of the test statistics which have been proposed for testing the equality of r variances have a simultaneous flavor to them. For the array $\{Y_{ij},\ i = 1, \ldots, r,\ j = 1, \ldots, n\}$ of independent random variables, let

$$E\{Y_{ij}\} = \mu_i \qquad \mathrm{Var}\,(Y_{ij}) = \sigma_i^2 \qquad i = 1, \ldots, r \qquad j = 1, \ldots, n. \tag{34}$$

Let the sample variance within population i be s_i^2, $i = 1, \ldots, r$; that is,

$$s_i^2 = \frac{1}{n-1}\sum_j (Y_{ij} - \bar{Y}_{i.})^2 \qquad i = 1, \ldots, r. \tag{35}$$

To test the null hypothesis

$$H_0\colon\ \sigma_1^2 = \sigma_2^2 = \cdots = \sigma_r^2 \tag{36}$$

Hartley (1950) proposed the statistic

$$F_{\max} = \frac{s^2_{\max}}{s^2_{\min}} \tag{37}$$

where

$$s^2_{\max} = \max\{s_1^2, \ldots, s_r^2\} \qquad s^2_{\min} = \min\{s_1^2, \ldots, s_r^2\}. \tag{38}$$

Under the assumption the Y_{ij} are normally distributed, Hartley tabulated the (approximate) upper $\alpha = .05$ percentile points of $F_{\max}$ for $r = 2(1)12$; $\nu = (n - 1) = 2(1)10, 12, 15, 20, 30, 60, \infty$. David (1952) improved upon the accuracy of Hartley's table and gave a corresponding table for $\alpha = .01$. These tables appear in Owen (1962), "Handbook of Statistical Tables," and Pearson and Hartley (1962), "Biometrika Tables for Statisticians," vol. I.

Another statistic in the same vein is Cochran's (1941) statistic

$$\frac{s^2_{\max}}{\sum_{i=1}^{r} s_i^2}. \tag{39}$$

Cochran gives upper $\alpha = .05$ percentage points of (39) for $r = 3(1)10$; $\nu = n - 1 = 1(1)6(2)10$.

The trouble with the Hartley and Cochran tests is the same ailment that afflicts Bartlett's (1937) M test—extreme sensitivity to nonnormality.

This was vividly demonstrated by Box (1953). The actual significance levels for these tests can be wildly different from the supposed levels if the distribution of the Y_{ij} is nonnormal (viz., the distribution has kurtosis γ_2 different from zero). This makes these tests rather dangerous to use in practice.

Box points out the reason for the nonrobustness of these variance tests. They do not utilize any evidence of variance variability within the samples. The sample variability is measured theoretically, and the theoretical variability changes as the underlying distribution changes.

To correct for this Box (1953) proposes an approximate test which transforms the dispersion problem into a location problem. Divide each population sample into c subsamples of size $m(n = cm)$, and compute the sample variance s_{ik}^2 for each subsample, $i = 1, \ldots, r$, $k = 1, \ldots, c$. Let

$$Z_{ik} = \log s_{ik}^2 \qquad i = 1, \ldots, r \qquad k = 1, \ldots, c. \tag{40}$$

Then, approximately, for $i = 1, \ldots, r$, $k = 1, \ldots, c$,

$$\begin{aligned} E\{Z_{ik}\} &= \eta_i = \log \sigma_i^2 \\ \operatorname{Var}(Z_{ik}) &= \frac{2}{m-1} + \frac{\gamma_2}{m}. \end{aligned} \tag{41}$$

[The kurtosis is assumed to be the same for all populations; this leads to a single γ_2 in (41) instead of one for each population.] The hypothesis (36) is equivalent to the hypothesis

$$H_0\colon \quad \eta_1 = \eta_2 = \cdots = \eta_r. \tag{42}$$

This is a location hypothesis for a one-way classification in the variables $\{Z_{ik}\}$, so Box proposes testing (42), and hence (36), by a one-way analysis of variance of the Z_{ik}. Simultaneous techniques could also be applied to the Z_{ik}.

No firm rules have been set down for the relative selection of c and m. Box (1953) studies several possibilities in his example. The statistician is left to rely on his own judgment in this matter.

A description and discussion of Box's test appears in Scheffé (1959, sec. 3.8).

4 PERIODOGRAM ANALYSIS

In the regression models discussed in this book, the model has always incorporated the unknown regression parameters in a linear fashion. This, obviously, will not cover all experimental situations. One type of data which cannot be analyzed by linear regression is data from trigonometric (i.e., stationary) time series.

No attempt is made in this section to discuss the analysis of stationary time series. The scope of the subject matter and research in this area is far too broad to be adequately covered even in one book. Rather the discussion here is concentrated solely on a classical model for periodic variation which leads to a statistic reminiscent of the ones treated in this book.

Let

$$Y_t = \sum_{i=1}^{p} (\alpha_i \cos t\lambda_i + \beta_i \sin t\lambda_i) + e_t \qquad t = 1, 2, \ldots \tag{43}$$

where $0 \leq \lambda_i \leq \pi$, $i = 1, \ldots, p$, and the process $\{e_t\}$ is white noise; that is,

$$e_t \text{ are independent } N(0,\sigma^2). \tag{44}$$

The observations Y_t are assumed to be composed of p periodic sine and cosine functions with unknown amplitudes α_i, β_i and frequencies λ_i, and of random noise e_t. For frequency λ_i the period of the sine (cosine) function is $2\pi/\lambda_i$. The model could alternatively be written in terms of sine (or cosine) functions alone with different phases φ_i; that is,

$$Y_t = \sum_{i=1}^{p} \gamma_i \sin (t\lambda_i + \varphi_i) + e_t \qquad t = 1, 2, \ldots. \tag{45}$$

The model (43) is linear in the amplitudes $\{\alpha_i,\beta_i\}$, but it is nonlinear in the frequencies $\{\lambda_i\}$. If the λ_i were known, then it would be a simple matter to estimate and treat the unknown amplitudes by linear techniques. When the frequencies are unknown, an analysis, called *periodogram analysis*, was proposed by Schuster (1898).

For n observations $Y_1, \ldots, Y_n$ let

$$I_n(\lambda) = \frac{1}{2\pi n} \left| \sum_{t=1}^{n} Y_t e^{-it\lambda} \right|^2 = \frac{n}{2\pi} \{A^2(\lambda) + B^2(\lambda)\} \tag{46}$$

where

$$A(\lambda) = \frac{1}{n} \sum_{t=1}^{n} Y_t \cos t\lambda$$
$$B(\lambda) = \frac{1}{n} \sum_{t=1}^{n} Y_t \sin t\lambda. \tag{47}$$

The function $I_n(\lambda)$ is called the *periodogram*. It is the inverse Fourier transform of the sample covariance sequence. For large n the periodogram will peak up around any true frequencies λ_i and remain negligible for other λ. If one were to compute and plot $I_n(\lambda)$ over the interval

$[0,\pi]$, one would be tempted to guess that the peaks of the function correspond to true frequencies.[1] Periodograms tend to be rather jagged, so there is the problem of evaluating which peak (if any) is a real peak and which is due to random fluctuation.

Fisher (1929) provided a rigorous statistical analysis of a periodogram by devising a test of whether the values of $I_n(\lambda)$ at selected frequencies are significantly large. For simplicity, let $n = 2m + 1$, m an integer; the analysis for n even is identical except for the end frequencies $(\lambda = 0, \pi)$. Define

$$\lambda_k^* = \frac{2\pi k}{2m+1} \qquad k = 1, \ldots, m. \tag{48}$$

By examining the coefficients in the linear combinations (47) for $\lambda = \lambda_k^*$, $k = 1, \ldots, m$, it can be straightforwardly demonstrated that if $\alpha_i = \beta_i = 0$, $i = 1, \ldots, p$ (that is, the process consists purely of white noise with no periodic variation), then the $2m$ random variables

$$A(\lambda_k^*) \qquad B(\lambda_k^*) \qquad k = 1, \ldots, m \tag{49}$$

are independently, normally distributed with zero means and variances $\sigma^2/2n$. The independence of the random variables follows from the trigonometric identities

$$\begin{aligned} &\sum_{t=1}^{2m+1} \cos \lambda_k^* t \cos \lambda_{k'}^* t = 0 \qquad k \neq k' \\ &\sum_{t=1}^{2m+1} \cos \lambda_k^* t \sin \lambda_{k'}^* t = 0 \qquad k = k' \quad \text{or} \quad k \neq k' \\ &\sum_{t=1}^{2m+1} \sin \lambda_k^* t \sin \lambda_{k'}^* t = 0 \qquad k \neq k' \end{aligned} \tag{50}$$

which impart zero covariances to the random variables. Thus, when there is no periodic variation, the values of the periodogram at the frequencies λ_k^*, that is,

$$I_n(\lambda_k^*) = \frac{n}{2\pi} \{A^2(\lambda_k^*) + B^2(\lambda_k^*)\} \qquad k = 1, \ldots, m \tag{51}$$

form a sequence of *independent* random variables, each distributed as $\sigma^2/4\pi$ times a χ^2 variable with 2 d.f. When, in fact, there is periodic variation, the value of $I_n(\lambda_k^*)$ will be inflated for λ_k^* near a true frequency λ_i.

[1] The spectrum of a discrete time stationary process is $[-\pi,\pi]$. However, since the process is real-valued, the spectral distribution function is symmetric about zero and $I_n(\lambda) = I_n(-\lambda)$.

Fisher proposed the statistic

$$\frac{\max_k \{I_n(\lambda_k^*)\}}{\sum_{k=1}^{m} I_n(\lambda_k^*)} \tag{52}$$

for testing the presence of periodic variation. In his 1929 paper he tabulated the upper 5 percent and 1 percent points of the null distribution for $m = 5(1)50$.

Fisher (1940) also studied the distribution of the rth largest $I_n(\lambda_k^*)$ divided by the sum. A different derivation of this latter distribution is given by Whittle (1951). This distribution can be found in Grenander and Rosenblatt (1957), E. J. Hannan (1960), Wilks (1962), and elsewhere.

Let $I_n(\lambda_{(1)}^*) < \cdots < I_n(\lambda_{(m)}^*)$ be the ordered values of $I_n(\lambda_k^*)$, $k = 1, \ldots, m$. If $I_n(\lambda_{(m)}^*)$ is judged to be significantly large by Fisher's test, then $\lambda_{(m)}^*$ is inferred to be a true frequency. In practice, it makes more sense to infer that the frequency for the peak of the periodogram nearest to $I_n(\lambda_{(m)}^*)$ is a true frequency. Of course, for large n the difference between these two frequencies is negligible. The actual maximum value of the periodogram cannot be tested for significance (unless by accident it occurs at some λ_k^*) because for $\lambda \neq \lambda_k^*$ the random variables $A(\lambda)$ and $B(\lambda)$ are not independent of each other or the other variables $A(\lambda_k^*)$, $B(\lambda_k^*)$. For small n this is unfortunate, but it makes little difference for large n.

The significance of the frequency with the second largest value $I_n(\lambda_{(m-1)}^*)$ can be judged from the appropriate critical value of the null distribution of $I_n(\lambda_{(m-1)}^*)/\sum_{k=1}^{m} I_n(\lambda_k^*)$. The next lower value $I_n(\lambda_{(m-2)}^*)$ can be tested similarly, and so on.

The periodogram $I_n(\lambda)$ is an estimate of the spectral density function, but it is well known to be a poor one [see Grenander and Rosenblatt (1957) or E. J. Hannan (1960)]. The field of spectral density estimation has received much consideration and is undergoing constant improvement. No discussion is given here of this area of research because it is not directly relevant to simultaneous inference or vice versa.

5 ALTERNATIVE APPROACHES: SELECTION, RANKING, SLIPPAGE

The techniques discussed in this book provide a means of making more detailed inferences than the *significant* or *nonsignificant* conclusions of classical analysis of variance. Their aim is to provide confidence intervals or tests for any and all comparisons involving the model parameters (e.g., means) which are of interest. There have been numerous other procedures proposed in the literature as alternatives to the classical analysis-of-variance tests. Among these other procedures are a variety of multiple decision rules referred to as selection, ranking, or slippage rules.

The aim of selection, ranking, or slippage rules is different from that of simultaneous confidence intervals or multiple comparisons tests. Basically, these procedures attempt to pick out the best population (or better populations) from a set of populations if the set is not homogeneous. The best population is ordinarily interpreted to be the population with the largest (or smallest) value of an unknown parameter such as the mean, variance, etc. Analogously, the better populations are those with the larger (or smaller) values of the parameter.

Gupta (1956) pioneered one type of selection rule. The goal of this rule is to choose a subset of the populations which contains the best population (or better populations) with probability equalling or exceeding a prescribed level P^*. The sample size for each population is fixed, so the only question is how large a subset to select in order to include the best population (or better populations) with the specified probability. The subset should be kept as small as possible while still maintaining the prescribed probability P^*. To illustrate, consider the model structure for many-one t statistics (Sec. 5 of Chap. 2), i.e., k treatment populations ($i = 1, \ldots, k$) and one control population ($i = 0$). Let there be n_i observations Y_{ij} from population i. The rule is to select for the subset those populations whose means satisfy

$$\bar{Y}_{i.} \geq \bar{Y}_{0.} - d \frac{s}{\sqrt{n_i}} \tag{53}$$

where s is the common estimate of the standard deviation based on $\sum_1^k n_i - k$ d.f. All populations whose sample means satisfy (53) are inferred to be better than the control population. The constant d should be so chosen that with probability at least P^* all populations with $\mu_i > \mu_0$ are included in the subset. It is easily verified that the probability of correctly selecting all populations with $\mu_i > \mu_0$ has an infimum at $\mu_0 = \mu_1 = \cdots = \mu_k$. For $n_i \equiv n$, $i = 0, 1, \ldots, k$, the choice of d thus reduces to finding the solution of

$$P\{\min_{i=1,\ldots,k} \{\sqrt{n}\,(\bar{Y}_{i.} - \mu_i)\} \geq \sqrt{n}\,(\bar{Y}_{0.} - \mu_0) - ds | \mu_0 = \mu_1 = \cdots = \mu_k\} = P^*. \tag{54}$$

Gupta and Sobel (1957) give tables of d for various values of P^*, k, and d.f. for normally distributed observations.

Gupta and Sobel (1958) discuss in detail the rule (53) and its analogs where μ_0 and/or σ^2 are known. They also consider the scale parameter case, and the analogous problem for binomial populations. In addition, Gupta and Sobel (1960, 1962) treat the selection of the best of k binomial populations and the selection of the smallest of k variances. Gupta (1962) discusses the selection of the best of k gamma populations.

A good survey article on this approach is Gupta (1965).

Prior to the Gupta and Sobel formulation of the selection problem, Paulson (1949, 1952a) suggested a similar approach in which the sample size is controlled to achieve a specific aim. In the comparison of k treatment populations with a control population, Paulson's rule decides that the treatment population with the largest sample mean $\bar{Y}_{(k)\cdot}$ is better than the control population if

$$\bar{Y}_{(k)\cdot} > \bar{Y}_{0\cdot} + \lambda_n \frac{s}{\sqrt{n}} \tag{55}$$

where λ_n is chosen so that

$$P\left\{\bar{Y}_{(k)\cdot} > \bar{Y}_{0\cdot} + \lambda_n \frac{s}{\sqrt{n}} \,\middle|\, \mu_0 = \mu_1 = \cdots = \mu_k\right\} = \alpha. \tag{56}$$

This choice of λ_n fixes the rule (55) so that if no treatment population is better than the control population, then with probability $1 - \alpha$, no treatment will be inferred to be better than the control. The sample size n is chosen so that if one treatment is better than all the others and the control by an amount $\Delta > 0$, it will be selected with probability $1 - \beta$; that is,

$$P\left\{\bar{Y}_{k\cdot} > \bar{Y}_{0\cdot} + \lambda_n \frac{s}{\sqrt{n}} \text{ and } \bar{Y}_{k\cdot} > \max\{\bar{Y}_{1\cdot}, \ldots, \bar{Y}_{k-1\cdot}\} \,\middle|\, \mu_0 = \mu_1 = \cdots = \mu_{k-1};\ \mu_k = \mu_0 + \Delta\right\} = 1 - \beta. \tag{57}$$

The interval $(\mu_0, \mu_0 + \Delta)$ is referred to as an *indifference zone* with the connotation that the statistician is not too concerned with picking out treatments whose means do not exceed μ_0 by more than Δ.

Bechhofer (1954) introduced ranking procedures. The goal of ranking procedures is a generalization of the two previous selection procedures. For a one-way classification the goal is to correctly divide the k populations into the k_s best populations, the k_{s-1} second best populations, and so on down to the k_1 worst populations $\left(\sum_{i=1}^{s} k_i = k\right)$. The k_s best populations are those with the largest means, etc. The special case $s = 2$, $k_2 = 1$ is equivalent to selecting the best population, and the special case $s = k$, $k_i \equiv 1$ is equivalent to ranking the population means.

The probability of a correct ranking depends on the configuration of the true population means, the unknown variance, and the sample sizes (assuming normality). If there is little difference between the s groups, then it is usually not important to have the probability of a correct ranking at a high level. But if the gaps between the groups are large, it is likely important to guarantee a correct ranking with high probability.

Bechhofer formalizes this philosophy by creating indifference zones similar to the one in the Paulson selection rule. If the gap between the k_s best populations and k_{s-1} second-best populations is at least $\Delta_{s-1} > 0$, the gap between the k_{s-1} second-best and k_{s-2} third-best populations is at least $\Delta_{s-2} > 0$, etc., then the probability of correctly dividing the means into s groups should be at least γ. This probability achieves a minimum when the means are equal within a group and the gaps between the groups are exactly $\Delta_{s-1}, \Delta_{s-2}, \ldots, \Delta_1$.

In his 1954 paper Bechhofer gives tables for determining the sample sizes necessary to achieve the probability γ when $s = 2$ and the variance is known. Bechhofer, Dunnett, and Sobel (1954) treat the case with unknown variance. The analogous ranking problem for variances is considered in Bechhofer and Sobel (1954). Selection of the multinomial event with the highest probability is handled in Bechhofer, Elmaghraby, and Morse (1959).

The three selection and ranking procedures just discussed were preceded chronologically, and were more or less inspired, by a slippage rule of Mosteller (1948). Mosteller formulated a nonparametric rule for testing whether the location of one population had slipped to the right (or left) of the other $k - 1$ populations in a one-way classification. Calculation of the test critical point depended on equal sample sizes, but this was relaxed in Mosteller and Tukey (1950).

Recently sequential selection procedures have been developed. The relevant references are Bechhofer (1958), Bechhofer and Blumenthal (1962), Paulson (1962, 1964).

There have been some theoretical papers considering the optimality of selection, ranking, and slippage procedures of the type discussed above. The reader is referred in particular to Paulson (1952b), Truax (1953), Karlin and Truax (1960), and Lehmann (1961).

In addition to the papers cited in this section there are other papers on selection, ranking, and slippage techniques. Gupta (1965) contains a good bibliography of these papers.

APPENDIX A

Strong Law for the Expected Error Rate

Let P_i, $i = 1, 2, \ldots$, be a sequence of independent random proportions with range spaces $\{0 = 0/N_i, 1/N_i, \ldots, N_i/N_i = 1\}$ and expectations $E\{P_i\} = p$, $i = 1, 2, \ldots$.

Lemma Var $(P_i) \leq p(1 - p)$, with equality when $P\{P_i = 1\} = p$.

Proof $E\{P_i\} = p$ by assumption.

$$E\{P_i^2\} = \sum_0^{N_i} \left(\frac{k}{N_i}\right)^2 P\left\{P_i = \frac{k}{N_i}\right\}$$
$$\leq \sum_0^{N_i} \left(k \frac{N_i}{N_i^2}\right) P\left\{P_i = \frac{k}{N_i}\right\} = E\{P_i\} = p.$$

Hence, $\quad \text{Var}\,(P_i) \leq p - p^2 = p(1 - p).$

When $P\{P_i = 1\} = p$, then $E\{P_i^2\} = p$. $\quad \|$

The theorem will use the following well-known [Loève (1963, p. 238)] result:

Kolmogorov's Criterion If $\{Y_n\}$ is a sequence of independent random variables, then for $b_n \uparrow \infty$, $\sum_1^\infty \text{Var}\,(Y_n)/b_n^2 < \infty$ implies that

$$(S_n - E\{S_n\})/b_n \to 0, \text{ a.s., where } S_n = \sum_1^n Y_i.$$

Theorem (1) $\bar{P} = \left(\sum_1^n P_i\right)/n \to p$, a.s.

(2) If there exists an N for which $N_i \leq N$, $i = 1, 2, \ldots$, then

$$\bar{\bar{P}} = \left(\sum_1^n N_i P_i\right) \Big/ \sum_1^n N_i \to p, \text{ a.s.}$$

Proof (1) $\displaystyle\sum_1^\infty \frac{\operatorname{Var}(P_n)}{n^2} \leq \sum_1^\infty \frac{p(1-p)}{n^2} < \infty.$

(2)
$$\sum_1^\infty \frac{\operatorname{Var}(N_n P_n)}{\left(\sum_1^n N_i\right)^2} \leq \sum_1^\infty \frac{N_n^2 p(1-p)}{\left(\sum_1^n N_i\right)^2}$$

$$\leq N^2 p(1-p) \sum_1^\infty \frac{1}{n^2} < \infty. \qquad \|$$

Without the condition $N_i \leq N$ in part (2), it is easy to construct sequences for which the convergence does not hold, but this is not true of all sequences not satisfying the condition.

APPENDIX B

Tables

Table I: Percentage points of the studentized range Reproduced from H. L. Harter, Tables of range and studentized range, *Ann. Math. Statist.*, **31**:1122–1147 (1960), with permission of author and editor.

Table II: Percentage points of the Bonferroni t statistic Reproduced from O. J. Dunn, Multiple comparisons among means, *J. Am. Statist. Assoc.*, **56**:52–64 (1961), with permission of author and editor.

Table III: Percentage points of the studentized maximum modulus Reproduced from K. C. S. Pillai and K. V. Ramachandran, On the distribution of the ratio of the ith observation in an ordered sample from a normal population to an independent estimate of the standard deviation, *Ann. Math. Statist.*, **25**:565–572 (1954), with permission of authors and editor.

Table IV: Percentage points of the many-one t statistics Reproduced from C. W. Dunnett, A multiple comparisons procedure for comparing several treatments with a control, *J. Am. Statist. Assoc.*, **50**:1096–1121 (1955), and C. W. Dunnett, New tables for multiple comparisons with a control, *Biometrics*, **20**:482–491 (1964), with permission of author and editors.

Table V: Percentage points of the Duncan multiple range test Reproduced from H. L. Harter, Critical values for Duncan's new multiple range test, *Biometrics*, **16**:671–685 (1960), with permission of author and editor.

Table VI: Percentage points of the many-one sign statistics One-tailed table computed by S. L. Boyle on the Stanford University IBM 7090. The two-tailed table is adapted from R. G. D. Steel, A multiple comparison sign test: treatments versus control, *J. Am. Statist. Assoc.*, **54**:767–775 (1959), with permission of author and editor.

Table VII: Percentage points of the k-sample sign statistics Computed by S. L. Boyle on the Stanford University IBM 7090.

Table VIII: Percentage points of the many-one rank statistics One-tailed table computed by S. L. Boyle on the Stanford University IBM 7090. The two-tailed table is adapted from R. G. D. Steel, A multiple comparison rank sum test: treatments vs. control, *Biometrics*, **15**:560–572 (1959), with permission of author and editor.

Table IX: Percentage points of the k-sample rank statistics Computed by S. L. Boyle on the Stanford University IBM 7090.

Table I

PERCENTAGE POINTS OF THE STUDENTIZED RANGE $q^{\alpha}_{r,\nu}$
(For reference, see Chap. 2, Sec. 1)

$\alpha = .05$

ν \ r	2	3	4	5	6	7	8	9	10
1	17.97	26.98	32.82	37.08	40.41	43.12	45.40	47.36	49.07
2	6.085	8.331	9.798	10.88	11.74	12.44	13.03	13.54	13.99
3	4.501	5.910	6.825	7.502	8.037	8.478	8.853	9.177	9.462
4	3.927	5.040	5.757	6.287	6.707	7.053	7.347	7.602	7.826
5	3.635	4.602	5.218	5.673	6.033	6.330	6.582	6.802	6.995
6	3.461	4.339	4.896	5.305	5.628	5.895	6.122	6.319	6.493
7	3.344	4.165	4.681	5.060	5.359	5.606	5.815	5.998	6.158
8	3.261	4.041	4.529	4.886	5.167	5.399	5.597	5.767	5.918
9	3.199	3.949	4.415	4.756	5.024	5.244	5.432	5.595	5.739
10	3.151	3.877	4.327	4.654	4.912	5.124	5.305	5.461	5.599
11	3.113	3.820	4.256	4.574	4.823	5.028	5.202	5.353	5.487
12	3.082	3.773	4.199	4.508	4.751	4.950	5.119	5.265	5.395
13	3.055	3.735	4.151	4.453	4.690	4.885	5.049	5.192	5.318
14	3.033	3.702	4.111	4.407	4.639	4.829	4.990	5.131	5.254
15	3.014	3.674	4.076	4.367	4.595	4.782	4.940	5.077	5.198
16	2.998	3.649	4.046	4.333	4.557	4.741	4.897	5.031	5.150
17	2.984	3.628	4.020	4.303	4.524	4.705	4.858	4.991	5.108
18	2.971	3.609	3.997	4.277	4.495	4.673	4.824	4.956	5.071
19	2.960	3.593	3.977	4.253	4.469	4.645	4.794	4.924	5.038
20	2.950	3.578	3.958	4.232	4.445	4.620	4.768	4.896	5.008
24	2.919	3.532	3.901	4.166	4.373	4.541	4.684	4.807	4.915
30	2.888	3.486	3.845	4.102	4.302	4.464	4.602	4.720	4.824
40	2.858	3.442	3.791	4.039	4.232	4.389	4.521	4.635	4.735
60	2.829	3.399	3.737	3.977	4.163	4.314	4.441	4.550	4.646
120	2.800	3.356	3.685	3.917	4.096	4.241	4.363	4.468	4.560
∞	2.772	3.314	3.633	3.858	4.030	4.170	4.286	4.387	4.474

ν \ r	11	12	13	14	15	16	17	18	19
1	50.59	51.96	53.20	54.33	55.36	56.32	57.22	58.04	58.83
2	14.39	14.75	15.08	15.38	15.65	15.91	16.14	16.37	16.57
3	9.717	9.946	10.15	10.35	10.53	10.69	10.84	10.98	11.11
4	8.027	8.208	8.373	8.525	8.664	8.794	8.914	9.028	9.134
5	7.168	7.324	7.466	7.596	7.717	7.828	7.932	8.030	8.122
6	6.649	6.789	6.917	7.034	7.143	7.244	7.338	7.426	7.508
7	6.302	6.431	6.550	6.658	6.759	6.852	6.939	7.020	7.097
8	6.054	6.175	6.287	6.389	6.483	6.571	6.653	6.729	6.802
9	5.867	5.983	6.089	6.186	6.276	6.359	6.437	6.510	6.579
10	5.722	5.833	5.935	6.028	6.114	6.194	6.269	6.339	6.405
11	5.605	5.713	5.811	5.901	5.984	6.062	6.134	6.202	6.265
12	5.511	5.615	5.710	5.798	5.878	5.953	6.023	6.089	6.151
13	5.431	5.533	5.625	5.711	5.789	5.862	5.931	5.995	6.055
14	5.364	5.463	5.554	5.637	5.714	5.786	5.852	5.915	5.974
15	5.306	5.404	5.493	5.574	5.649	5.720	5.785	5.846	5.904
16	5.256	5.352	5.439	5.520	5.593	5.662	5.727	5.786	5.843
17	5.212	5.307	5.392	5.471	5.544	5.612	5.675	5.734	5.790
18	5.174	5.267	5.352	5.429	5.501	5.568	5.630	5.688	5.743
19	5.140	5.231	5.315	5.391	5.462	5.528	5.589	5.647	5.701
20	5.108	5.199	5.282	5.357	5.427	5.493	5.553	5.610	5.663
24	5.012	5.099	5.179	5.251	5.319	5.381	5.439	5.494	5.545
30	4.917	5.001	5.077	5.147	5.211	5.271	5.327	5.379	5.429
40	4.824	4.904	4.977	5.044	5.106	5.163	5.216	5.266	5.313
60	4.732	4.808	4.878	4.942	5.001	5.056	5.107	5.154	5.199
120	4.641	4.714	4.781	4.842	4.898	4.950	4.998	5.044	5.086
∞	4.552	4.622	4.685	4.743	4.796	4.845	4.891	4.934	4.974

Table I (*Continued*)

$\alpha = .05$

ν \ r	20	22	24	26	28	30	32	34	36
1	59.56	60.91	62.12	63.22	64.23	65.15	66.01	66.81	67.56
2	16.77	17.13	17.45	17.75	18.02	18.27	18.50	18.72	18.92
3	11.24	11.47	11.68	11.87	12.05	12.21	12.36	12.50	12.63
4	9.233	9.418	9.584	9.736	9.875	10.00	10.12	10.23	10.34
5	8.208	8.368	8.512	8.643	8.764	8.875	8.979	9.075	9.165
6	7.587	7.730	7.861	7.979	8.088	8.189	8.283	8.370	8.452
7	7.170	7.303	7.423	7.533	7.634	7.728	7.814	7.895	7.972
8	6.870	6.995	7.109	7.212	7.307	7.395	7.477	7.554	7.625
9	6.644	6.763	6.871	6.970	7.061	7.145	7.222	7.295	7.363
10	6.467	6.582	6.686	6.781	6.868	6.948	7.023	7.093	7.159
11	6.326	6.436	6.536	6.628	6.712	6.790	6.863	6.930	6.994
12	6.209	6.317	6.414	6.503	6.585	6.660	6.731	6.796	6.858
13	6.112	6.217	6.312	6.398	6.478	6.551	6.620	6.684	6.744
14	6.029	6.132	6.224	6.309	6.387	6.459	6.526	6.588	6.647
15	5.958	6.059	6.149	6.233	6.309	6.379	6.445	6.506	6.564
16	5.897	5.995	6.084	6.166	6.241	6.310	6.374	6.434	6.491
17	5.842	5.940	6.027	6.107	6.181	6.249	6.313	6.372	6.427
18	5.794	5.890	5.977	6.055	6.128	6.195	6.258	6.316	6.371
19	5.752	5.846	5.932	6.009	6.081	6.147	6.209	6.267	6.321
20	5.714	5.807	5.891	5.968	6.039	6.104	6.165	6.222	6.275
24	5.594	5.683	5.764	5.838	5.906	5.968	6.027	6.081	6.132
30	5.475	5.561	5.638	5.709	5.774	5.833	5.889	5.941	5.990
40	5.358	5.439	5.513	5.581	5.642	5.700	5.753	5.803	5.849
60	5.241	5.319	5.389	5.453	5.512	5.566	5.617	5.664	5.708
120	5.126	5.200	5.266	5.327	5.382	5.434	5.481	5.526	5.568
∞	5.012	5.081	5.144	5.201	5.253	5.301	5.346	5.388	5.427

ν \ r	38	40	50	60	70	80	90	100
1	68.26	68.92	71.73	73.97	75.82	77.40	78.77	79.98
2	19.11	19.28	20.05	20.66	21.16	21.59	21.96	22.29
3	12.75	12.87	13.36	13.76	14.08	14.36	14.61	14.82
4	10.44	10.53	10.93	11.24	11.51	11.73	11.92	12.09
5	9.250	9.330	9.674	9.949	10.18	10.38	10.54	10.69
6	8.529	8.601	8.913	9.163	9.370	9.548	9.702	9.839
7	8.043	8.110	8.400	8.632	8.824	8.989	9.133	9.261
8	7.693	7.756	8.029	8.248	8.430	8.586	8.722	8.843
9	7.428	7.488	7.749	7.958	8.132	8.281	8.410	8.526
10	7.220	7.279	7.529	7.730	7.897	8.041	8.166	8.276
11	7.053	7.110	7.352	7.546	7.708	7.847	7.968	8.075
12	6.916	6.970	7.205	7.394	7.552	7.687	7.804	7.909
13	6.800	6.854	7.083	7.267	7.421	7.552	7.667	7.769
14	6.702	6.754	6.979	7.159	7.309	7.438	7.550	7.650
15	6.618	6.669	6.888	7.065	7.212	7.339	7.449	7.546
16	6.544	6.594	6.810	6.984	7.128	7.252	7.360	7.457
17	6.479	6.529	6.741	6.912	7.054	7.176	7.283	7.377
18	6.422	6.471	6.680	6.848	6.989	7.109	7.213	7.307
19	6.371	6.419	6.626	6.792	6.930	7.048	7.152	7.244
20	6.325	6.373	6.576	6.740	6.877	6.994	7.097	7.187
24	6.181	6.226	6.421	6.579	6.710	6.822	6.920	7.008
30	6.037	6.080	6.267	6.417	6.543	6.650	6.744	6.827
40	5.893	5.934	6.112	6.255	6.375	6.477	6.566	6.645
60	5.750	5.789	5.958	6.093	6.206	6.303	6.387	6.462
120	5.607	5.644	5.802	5.929	6.035	6.126	6.205	6.275
∞	5.463	5.498	5.646	5.764	5.863	5.947	6.020	6.085

Table I (*Continued*)

$\alpha = .01$

ν \ r	2	3	4	5	6	7	8	9	10
1	90.03	135.0	164.3	185.6	202.2	215.8	227.2	237.0	245.6
2	14.04	19.02	22.29	24.72	26.63	28.20	29.53	30.68	31.69
3	8.261	10.62	12.17	13.33	14.24	15.00	15.64	16.20	16.69
4	6.512	8.120	9.173	9.958	10.58	11.10	11.55	11.93	12.27
5	5.702	6.976	7.804	8.421	8.913	9.321	9.669	9.972	10.24
6	5.243	6.331	7.033	7.556	7.973	8.318	8.613	8.869	9.097
7	4.949	5.919	6.543	7.005	7.373	7.679	7.939	8.166	8.368
8	4.746	5.635	6.204	6.625	6.960	7.237	7.474	7.681	7.863
9	4.596	5.428	5.957	6.348	6.658	6.915	7.134	7.325	7.495
10	4.482	5.270	5.769	6.136	6.428	6.669	6.875	7.055	7.213
11	4.392	5.146	5.621	5.970	6.247	6.476	6.672	6.842	6.992
12	4.320	5.046	5.502	5.836	6.101	6.321	6.507	6.670	6.814
13	4.260	4.964	5.404	5.727	5.981	6.192	6.372	6.528	6.667
14	4.210	4.895	5.322	5.634	5.881	6.085	6.258	6.409	6.543
15	4.168	4.836	5.252	5.556	5.796	5.994	6.162	6.309	6.439
16	4.131	4.786	5.192	5.489	5.722	5.915	6.079	6.222	6.349
17	4.099	4.742	5.140	5.430	5.659	5.847	6.007	6.147	6.270
18	4.071	4.703	5.094	5.379	5.603	5.788	5.944	6.081	6.201
19	4.046	4.670	5.054	5.334	5.554	5.735	5.889	6.022	6.141
20	4.024	4.639	5.018	5.294	5.510	5.688	5.839	5.970	6.087
24	3.956	4.546	4.907	5.168	5.374	5.542	5.685	5.809	5.919
30	3.889	4.455	4.799	5.048	5.242	5.401	5.536	5.653	5.756
40	3.825	4.367	4.696	4.931	5.114	5.265	5.392	5.502	5.599
60	3.762	4.282	4.595	4.818	4.991	5.133	5.253	5.356	5.447
120	3.702	4.200	4.497	4.709	4.872	5.005	5.118	5.214	5.299
∞	3.643	4.120	4.403	4.603	4.757	4.882	4.987	5.078	5.157

ν \ r	11	12	13	14	15	16	17	18	19
1	253.2	260.0	266.2	271.8	277.0	281.8	286.3	290.4	294.3
2	32.59	33.40	34.13	34.81	35.43	36.00	36.53	37.03	37.50
3	17.13	17.53	17.89	18.22	18.52	18.81	19.07	19.32	19.55
4	12.57	12.84	13.09	13.32	13.53	13.73	13.91	14.08	14.24
5	10.48	10.70	10.89	11.08	11.24	11.40	11.55	11.68	11.81
6	9.301	9.485	9.653	9.808	9.951	10.08	10.21	10.32	10.43
7	8.548	8.711	8.860	8.997	9.124	9.242	9.353	9.456	9.554
8	8.027	8.176	8.312	8.436	8.552	8.659	8.760	8.854	8.943
9	7.647	7.784	7.910	8.025	8.132	8.232	8.325	8.412	8.495
10	7.356	7.485	7.603	7.712	7.812	7.906	7.993	8.076	8.153
11	7.128	7.250	7.362	7.465	7.560	7.649	7.732	7.809	7.883
12	6.943	7.060	7.167	7.265	7.356	7.441	7.520	7.594	7.665
13	6.791	6.903	7.006	7.101	7.188	7.269	7.345	7.417	7.485
14	6.664	6.772	6.871	6.962	7.047	7.126	7.199	7.268	7.333
15	6.555	6.660	6.757	6.845	6.927	7.003	7.074	7.142	7.204
16	6.462	6.564	6.658	6.744	6.823	6.898	6.967	7.032	7.093
17	6.381	6.480	6.572	6.656	6.734	6.806	6.873	6.937	6.997
18	6.310	6.407	6.497	6.579	6.655	6.725	6.792	6.854	6.912
19	6.247	6.342	6.430	6.510	6.585	6.654	6.719	6.780	6.837
20	6.191	6.285	6.371	6.450	6.523	6.591	6.654	6.714	6.771
24	6.017	6.106	6.186	6.261	6.330	6.394	6.453	6.510	6.563
30	5.849	5.932	6.008	6.078	6.143	6.203	6.259	6.311	6.361
40	5.686	5.764	5.835	5.900	5.961	6.017	6.069	6.119	6.165
60	5.528	5.601	5.667	5.728	5.785	5.837	5.886	5.931	5.974
120	5.375	5.443	5.505	5.562	5.614	5.662	5.708	5.750	5.790
∞	5.227	5.290	5.348	5.400	5.448	5.493	5.535	5.574	5.611

Table I (*Continued*)

$\alpha = .01$

ν \ r	20	22	24	26	28	30	32	34	36
1	298.0	304.7	310.8	316.3	321.3	326.0	330.3	334.3	338.0
2	37.95	38.76	39.49	40.15	40.76	41.32	41.84	42.33	42.78
3	19.77	20.17	20.53	20.86	21.16	21.44	21.70	21.95	22.17
4	14.40	14.68	14.93	15.16	15.37	15.57	15.75	15.92	16.08
5	11.93	12.16	12.36	12.54	12.71	12.87	13.02	13.15	13.28
6	10.54	10.73	10.91	11.06	11.21	11.34	11.47	11.58	11.69
7	9.646	9.815	9.970	10.11	10.24	10.36	10.47	10.58	10.67
8	9.027	9.182	9.322	9.450	9.569	9.678	9.779	9.874	9.964
9	8.573	8.717	8.847	8.966	9.075	9.177	9.271	9.360	9.443
10	8.226	8.361	8.483	8.595	8.698	8.794	8.883	8.966	9.044
11	7.952	8.080	8.196	8.303	8.400	8.491	8.575	8.654	8.728
12	7.731	7.853	7.964	8.066	8.159	8.246	8.327	8.402	8.473
13	7.548	7.665	7.772	7.870	7.960	8.043	8.121	8.193	8.262
14	7.395	7.508	7.611	7.705	7.792	7.873	7.948	8.018	8.084
15	7.264	7.374	7.474	7.566	7.650	7.728	7.800	7.869	7.932
16	7.152	7.258	7.356	7.445	7.527	7.602	7.673	7.739	7.802
17	7.053	7.158	7.253	7.340	7.420	7.493	7.563	7.627	7.687
18	6.968	7.070	7.163	7.247	7.325	7.398	7.465	7.528	7.587
19	6.891	6.992	7.082	7.166	7.242	7.313	7.379	7.440	7.498
20	6.823	6.922	7.011	7.092	7.168	7.237	7.302	7.362	7.419
24	6.612	6.705	6.789	6.865	6.936	7.001	7.062	7.119	7.173
30	6.407	6.494	6.572	6.644	6.710	6.772	6.828	6.881	6.932
40	6.209	6.289	6.362	6.429	6.490	6.547	6.600	6.650	6.697
60	6.015	6.090	6.158	6.220	6.277	6.330	6.378	6.424	6.467
120	5.827	5.897	5.959	6.016	6.069	6.117	6.162	6.204	6.244
∞	5.645	5.709	5.766	5.818	5.866	5.911	5.952	5.990	6.026

ν \ r	38	40	50	60	70	80	90	100
1	341.5	344.8	358.9	370.1	379.4	387.3	394.1	400.1
2	43.21	43.61	45.33	46.70	47.83	48.80	49.64	50.38
3	22.39	22.59	23.45	24.13	24.71	25.19	25.62	25.99
4	16.23	16.37	16.98	17.46	17.86	18.20	18.50	18.77
5	13.40	13.52	14.00	14.39	14.72	14.99	15.23	15.45
6	11.80	11.90	12.31	12.65	12.92	13.16	13.37	13.55
7	10.77	10.85	11.23	11.52	11.77	11.99	12.17	12.34
8	10.05	10.13	10.47	10.75	10.97	11.17	11.34	11.49
9	9.521	9.594	9.912	10.17	10.38	10.57	10.73	10.87
10	9.117	9.187	9.486	9.726	9.927	10.10	10.25	10.39
11	8.798	8.864	9.148	9.377	9.568	9.732	9.875	10.00
12	8.539	8.603	8.875	9.094	9.277	9.434	9.571	9.693
13	8.326	8.387	8.648	8.859	9.035	9.187	9.318	9.436
14	8.146	8.204	8.457	8.661	8.832	8.978	9.106	9.219
15	7.992	8.049	8.295	8.492	8.658	8.800	8.924	9.035
16	7.860	7.916	8.154	8.347	8.507	8.646	8.767	8.874
17	7.745	7.799	8.031	8.219	8.377	8.511	8.630	8.735
18	7.643	7.696	7.924	8.107	8.261	8.393	8.508	8.611
19	7.553	7.605	7.828	8.008	8.159	8.288	8.401	8.502
20	7.473	7.523	7.742	7.919	8.067	8.194	8.305	8.404
24	7.223	7.270	7.476	7.642	7.780	7.900	8.004	8.097
30	6.978	7.023	7.215	7.370	7.500	7.611	7.709	7.796
40	6.740	6.782	6.960	7.104	7.225	7.328	7.419	7.500
60	6.507	6.546	6.710	6.843	6.954	7.050	7.133	7.207
120	6.281	6.316	6.467	6.588	6.689	6.776	6.852	6.919
∞	6.060	6.092	6.228	6.338	6.429	6.507	6.575	6.636

Table II

PERCENTAGE POINTS OF THE BONFERRONI t STATISTIC $t_{\nu}^{\alpha/2k}$

(For reference, see Chap. 2, Sec. 3)

$\alpha = .05$

ν \ k	2	3	4	5	6	7	8	9	10	15	20	25	30	35	40	45	50
5	3.17	3.54	3.81	4.04	4.22	4.38	4.53	4.66	4.78	5.25	5.60	5.89	6.15	6.36	6.56	6.70	6.86
7	2.84	3.13	3.34	3.50	3.64	3.76	3.86	3.95	4.03	4.36	4.59	4.78	4.95	5.09	5.21	5.31	5.40
10	2.64	2.87	3.04	3.17	3.28	3.37	3.45	3.52	3.58	3.83	4.01	4.15	4.27	4.37	4.45	4.53	4.59
12	2.56	2.78	2.94	3.06	3.15	3.24	3.31	3.37	3.43	3.65	3.80	3.93	4.04	4.13	4.20	4.26	4.32
15	2.49	2.69	2.84	2.95	3.04	3.11	3.18	3.24	3.29	3.48	3.62	3.74	3.82	3.90	3.97	4.02	4.07
20	2.42	2.61	2.75	2.85	2.93	3.00	3.06	3.11	3.16	3.33	3.46	3.55	3.63	3.70	3.76	3.80	3.85
24	2.39	2.58	2.70	2.80	2.88	2.94	3.00	3.05	3.09	3.26	3.38	3.47	3.54	3.61	3.66	3.70	3.74
30	2.36	2.54	2.66	2.75	2.83	2.89	2.94	2.99	3.03	3.19	3.30	3.39	3.46	3.52	3.57	3.61	3.65
40	2.33	2.50	2.62	2.71	2.78	2.84	2.89	2.93	2.97	3.12	3.23	3.31	3.38	3.43	3.48	3.51	3.55
60	2.30	2.47	2.58	2.66	2.73	2.79	2.84	2.88	2.92	3.06	3.16	3.24	3.30	3.34	3.39	3.42	3.46
120	2.27	2.43	2.54	2.62	2.68	2.74	2.79	2.83	2.86	2.99	3.09	3.16	3.22	3.27	3.31	3.34	3.37
∞	2.24	2.39	2.50	2.58	2.64	2.69	2.74	2.77	2.81	2.94	3.02	3.09	3.15	3.19	3.23	3.26	3.29

$\alpha = .01$

ν \ k	2	3	4	5	6	7	8	9	10	15	20	25	30	35	40	45	50
5	4.78	5.25	5.60	5.89	6.15	6.36	6.56	6.70	6.86	7.51	8.00	8.37	8.68	8.95	9.19	9.41	9.68
7	4.03	4.36	4.59	4.78	4.95	5.09	5.21	5.31	5.40	5.79	6.08	6.30	6.49	6.67	6.83	6.93	7.06
10	3.58	3.83	4.01	4.15	4.27	4.37	4.45	4.53	4.59	4.86	5.06	5.20	5.33	5.44	5.52	5.60	5.70
12	3.43	3.65	3.80	3.93	4.04	4.13	4.20	4.26	4.32	4.56	4.73	4.86	4.95	5.04	5.12	5.20	5.27
15	3.29	3.48	3.62	3.74	3.82	3.90	3.97	4.02	4.07	4.29	4.42	4.53	4.61	4.71	4.78	4.84	4.90
20	3.16	3.33	3.46	3.55	3.63	3.70	3.76	3.80	3.85	4.03	4.15	4.25	4.33	4.39	4.46	4.52	4.56
24	3.09	3.26	3.38	3.47	3.54	3.61	3.66	3.70	3.74	3.91	4.04	4.1†	4.2†	4.3†	4.3†	4.3†	4.4†
30	3.03	3.19	3.30	3.39	3.46	3.52	3.57	3.61	3.65	3.80	3.90	3.98	4.13	4.26	4.1†	4.2†	4.2†
40	2.97	3.12	3.23	3.31	3.38	3.43	3.48	3.51	3.55	3.70	3.79	3.88	3.93	3.97	4.01	4.1†	4.1†
60	2.92	3.06	3.16	3.24	3.30	3.34	3.39	3.42	3.46	3.59	3.69	3.76	3.81	3.84	3.89	3.93	3.97
120	2.86	2.99	3.09	3.16	3.22	3.27	3.31	3.34	3.37	3.50	3.58	3.64	3.69	3.73	3.77	3.80	3.83
∞	2.81	2.94	3.02	3.09	3.15	3.19	3.23	3.26	3.29	3.40	3.48	3.54	3.59	3.63	3.66	3.69	3.72

† Obtained by graphical interpolation.

Table III†

PERCENTAGE POINTS OF THE STUDENTIZED MAXIMUM MODULUS $|m|_{k,\nu}^{\alpha}$

(For reference, see Chap. 2, Sec. 4)

$\alpha = .05$

ν \ k	1	2	3	4	5	6	7	8
5	2.57	3.09	3.40	3.62	3.78	3.92	4.04	4.14
10	2.23	2.61	2.83	2.98	3.10	3.19	3.28	3.35
15	2.13	2.47	2.67	2.81	2.91	2.99	3.06	3.12
20	2.09	2.41	2.59	2.72	2.82	2.90	2.97	3.02
24	2.06	2.38	2.56	2.68	2.78	2.84	2.91	2.96
30	2.04	2.35	2.52	2.64	2.73	2.80	2.86	2.91
40	2.02	2.32	2.49	2.60	2.69	2.76	2.82	2.86
60	2.00	2.29	2.46	2.56	2.65	2.72	2.77	2.82
120	1.98	2.26	2.43	2.53	2.61	2.68	2.73	2.77
∞	1.96	2.23	2.39	2.49	2.57	2.64	2.69	2.73

† Footnote to second edition: See the new table of the studentized maximum modulus in the Addendum.

Table IV

PERCENTAGE POINTS OF THE MANY-ONE t STATISTICS
(For reference, see Chap. 2, Sec. 5)

(One-tailed) $d^{\alpha}_{k,\nu}$

	$\alpha = .05$									$\alpha = .01$								
ν \ k	1	2	3	4	5	6	7	8	9	1	2	3	4	5	6	7	8	9
5	2.02	2.44	2.68	2.85	2.98	3.08	3.16	3.24	3.30	3.37	3.90	4.21	4.43	4.60	4.73	4.85	4.94	5.03
6	1.94	2.34	2.56	2.71	2.83	2.92	3.00	3.07	3.12	3.14	3.61	3.88	4.07	4.21	4.33	4.43	4.51	4.59
7	1.89	2.27	2.48	2.62	2.73	2.82	2.89	2.95	3.01	3.00	3.42	3.66	3.83	3.96	4.07	4.15	4.23	4.30
8	1.86	2.22	2.42	2.55	2.66	2.74	2.81	2.87	2.92	2.90	3.29	3.51	3.67	3.79	3.88	3.96	4.03	4.09
9	1.83	2.18	2.37	2.50	2.60	2.68	2.75	2.81	2.86	2.82	3.19	3.40	3.55	3.66	3.75	3.82	3.89	3.94
10	1.81	2.15	2.34	2.47	2.56	2.64	2.70	2.76	2.81	2.76	3.11	3.31	3.45	3.56	3.64	3.71	3.78	3.83
11	1.80	2.13	2.31	2.44	2.53	2.60	2.67	2.72	2.77	2.72	3.06	3.25	3.38	3.48	3.56	3.63	3.69	3.74
12	1.78	2.11	2.29	2.41	2.50	2.58	2.64	2.69	2.74	2.68	3.01	3.19	3.32	3.42	3.50	3.56	3.62	3.67
13	1.77	2.09	2.27	2.39	2.48	2.55	2.61	2.66	2.71	2.65	2.97	3.15	3.27	3.37	3.44	3.51	3.56	3.61
14	1.76	2.08	2.25	2.37	2.46	2.53	2.59	2.64	2.69	2.62	2.94	3.11	3.23	3.32	3.40	3.46	3.51	3.56
15	1.75	2.07	2.24	2.36	2.44	2.51	2.57	2.62	2.67	2.60	2.91	3.08	3.20	3.29	3.36	3.42	3.47	3.52
16	1.75	2.06	2.23	2.34	2.43	2.50	2.56	2.61	2.65	2.58	2.88	3.05	3.17	3.26	3.33	3.39	3.44	3.48
17	1.74	2.05	2.22	2.33	2.42	2.49	2.54	2.59	2.64	2.57	2.86	3.03	3.14	3.23	3.30	3.36	3.41	3.45
18	1.73	2.04	2.21	2.32	2.41	2.48	2.53	2.58	2.62	2.55	2.84	3.01	3.12	3.21	3.27	3.33	3.38	3.42
19	1.73	2.03	2.20	2.31	2.40	2.47	2.52	2.57	2.61	2.54	2.83	2.99	3.10	3.18	3.25	3.31	3.36	3.40
20	1.72	2.03	2.19	2.30	2.39	2.46	2.51	2.56	2.60	2.53	2.81	2.97	3.08	3.17	3.23	3.29	3.34	3.38
24	1.71	2.01	2.17	2.28	2.36	2.43	2.48	2.53	2.57	2.49	2.77	2.92	3.03	3.11	3.17	3.22	3.27	3.31
30	1.70	1.99	2.15	2.25	2.33	2.40	2.45	2.50	2.54	2.46	2.72	2.87	2.97	3.05	3.11	3.16	3.21	3.24
40	1.68	1.97	2.13	2.23	2.31	2.37	2.42	2.47	2.51	2.42	2.68	2.82	2.92	2.99	3.05	3.10	3.14	3.18
60	1.67	1.95	2.10	2.21	2.28	2.35	2.39	2.44	2.48	2.39	2.64	2.78	2.87	2.94	3.00	3.04	3.08	3.12
120	1.66	1.93	2.08	2.18	2.26	2.32	2.37	2.41	2.45	2.36	2.60	2.73	2.82	2.89	2.94	2.99	3.03	3.06
∞	1.64	1.92	2.06	2.16	2.23	2.29	2.34	2.38	2.42	2.33	2.56	2.68	2.77	2.84	2.89	2.93	2.97	3.00

Table IV (Continued)

(Two-tailed) $|d|^{\alpha}_{k,\nu}$

$\alpha = .05$

ν \ k	1	2	3	4	5	6	7	8	9	10	11	12	15	20
5	2.57	3.03	3.29	3.48	3.62	3.73	3.82	3.90	3.97	4.03	4.09	4.14	4.26	4.42
6	2.45	2.86	3.10	3.26	3.39	3.49	3.57	3.64	3.71	3.76	3.81	3.86	3.97	4.11
7	2.36	2.75	2.97	3.12	3.24	3.33	3.41	3.47	3.53	3.58	3.63	3.67	3.78	3.91
8	2.31	2.67	2.88	3.02	3.13	3.22	3.29	3.35	3.41	3.46	3.50	3.54	3.64	3.76
9	2.26	2.61	2.81	2.95	3.05	3.14	3.20	3.26	3.32	3.36	3.40	3.44	3.53	3.65
10	2.23	2.57	2.76	2.89	2.99	3.07	3.14	3.19	3.24	3.29	3.33	3.36	3.45	3.57
11	2.20	2.53	2.72	2.84	2.94	3.02	3.08	3.14	3.19	3.23	3.27	3.30	3.39	3.50
12	2.18	2.50	2.68	2.81	2.90	2.98	3.04	3.09	3.14	3.18	3.22	3.25	3.34	3.45
13	2.16	2.48	2.65	2.78	2.87	2.94	3.00	3.06	3.10	3.14	3.18	3.21	3.29	3.40
14	2.14	2.46	2.63	2.75	2.84	2.91	2.97	3.02	3.07	3.11	3.14	3.18	3.26	3.36
15	2.13	2.44	2.61	2.73	2.82	2.89	2.95	3.00	3.04	3.08	3.12	3.15	3.23	3.33
16	2.12	2.42	2.59	2.71	2.80	2.87	2.92	2.97	3.02	3.06	3.09	3.12	3.20	3.30
17	2.11	2.41	2.58	2.69	2.78	2.85	2.90	2.95	3.00	3.03	3.07	3.10	3.18	3.27
18	2.10	2.40	2.56	2.68	2.76	2.83	2.89	2.94	2.98	3.01	3.05	3.08	3.16	3.25
19	2.09	2.39	2.55	2.66	2.75	2.81	2.87	2.92	2.96	3.00	3.03	3.06	3.14	3.23
20	2.09	2.38	2.54	2.65	2.73	2.80	2.86	2.90	2.95	2.98	3.02	3.05	3.12	3.22
24	2.06	2.35	2.51	2.61	2.70	2.76	2.81	2.86	2.90	2.94	2.97	3.00	3.07	3.16
30	2.04	2.32	2.47	2.58	2.66	2.72	2.77	2.82	2.86	2.89	2.92	2.95	3.02	3.11
40	2.02	2.29	2.44	2.54	2.62	2.68	2.73	2.77	2.81	2.85	2.87	2.90	2.97	3.06
60	2.00	2.27	2.41	2.51	2.58	2.64	2.69	2.73	2.77	2.80	2.83	2.86	2.92	3.00
120	1.98	2.24	2.38	2.47	2.55	2.60	2.65	2.69	2.73	2.76	2.79	2.81	2.87	2.95
∞	1.96	2.21	2.35	2.44	2.51	2.57	2.61	2.65	2.69	2.72	2.74	2.77	2.83	2.91

Table IV (Continued)

$\alpha = .01$

ν \ k	1	2	3	4	5	6	7	8	9	10	11	12	15	20
5	4.03	4.63	4.98	5.22	5.41	5.56	5.69	5.80	5.89	5.98	6.05	6.12	6.30	6.52
6	3.71	4.21	4.51	4.71	4.87	5.00	5.10	5.20	5.28	5.35	5.41	5.47	5.62	5.81
7	3.50	3.95	4.21	4.39	4.53	4.64	4.74	4.82	4.89	4.95	5.01	5.06	5.19	5.36
8	3.36	3.77	4.00	4.17	4.29	4.40	4.48	4.56	4.62	4.68	4.73	4.78	4.90	5.05
9	3.25	3.63	3.85	4.01	4.12	4.22	4.30	4.37	4.43	4.48	4.53	4.57	4.68	4.82
10	3.17	3.53	3.74	3.88	3.99	4.08	4.16	4.22	4.28	4.33	4.37	4.42	4.52	4.65
11	3.11	3.45	3.65	3.79	3.89	3.98	4.05	4.11	4.16	4.21	4.25	4.29	4.39	4.52
12	3.05	3.39	3.58	3.71	3.81	3.89	3.96	4.02	4.07	4.12	4.16	4.19	4.29	4.41
13	3.01	3.33	3.52	3.65	3.74	3.82	3.89	3.94	3.99	4.04	4.08	4.11	4.20	4.32
14	2.98	3.29	3.47	3.59	3.69	3.76	3.83	3.88	3.93	3.97	4.01	4.05	4.13	4.24
15	2.95	3.25	3.43	3.55	3.64	3.71	3.78	3.83	3.88	3.92	3.95	3.99	4.07	4.18
16	2.92	3.22	3.39	3.51	3.60	3.67	3.73	3.78	3.83	3.87	3.91	3.94	4.02	4.13
17	2.90	3.19	3.36	3.47	3.56	3.63	3.69	3.74	3.79	3.83	3.86	3.90	3.98	4.08
18	2.88	3.17	3.33	3.44	3.53	3.60	3.66	3.71	3.75	3.79	3.83	3.86	3.94	4.04
19	2.86	3.15	3.31	3.42	3.50	3.57	3.63	3.68	3.72	3.76	3.79	3.83	3.90	4.00
20	2.85	3.13	3.29	3.40	3.48	3.55	3.60	3.65	3.69	3.73	3.77	3.80	3.87	3.97
24	2.80	3.07	3.22	3.32	3.40	3.47	3.52	3.57	3.61	3.64	3.68	3.70	3.78	3.87
30	2.75	3.01	3.15	3.25	3.33	3.39	3.44	3.49	3.52	3.56	3.59	3.62	3.69	3.78
40	2.70	2.95	3.09	3.19	3.26	3.32	3.37	3.41	3.44	3.48	3.51	3.53	3.60	3.68
60	2.66	2.90	3.03	3.12	3.19	3.25	3.29	3.33	3.37	3.40	3.42	3.45	3.51	3.59
120	2.62	2.85	2.97	3.06	3.12	3.18	3.22	3.26	3.29	3.32	3.35	3.37	3.43	3.51
∞	2.58	2.79	2.92	3.00	3.06	3.11	3.15	3.19	3.22	3.25	3.27	3.29	3.35	3.42

Table V

PERCENTAGE POINTS OF THE DUNCAN MULTIPLE RANGE TEST $q_{p,\nu}^{\alpha_p}$ $[\alpha_p = 1 - (1 - \alpha)^{p-1}]$

(For reference, see Chap. 2, Sec. 6)

$\alpha = .05$

ν \ p	2	3	4	5	6	7	8	9	10	11	12	13	14	15	16	17	18	19
1	17.97	17.97	17.97	17.97	17.97	17.97	17.97	17.97	17.97	17.97	17.97	17.97	17.97	17.97	17.97	17.97	17.97	17.97
2	6.085	6.085	6.085	6.085	6.085	6.085	6.085	6.085	6.085	6.085	6.085	6.085	6.085	6.085	6.085	6.085	6.085	6.085
3	4.501	4.516	4.516	4.516	4.516	4.516	4.516	4.516	4.516	4.516	4.516	4.516	4.516	4.516	4.516	4.516	4.516	4.516
4	3.927	4.013	4.033	4.033	4.033	4.033	4.033	4.033	4.033	4.033	4.033	4.033	4.033	4.033	4.033	4.033	4.033	4.033
5	3.635	3.749	3.797	3.814	3.814	3.814	3.814	3.814	3.814	3.814	3.814	3.814	3.814	3.814	3.814	3.814	3.814	3.814
6	3.461	3.587	3.649	3.680	3.694	3.697	3.697	3.697	3.697	3.697	3.697	3.697	3.697	3.697	3.697	3.697	3.697	3.697
7	3.344	3.477	3.548	3.588	3.611	3.622	3.626	3.626	3.626	3.626	3.626	3.626	3.626	3.626	3.626	3.626	3.626	3.626
8	3.261	3.399	3.475	3.521	3.549	3.566	3.575	3.579	3.579	3.579	3.579	3.579	3.579	3.579	3.579	3.579	3.579	3.579
9	3.199	3.339	3.420	3.470	3.502	3.523	3.536	3.544	3.547	3.547	3.547	3.547	3.547	3.547	3.547	3.547	3.547	3.547
10	3.151	3.293	3.376	3.430	3.465	3.489	3.505	3.516	3.522	3.525	3.526	3.526	3.526	3.526	3.526	3.526	3.526	3.526
11	3.113	3.256	3.342	3.397	3.435	3.462	3.480	3.493	3.501	3.506	3.509	3.510	3.510	3.510	3.510	3.510	3.510	3.510
12	3.082	3.225	3.313	3.370	3.410	3.439	3.459	3.474	3.484	3.491	3.496	3.498	3.499	3.499	3.499	3.499	3.499	3.499
13	3.055	3.200	3.289	3.348	3.389	3.419	3.442	3.458	3.470	3.478	3.484	3.488	3.490	3.490	3.490	3.490	3.490	3.490
14	3.033	3.178	3.268	3.329	3.372	3.403	3.426	3.444	3.457	3.467	3.474	3.479	3.482	3.484	3.484	3.485	3.485	3.485
15	3.014	3.160	3.250	3.312	3.356	3.389	3.413	3.432	3.446	3.457	3.465	3.471	3.476	3.478	3.480	3.481	3.481	3.481
16	2.998	3.144	3.235	3.298	3.343	3.376	3.402	3.422	3.437	3.449	3.458	3.465	3.470	3.473	3.477	3.478	3.478	3.478
17	2.984	3.130	3.222	3.285	3.331	3.366	3.392	3.412	3.429	3.441	3.451	3.459	3.465	3.469	3.473	3.475	3.476	3.476
18	2.971	3.118	3.210	3.274	3.321	3.356	3.383	3.405	3.421	3.435	3.445	3.454	3.460	3.465	3.470	3.472	3.474	3.474
19	2.960	3.107	3.199	3.264	3.311	3.347	3.375	3.397	3.415	3.429	3.440	3.449	3.456	3.462	3.467	3.470	3.472	3.473
20	2.950	3.097	3.190	3.255	3.303	3.339	3.368	3.391	3.409	3.424	3.436	3.445	3.453	3.459	3.464	3.467	3.470	3.472
24	2.919	3.066	3.160	3.226	3.276	3.315	3.345	3.370	3.390	3.406	3.420	3.432	3.441	3.449	3.456	3.461	3.465	3.469
30	2.888	3.035	3.131	3.199	3.250	3.290	3.322	3.349	3.371	3.389	3.405	3.418	3.430	3.439	3.447	3.454	3.460	3.466
40	2.858	3.006	3.102	3.171	3.224	3.266	3.300	3.328	3.352	3.373	3.390	3.405	3.418	3.429	3.439	3.448	3.456	3.463
60	2.829	2.976	3.073	3.143	3.198	3.241	3.277	3.307	3.333	3.355	3.374	3.391	3.406	3.419	3.431	3.442	3.451	3.460
120	2.800	2.947	3.045	3.116	3.172	3.217	3.254	3.287	3.314	3.337	3.359	3.377	3.394	3.409	3.423	3.435	3.446	3.457
∞	2.772	2.918	3.017	3.089	3.146	3.193	3.232	3.265	3.294	3.320	3.343	3.363	3.382	3.399	3.414	3.428	3.442	3.454

Table V (Continued): $\alpha = .05$

ν \ p	20	22	24	26	28	30	32	34	36	38	40	50	60	70	80	90	100
1	17.97	17.97	17.97	17.97	17.97	17.97	17.97	17.97	17.97	17.97	17.97	17.97	17.97	17.97	17.97	17.97	17.97
2	6.085	6.085	6.085	6.085	6.085	6.085	6.085	6.085	6.085	6.085	6.085	6.085	6.085	6.085	6.085	6.085	6.085
3	4.516	4.516	4.516	4.516	4.516	4.516	4.516	4.516	4.516	4.516	4.516	4.516	4.516	4.516	4.516	4.516	4.516
4	4.033	4.033	4.033	4.033	4.033	4.033	4.033	4.033	4.033	4.033	4.033	4.033	4.033	4.033	4.033	4.033	4.033
5	3.814	3.814	3.814	3.814	3.814	3.814	3.814	3.814	3.814	3.814	3.814	3.814	3.814	3.814	3.814	3.814	3.814
6	3.697	3.697	3.697	3.697	3.697	3.697	3.697	3.697	3.697	3.697	3.697	3.697	3.697	3.697	3.697	3.697	3.697
7	3.626	3.626	3.626	3.626	3.626	3.626	3.626	3.626	3.626	3.626	3.626	3.626	3.626	3.626	3.626	3.626	3.626
8	3.579	3.579	3.579	3.579	3.579	3.579	3.579	3.579	3.579	3.579	3.579	3.579	3.579	3.579	3.579	3.579	3.579
9	3.547	3.547	3.547	3.547	3.547	3.547	3.547	3.547	3.547	3.547	3.547	3.547	3.547	3.547	3.547	3.547	3.547
10	3.526	3.526	3.526	3.526	3.526	3.526	3.526	3.526	3.526	3.526	3.526	3.526	3.526	3.526	3.526	3.526	3.526
11	3.510	3.510	3.510	3.510	3.510	3.510	3.510	3.510	3.510	3.510	3.510	3.510	3.510	3.510	3.510	3.510	3.510
12	3.499	3.499	3.499	3.499	3.499	3.499	3.499	3.499	3.499	3.499	3.499	3.499	3.499	3.499	3.499	3.499	3.499
13	3.490	3.490	3.490	3.490	3.490	3.490	3.490	3.490	3.490	3.490	3.490	3.490	3.490	3.490	3.490	3.490	3.490
14	3.485	3.485	3.485	3.485	3.485	3.485	3.485	3.485	3.485	3.485	3.485	3.485	3.485	3.485	3.485	3.485	3.485
15	3.481	3.481	3.481	3.481	3.481	3.481	3.481	3.481	3.481	3.481	3.481	3.481	3.481	3.481	3.481	3.481	3.481
16	3.478	3.478	3.478	3.478	3.478	3.478	3.478	3.478	3.478	3.478	3.478	3.478	3.478	3.478	3.478	3.478	3.478
17	3.476	3.476	3.476	3.476	3.476	3.476	3.476	3.476	3.476	3.476	3.476	3.476	3.476	3.476	3.476	3.476	3.476
18	3.474	3.474	3.474	3.474	3.474	3.474	3.474	3.474	3.474	3.474	3.474	3.474	3.474	3.474	3.474	3.474	3.474
19	3.474	3.474	3.474	3.474	3.474	3.474	3.474	3.474	3.474	3.474	3.474	3.474	3.474	3.474	3.474	3.474	3.474
20	3.473	3.474	3.474	3.474	3.474	3.474	3.474	3.474	3.474	3.474	3.474	3.474	3.474	3.474	3.474	3.474	3.474
24	3.471	3.475	3.477	3.477	3.477	3.477	3.477	3.477	3.477	3.477	3.477	3.477	3.477	3.477	3.477	3.477	3.477
30	3.470	3.477	3.481	3.484	3.486	3.486	3.486	3.486	3.486	3.486	3.486	3.486	3.486	3.486	3.486	3.486	3.486
40	3.469	3.479	3.486	3.492	3.497	3.500	3.503	3.504	3.504	3.504	3.504	3.504	3.504	3.504	3.504	3.504	3.504
60	3.467	3.481	3.492	3.501	3.509	3.515	3.521	3.525	3.529	3.531	3.534	3.537	3.537	3.537	3.537	3.537	3.537
120	3.466	3.483	3.498	3.511	3.522	3.532	3.541	3.548	3.555	3.561	3.566	3.585	3.596	3.600	3.601	3.601	3.601
∞	3.466	3.486	3.505	3.522	3.536	3.550	3.562	3.574	3.584	3.594	3.603	3.640	3.668	3.690	3.708	3.722	3.735

Table V (Continued)

$\alpha = .01$

ν \ p	2	3	4	5	6	7	8	9	10	11	12	13	14	15	16	17	18	19
1	90.03	90.03	90.03	90.03	90.03	90.03	90.03	90.03	90.03	90.03	90.03	90.03	90.03	90.03	90.03	90.03	90.03	90.03
2	14.04	14.04	14.04	14.04	14.04	14.04	14.04	14.04	14.04	14.04	14.04	14.04	14.04	14.04	14.04	14.04	14.04	14.04
3	8.261	8.321	8.321	8.321	8.321	8.321	8.321	8.321	8.321	8.321	8.321	8.321	8.321	8.321	8.321	8.321	8.321	8.321
4	6.512	6.677	6.740	6.756	6.756	6.756	6.756	6.756	6.756	6.756	6.756	6.756	6.756	6.756	6.756	6.756	6.756	6.756
5	5.702	5.893	5.989	6.040	6.065	6.074	6.074	6.074	6.074	6.074	6.074	6.074	6.074	6.074	6.074	6.074	6.074	6.074
6	5.243	5.439	5.549	5.614	5.655	5.680	5.694	5.701	5.703	5.703	5.703	5.703	5.703	5.703	5.703	5.703	5.703	5.703
7	4.949	5.145	5.260	5.334	5.383	5.416	5.439	5.454	5.464	5.470	5.472	5.472	5.472	5.472	5.472	5.472	5.472	5.472
8	4.746	4.939	5.057	5.135	5.189	5.227	5.256	5.276	5.291	5.302	5.309	5.314	5.316	5.317	5.317	5.317	5.317	5.317
9	4.596	4.787	4.906	4.986	5.043	5.086	5.118	5.142	5.160	5.174	5.185	5.193	5.199	5.203	5.205	5.206	5.206	5.206
10	4.482	4.671	4.790	4.871	4.931	4.975	5.010	5.037	5.058	5.074	5.088	5.098	5.106	5.112	5.117	5.120	5.122	5.124
11	4.392	4.579	4.697	4.780	4.841	4.887	4.924	4.952	4.975	4.994	5.009	5.021	5.031	5.039	5.045	5.050	5.054	5.057
12	4.320	4.504	4.622	4.706	4.767	4.815	4.852	4.883	4.907	4.927	4.944	4.958	4.969	4.978	4.986	4.993	4.998	5.002
13	4.260	4.442	4.560	4.644	4.706	4.755	4.793	4.824	4.850	4.872	4.889	4.904	4.917	4.928	4.937	4.944	4.950	4.956
14	4.210	4.391	4.508	4.591	4.654	4.704	4.743	4.775	4.802	4.824	4.843	4.859	4.872	4.884	4.894	4.902	4.910	4.916
15	4.168	4.347	4.463	4.547	4.610	4.660	4.700	4.733	4.760	4.783	4.803	4.820	4.834	4.846	4.857	4.866	4.874	4.881
16	4.131	4.309	4.425	4.509	4.572	4.622	4.663	4.696	4.724	4.748	4.768	4.786	4.800	4.813	4.825	4.835	4.844	4.851
17	4.099	4.275	4.391	4.475	4.539	4.589	4.630	4.664	4.693	4.717	4.738	4.756	4.771	4.785	4.797	4.807	4.816	4.824
18	4.071	4.246	4.362	4.445	4.509	4.560	4.601	4.635	4.664	4.689	4.711	4.729	4.745	4.759	4.772	4.783	4.792	4.801
19	4.046	4.220	4.335	4.419	4.483	4.534	4.575	4.610	4.639	4.665	4.686	4.705	4.722	4.736	4.749	4.761	4.771	4.780
20	4.024	4.197	4.312	4.395	4.459	4.510	4.552	4.587	4.617	4.642	4.664	4.684	4.701	4.716	4.729	4.741	4.751	4.761
24	3.956	4.126	4.239	4.322	4.386	4.437	4.480	4.516	4.546	4.573	4.596	4.616	4.634	4.651	4.665	4.678	4.690	4.700
30	3.889	4.056	4.168	4.250	4.314	4.366	4.409	4.445	4.477	4.504	4.528	4.550	4.569	4.586	4.601	4.615	4.628	4.640
40	3.825	3.988	4.098	4.180	4.244	4.296	4.339	4.376	4.408	4.436	4.461	4.483	4.503	4.521	4.537	4.553	4.566	4.579
60	3.762	3.922	4.031	4.111	4.174	4.226	4.270	4.307	4.340	4.368	4.394	4.417	4.438	4.456	4.474	4.490	4.504	4.518
120	3.702	3.858	3.965	4.044	4.107	4.158	4.202	4.239	4.272	4.301	4.327	4.351	4.372	4.392	4.410	4.426	4.442	4.456
∞	3.643	3.796	3.900	3.978	4.040	4.091	4.135	4.172	4.205	4.235	4.261	4.285	4.307	4.327	4.345	4.363	4.379	4.394

Table V (*Continued*): $\alpha = .01$

ν \ p	20	22	24	26	28	30	32	34	36	38	40	50	60	70	80	90	100
1	90.03	90.03	90.03	90.03	90.03	90.03	90.03	90.03	90.03	90.03	90.03	90.03	90.03	90.03	90.03	90.03	90.03
2	14.04	14.04	14.04	14.04	14.04	14.04	14.04	14.04	14.04	14.04	14.04	14.04	14.04	14.04	14.04	14.04	14.04
3	8.321	8.321	8.321	8.321	8.321	8.321	8.321	8.321	8.321	8.321	8.321	8.321	8.321	8.321	8.321	8.321	8.321
4	6.756	6.756	6.756	6.756	6.756	6.756	6.756	6.756	6.756	6.756	6.756	6.756	6.756	6.756	6.756	6.756	6.756
5	6.074	6.074	6.074	6.074	6.074	6.074	6.074	6.074	6.074	6.074	6.074	6.074	6.074	6.074	6.074	6.074	6.074
6	5.703	5.703	5.703	5.703	5.703	5.703	5.703	5.703	5.703	5.703	5.703	5.703	5.703	5.703	5.703	5.703	5.703
7	5.472	5.472	5.472	5.472	5.472	5.472	5.472	5.472	5.472	5.472	5.472	5.472	5.472	5.472	5.472	5.472	5.472
8	5.317	5.317	5.317	5.317	5.317	5.317	5.317	5.317	5.317	5.317	5.317	5.317	5.317	5.317	5.317	5.317	5.317
9	5.206	5.206	5.206	5.206	5.206	5.206	5.206	5.206	5.206	5.206	5.206	5.206	5.206	5.206	5.206	5.206	5.206
10	5.124	5.124	5.124	5.124	5.124	5.124	5.124	5.124	5.124	5.124	5.124	5.124	5.124	5.124	5.124	5.124	5.124
11	5.059	5.061	5.061	5.061	5.061	5.061	5.061	5.061	5.061	5.061	5.061	5.061	5.061	5.061	5.061	5.061	5.061
12	5.006	5.010	5.011	5.011	5.011	5.011	5.011	5.011	5.011	5.011	5.011	5.011	5.011	5.011	5.011	5.011	5.011
13	4.960	4.966	4.970	4.972	4.972	4.972	4.972	4.972	4.972	4.972	4.972	4.972	4.972	4.972	4.972	4.972	4.972
14	4.921	4.929	4.935	4.938	4.940	4.940	4.940	4.940	4.940	4.940	4.940	4.940	4.940	4.940	4.940	4.940	4.940
15	4.887	4.897	4.904	4.909	4.912	4.914	4.914	4.914	4.914	4.914	4.914	4.914	4.914	4.914	4.914	4.914	4.914
16	4.858	4.869	4.877	4.883	4.887	4.890	4.892	4.892	4.892	4.892	4.892	4.892	4.892	4.892	4.892	4.892	4.892
17	4.832	4.844	4.853	4.860	4.865	4.869	4.872	4.873	4.874	4.874	4.874	4.874	4.874	4.874	4.874	4.874	4.874
18	4.808	4.821	4.832	4.839	4.846	4.850	4.854	4.856	4.857	4.858	4.858	4.858	4.858	4.858	4.858	4.858	4.858
19	4.788	4.802	4.812	4.821	4.828	4.833	4.838	4.841	4.843	4.844	4.845	4.845	4.845	4.845	4.845	4.845	4.845
20	4.769	4.784	4.795	4.805	4.813	4.818	4.823	4.827	4.830	4.832	4.833	4.833	4.833	4.833	4.833	4.833	4.833
24	4.710	4.727	4.741	4.752	4.762	4.770	4.777	4.783	4.788	4.791	4.794	4.802	4.802	4.802	4.802	4.802	4.802
30	4.650	4.669	4.685	4.699	4.711	4.721	4.730	4.738	4.744	4.750	4.755	4.772	4.777	4.777	4.777	4.777	4.777
40	4.591	4.611	4.630	4.645	4.659	4.671	4.682	4.692	4.700	4.708	4.715	4.740	4.754	4.761	4.764	4.764	4.764
60	4.530	4.553	4.573	4.591	4.607	4.620	4.633	4.645	4.655	4.665	4.673	4.707	4.730	4.745	4.755	4.761	4.765
120	4.469	4.494	4.516	4.535	4.552	4.568	4.583	4.596	4.609	4.619	4.630	4.673	4.703	4.727	4.745	4.759	4.770
∞	4.408	4.434	4.457	4.478	4.497	4.514	4.530	4.545	4.559	4.572	4.584	4.635	4.675	4.707	4.734	4.756	4.776

Table VI

PERCENTAGE POINTS OF THE MANY-ONE SIGN STATISTICS
(For reference, see Chap. 4, Sec. 1)

(One-tailed) s_+^α

	$\alpha = .05$									$\alpha = .01$								
n \ k	2	3	4	5	6	7	8	9	10	2	3	4	5	6	7	8	9	10
5	-	-	-	-	-	-	-	-	-									
6	6	-	-	-	-	-	-	-	-									
7	7	7	7	-	-	-	-	-	-									
8	8	8	8	8	8	8	8	-	-									
9	8	9	9	9	9	9	9	9	9	9	-	-	-	-	-	-	-	-
10	9	9	9	10	10	10	10	10	10	10	10	10	-	-	-	-	-	-
11	10	10	10	10	10	10	11	11	11	11	11	11	11	11	11	11	-	-
12	10	11	11	11	11	11	11	11	11	11	12	12	12	12	12	12	12	12
13	11	11	11	12	12	12	12	12	12	12	12	13	13	13	13	13	13	13
14	12	12	12	12	12	12	13	13	13	13	13	13	13	13	14	14	14	14
15	12	13	13	13	13	13	13	13	13	13	14	14	14	14	14	14	14	14
16	13	13	13	14	14	14	14	14	14	14	14	15	15	15	15	15	15	15
17	13	14	14	14	14	14	15	15	15	15	15	15	15	16	16	16	16	16
18	14	14	15	15	15	15	15	15	15	15	16	16	16	16	16	16	16	17
19	15	15	15	15	16	16	16	16	16	16	16	17	17	17	17	17	17	17
20	15	16	16	16	16	16	16	17	17	17	17	17	17	18	18	18	18	18
25	18	19	19	19	19	19	20	20	20	20	20	20	21	21	21	21	21	21
30	21	22	22	22	22	23	23	23	23	23	23	24	24	24	24	24	24	24
35	24	25	25	25	25	26	26	26	26	26	26	27	27	27	27	27	27	28
40	27	28	28	28	28	29	29	29	29	29	30	30	30	30	30	30	31	31
45	30	31	31	31	31	32	32	32	32	32	33	33	33	33	33	34	34	34
50	33	33	34	34	34	34	35	35	35	35	36	36	36	36	36	37	37	37
100	61	61	62	62	63	63	63	63	64	64	65	65	65	66	66	66	66	66

Table VI (*Continued*)

(Two-tailed) $s_{\pm}^{\alpha}$																
	$\alpha = .05$								$\alpha = .01$							
$n \backslash k$	2	3	4	5	6	7	8	9	2	3	4	5	6	7	8	9
6	-	-	-	-	-	-	-	-								
7	7	-	-	-	-	-	-	-								
8	8	8	8	-	-	-	-	-								
9	9	9	9	9	9	-	-	-	-	-	-	-	-	-	-	-
10	9	10	10	10	10	10	10	10	10	-	-	-	-	-	-	-
11	10	10	11	11	11	11	11	11	11	11	11	-	-	-	-	-
12	11	11	11	11	12	12	12	12	12	12	12	12	12	-	-	-
13	11	12	12	12	12	12	12	12	13	13	13	13	13	13	13	13
14	12	12	13	13	13	13	13	13	13	13	14	14	14	14	14	14
15	13	13	13	13	14	14	14	14	14	14	14	14	15	15	15	15
16	13	14	14	14	14	14	14	14	15	15	15	15	15	15	15	15
17	14	14	15	15	15	15	15	15	15	15	16	16	16	16	16	16
18	15	15	15	15	16	16	16	16	16	16	16	17	17	17	17	17
19	15	16	16	16	16	16	16	16	17	17	17	17	17	17	18	18
20	16	16	17	17	17	17	17	17	17	18	18	18	18	18	18	18

Table VII

PERCENTAGE POINTS OF THE k-SAMPLE SIGN STATISTICS s^{α}
(For reference, see Chap. 4, Sec. 2)

	$\alpha = .05$									$\alpha = .01$								
n \ k	2	3	4	5	6	7	8	9	10	2	3	4	5	6	7	8	9	10
5	-	-	-	-	-	-	-	-	-									
6	6	-	-	-	-	-	-	-	-									
7	7	-	-	-	-	-	-	-	-									
8	8	8	-	-	-	-	-	-	-	-	-	-	-	-	-	-	-	-
9	8	9	9	-	-	-	-	-	-	9	-	-	-	-	-	-	-	-
10	9	10	10	10	-	-	-	-	-	10	-	-	-	-	-	-	-	-
11	10	10	11	11	11	11	-	-	-	11	11	-	-	-	-	-	-	-
12	10	11	11	12	12	12	12	12	12	11	12	12	-	-	-	-	-	-
13	11	12	12	12	13	13	13	13	13	12	13	13	13	-	-	-	-	-
14	12	12	13	13	13	14	14	14	14	13	13	14	14	14	14	-	-	-
15	12	13	13	14	14	14	14	15	15	13	14	15	15	15	15	15	15	-
16	13	14	14	14	15	15	15	15	15	14	15	15	16	16	16	16	16	16
17	14	14	15	15	15	16	16	16	16	15	16	16	16	16	17	17	17	17
18	14	15	15	16	16	16	16	17	17	15	16	17	17	17	17	17	18	18
19	15	16	16	16	17	17	17	17	17	16	17	17	18	18	18	18	18	18
20	15	16	17	17	17	18	18	18	18	17	18	18	18	19	19	19	19	19
25	18	19	20	20	21	21	21	21	21	20	21	21	22	22	22	22	22	23
30	21	22	23	23	24	24	24	24	25	23	24	25	25	25	25	26	26	26
35	24	25	26	27	27	27	27	28	28	26	27	28	28	28	29	29	29	29
40	27	28	29	30	30	30	31	31	31	29	30	31	31	32	32	32	32	33
45	30	31	32	33	33	33	34	34	34	32	33	34	34	35	35	35	36	36
50	33	34	35	36	36	36	37	37	37	35	36	37	38	38	38	38	39	39
100	61	63	64	65	65	66	66	67	67	64	66	67	67	68	68	69	69	69

Table VIII

PERCENTAGE POINTS OF THE MANY-ONE RANK STATISTICS
(For reference, see Chap. 4, Sec. 3)

(One-tailed) r^{α}

n \ k	$\alpha = .05$ 2	3	4	5	6	7	8	9	10	$\alpha = .01$ 2	3	4	5	6	7	8	9	10
6	52	53	54	54	54	55	55	55	55	56	57	57	-	-	-	-	-	-
7	69	70	70	71	72	72	72	73	73	74	75	75	76	76	77	77	77	77
8	87	89	90	90	91	91	92	92	92	93	95	95	96	97	97	97	98	98
9	108	110	111	112	113	113	114	114	114	116	117	118	119	119	120	120	121	121
10	131	133	135	136	136	137	138	138	139	140	142	143	144	144	145	145	146	146
11	157	159	161	162	163	163	164	164	165	167	168	170	171	172	172	173	173	174
12	184	187	189	190	191	192	192	193	194	195	198	199	200	201	202	203	203	204
13	214	217	219	220	221	222	223	224	224	226	229	231	232	233	234	235	235	236
14	246	249	251	253	254	255	256	257	258	260	263	264	266	267	268	269	269	270
15	280	283	286	288	289	290	291	292	293	295	298	300	302	303	304	305	306	307
16	316	320	322	324	326	327	328	329	330	333	336	339	340	342	343	344	345	346
17	354	359	361	364	365	367	368	369	370	373	377	379	381	382	384	385	386	387
18	395	399	402	405	407	408	409	411	412	415	419	422	424	425	427	428	429	430
19	437	442	446	448	450	452	453	455	456	459	464	467	469	471	472	473	474	476
20	482	487	491	494	496	498	499	501	502	506	510	514	516	518	520	521	522	523
25	737	745	750	754	757	759	761	763	765	771	777	782	785	788	790	792	793	795
30	1046	1056	1062	1067	1071	1075	1077	1080	1082	1089	1098	1104	1108	1112	1115	1117	1119	1121
35	1406	1419	1428	1434	1439	1443	1447	1450	1452	1462	1472	1480	1485	1490	1494	1497	1499	1502
40	1820	1836	1846	1854	1859	1865	1869	1873	1876	1887	1900	1909	1916	1922	1926	1930	1933	1936
45	2286	2304	2317	2326	2333	2339	2344	2349	2353	2366	2381	2392	2401	2407	2412	2417	2421	2425
50	2804	2826	2840	2851	2859	2866	2872	2877	2882	2898	2916	2929	2938	2945	2952	2957	2962	2966
100	10834	10896	10936	10966	10990	11010	11027	11041	11055	11099	11151	11186	11214	11234	11253	11268	11280	11293

Table VIII (*Continued*)

(Two-tailed) r_*^α																
	α = .05								α = .01							
n \ k	2	3	4	5	6	7	8	9	2	3	4	5	6	7	8	9
4	26	-	-	-	-	-	-	-	-	-	-	-	-	-	-	-
5	39	39	40	40	-	-	-	-	-	-	-	-	-	-	-	-
6	53	54	55	55	56	56	56	57	57	-	-	-	-	-	-	-
7	70	72	72	73	73	74	74	75	75	76	77	77	-	-	-	-
8	90	91	92	93	93	94	94	95	95	96	97	98	98	99	99	99
9	111	113	114	115	116	116	117	117	118	119	120	121	122	122	122	123
10	135	137	138	139	140	141	141	142	142	144	145	146	147	148	148	148
11	161	163	165	166	167	168	168	169	169	171	173	174	175	175	176	176
12	189	192	193	195	196	197	197	198	199	201	203	204	205	206	206	207
13	219	222	224	226	227	228	229	230	230	233	235	236	237	238	239	239
14	252	255	257	259	261	262	262	263	264	267	269	271	272	273	274	274
15	286	290	293	294	296	297	298	299	300	303	306	307	309	310	311	311
16	323	327	332	332	334	335	336	337	339	342	344	346	348	349	350	351
17	362	367	370	372	374	376	377	378	379	383	385	387	389	390	391	392
18	403	408	412	414	416	418	419	420	422	426	429	431	433	434	435	436
19	447	452	456	458	461	462	464	465	467	471	474	476	479	480	481	482
20	492	498	502	505	507	509	511	512	514	518	522	524	527	528	530	531

Table IX

PERCENTAGE POINTS OF THE k-SAMPLE RANK STATISTICS r_{**}^{α}
(For reference, see Chap. 4, Sec. 4)

	$\alpha = .05$									$\alpha = .01$								
n \ k	2	3	4	5	6	7	8	9	10	2	3	4	5	6	7	8	9	10
6	52	55	56	57	-	-	-	-	-	56	-	-	-	-	-	-	-	-
7	69	72	74	75	76	77	77	-	-	74	76	-	-	-	-	-	-	-
8	88	91	93	95	96	97	98	99	99	94	97	99	100	-	-	-	-	-
9	109	113	116	117	119	120	121	122	122	116	119	122	123	125	126	126	-	-
10	132	137	140	142	144	145	146	147	148	140	145	147	149	150	152	153	154	154
11	157	163	167	169	171	172	174	175	176	167	172	175	177	179	180	181	182	183
12	185	192	195	198	200	202	203	205	206	196	201	205	207	209	211	212	213	214
13	215	222	227	230	232	234	236	237	238	227	233	237	240	242	244	245	247	248
14	247	255	260	263	266	268	270	272	273	260	267	272	275	277	279	281	282	283
15	281	290	295	299	302	305	307	308	310	296	304	309	312	315	317	319	320	321
16	317	327	333	337	341	343	345	347	349	333	342	348	351	354	357	359	360	362
17	355	367	373	378	381	384	386	389	390	373	383	389	393	396	399	401	403	404
18	396	408	415	420	424	427	430	432	434	415	426	432	437	440	443	445	447	449
19	439	452	459	465	469	472	475	478	480	460	471	478	483	487	490	492	494	496
20	483	498	506	512	516	520	523	526	528	506	519	526	531	535	539	541	544	546
25	740	759	771	779	785	790	795	798	802	771	789	799	806	812	816	820	824	826
30	1049	1075	1090	1101	1109	1115	1121	1126	1130	1090	1113	1127	1136	1144	1149	1155	1159	1163
35	1410	1443	1462	1476	1486	1495	1502	1508	1513	1463	1492	1509	1521	1530	1537	1544	1549	1554
40	1825	1865	1888	1905	1917	1927	1936	1943	1950	1889	1924	1945	1959	1971	1980	1987	1994	2000
45	2291	2339	2367	2387	2402	2414	2424	2433	2441	2368	2410	2434	2452	2465	2476	2485	2493	2500
50	2810	2866	2899	2922	2939	2954	2966	2976	2985	2900	2949	2978	2998	3014	3027	3038	3047	3055
100	10853	11010	11102	11167	11217	11258	11291	11321	11346	11105	11243	11325	11383	11428	11464	11494	11521	11543

Developments in Multiple Comparisons 1966–1976

1 INTRODUCTION

In 1966, *Simultaneous Statistical Inference* [6] summarized what was then known about the theory and methods of multiple comparisons. Since that time research in this field has continued to yield a variety of new results. This article is an overview and bibliography of these developments.

The bibliography was compiled by scanning most statistical journals for the years 1966 through 1976, except for an occasional unavailable issue. A list of the journals covered in the scan is given in Section 5. In addition, I examined a number of statistical symposia and festschrifts and some psychological journals which publish statistical articles. A few articles in nonstatistical journals were brought to my attention by other statisticians. The papers collected in this search are listed in Section 4. A small collection of papers published prior to 1966 but missed in [6] is also included.

It was necessary to exclude certain topics. In particular, papers on ranking and selection were omitted. These are involved with multiple comparisons questions, but the general purpose of ranking and selection methods and the probabilities which they control are slightly different from the multiple comparisons techniques considered here. A separate bibliography is required for ranking and selection because of the large number of papers on this topic.

Also, papers on outlier detection have been excluded. Early techniques in this field employed statistics (e.g., the studentized extreme deviate from the sample mean) which are clearly multiple comparisons statistics. How-

Reprinted from: © Journal of the American Statistical Association, December 1977, Volume 72, Number 360. Invited Paper, Pages 779–788.

ever, recent work on outliers has shifted into other directions and more emphasis is placed on automatically handling outliers through robust estimators than in detecting them.

For a few papers, it was difficult to decide whether they should be considered to be on multiple comparisons or not. Primarily, these papers involved distribution theory in multivariate analysis where the statistics could be used for multiple comparisons. If a paper was concerned exclusively with analytic distribution theory and did not mention the multiple comparisons problem or include a table for use with the multiple comparisons test, then it was excluded.

2 PAPERS OF SPECIAL INTEREST

The papers in the bibliography have been grouped according to their primary topic. Research has been more active on some of these topics than on others, and some topics are of greater interest to me than others. There are many worthwhile papers in the bibliography, but in this overview I concentrate on those topics and papers which, in my opinion, have the greatest interest and potential usefulness.

2.1 Probability inequalities

At the time I wrote my book, I knew the Bonferroni inequality $P\{\bigcap_1^r A_i\} \geq 1 - \sum_1^r P\{A_i^c\}$ was very useful, but over the course of the past ten years, I have become even more impressed with the tightness of the bound if the values of $P\{A_i^c\}$ are small. Although special techniques and distribution theory can improve on it, the improvement is very often only minor. This is true whether the Bonferroni inequality is used simply to combine multiple t tests (see [6, pp. 67–70]) or to weld together separate confidence intervals on different parameters for a confidence interval on a function of all the parameters, as in tolerance intervals about regression lines [5].

Nonetheless, Šidák [24] proved a general inequality which gives a slight improvement over the Bonferroni inequality when both are applicable (see [48, 62]) and which is useful in the proofs of some theorems concerned with multiple comparisons techniques (see, e.g., [49]). The inequality states that, if the random vector $\mathbf{Y} = (Y_1, \ldots, Y_r)$ has a multivariate normal distribution $N(\mathbf{0}, \boldsymbol{\Sigma})$ with zero mean vector and arbitrary covariance matrix $\boldsymbol{\Sigma}$, and if S^2 is an independent random variable distributed as χ_ν^2/ν, then

$$P\left\{\frac{|Y_1|}{S} \leq c_1, \ldots, \frac{|Y_r|}{S} \leq c_r\right\} \geq \prod_1^r P\left\{\frac{|Y_i|}{S} \leq c_i\right\}. \tag{2.1}$$

In words, the probability that a random normal vector (divided by S) with arbitrary covariances falls inside a rectangle centered at its mean is always at least as large as the corresponding probability for the case where the covariances are zero, i.e., where the coordinate variables are independent.

Application of this inequality to confidence intervals or tests for multiple t statistics which may have dependent numerators is immediate. It produces slightly sharper intervals or tests than the Bonferroni inequality, because $(1-\alpha)^r > 1 - r\alpha$. Whether for use with the Šidák inequality or the original Bonferroni inequality, recent charts by Moses [81] facilitate determination of the appropriate t critical value.

Earlier, Dunn [1] had proved this inequality for $k = 2, 3$ or for general k but special $\boldsymbol{\Sigma}$. Also, in earlier work, Slepian [7] had obtained a similar one-sided inequality. Slepian's inequality states that, if $\sigma_{ij}^1 \geq \sigma_{ij}^2$ for $i \neq j$, and $\sigma_{ii}^1 = \sigma_{ii}^2$ for $i = 1, \ldots, r$, then

$$P_1\{Y_1 \leq c_1, \ldots, Y_r \leq c_r\} \geq P_2\{Y_1 \leq c_1, \ldots, Y_r \leq c_r\}, \tag{2.2}$$

where P_k, $k = 1, 2$, is computed under the multivariate distribution $N(\mathbf{0}, \boldsymbol{\Sigma}^k)$, with $\boldsymbol{\Sigma}^k = (\sigma_{ij}^k)$.

For a summary of results on the case where the denominator variable S in (2.1) is allowed to be different for different Y_i, the reader is referred to Šidák [26]. Khatri [19, 20] considered generalizations of $P\{\bigcap_1^r [|Y_i| \leq c_i]\} \geq \prod_1^r P\{|Y_i| \leq c_i\}$ to symmetric convex regions about the origin. For similar inequalities involving more general densities, the reader is referred to Das Gupta et al. [254].

2.2 Methods for unbalanced ANOVA

When [6] was written, the Tukey studentized range technique was applicable only for the case of equal sample sizes. For one-way classifications where different populations had different sample sizes, the only techniques available were Bonferroni t statistics and Scheffé F projection intervals. In practice, the fearless statistician used the studentized range with a modal or median sample size for slightly imbalanced designs.

Recently, two new methods have been proposed. For Y_{ij} distributed independently as $N(\mu_i, \sigma^2)$, $i = 1, \ldots, r$, $j = 1, \ldots, n_i$, Spjøtvoll and Stoline [53] gave the intervals

$$\mu_i - \mu_j \in \bar{Y}_{i\cdot} - \bar{Y}_{j\cdot} \pm q'_{\alpha, r, \nu} S \max(1/\sqrt{n_i}, 1/\sqrt{n_j}), \tag{2.3}$$

where

$$\bar{Y}_{i\cdot} = \sum_1^{n_i} Y_{ij}/n_i, \quad S^2 = \sum_1^r \sum_1^{n_i} (Y_{ij} - \bar{Y}_{i\cdot})^2/\nu, \quad \nu = \sum_1^r (n_i - 1),$$

and $q'_{\alpha, r, \nu}$ is the $1 - \alpha$ percentile point of the studentized augmented range distribution (see [6, p. 40]). These intervals are conservative in the sense that the probability they simultaneously contain the true mean differences for all i, j is greater than or equal to $1 - \alpha$. Tables of the studentized augmented range distribution are not available, but the critical point $q_{\alpha, r, v}$, from a studentized range distribution (see [6, p. 38]), gives a good approximation to $q'_{\alpha, r, v}$, provided $r > 2$ and $\alpha \leq .05$.

Hochberg [49] gave the different intervals

$$\mu_i - \mu_j \in \bar{Y}_{i\cdot} - \bar{Y}_{j\cdot} \pm |m|_{\alpha, c, v} S(1/n_i + 1/n_j)^{1/2}, \tag{2.4}$$

where $c = \binom{r}{2}$ and $|m|_{\alpha, c, v}$ is the $1 - \alpha$ percentile point of the studentized maximum modulus distribution (see [6, p. 71]). Use of this technique is facilitated by improved tables of critical points for the studentized maximum modulus distribution computed by Hahn and Hendrickson [34].†

There arises the natural question of which technique gives the shortest intervals. Ury [62] has made a valuable study of the Spjøtvoll-Stoline and Hochberg procedures as well as the Scheffé S method and the Bonferroni intervals as modified by Šidák [24] and Dunn [48]. Unfortunately, the answer is not simple. No one of the four techniques dominates the others. Which type of interval is best depends upon the particular combination of sample sizes, significance level, number of populations, and degrees of freedom. Roughly, the Spjøtvoll-Stoline method (2.3) is preferable for very mildly imbalanced designs. For designs with greater imbalance, the Hochberg procedure (2.4) is best when tables for it are available (i.e., when the number of populations is six or smaller) and the Bonferroni-Dunn-Šidák intervals are best otherwise. Ury's paper contains two tables which are very helpful in determining the region of optimality for each technique.

Carmer and Swanson [56] and Einot and Gabriel [57] compare some of the aforementioned procedures and other multiple comparisons methods for the one-way classification, such as multiple range tests, least significance difference tests, and empirical Bayes tests.

For recent work on the companion problem of unequal population variances σ_i^2, $i = 1, \ldots, r$, the reader is referred to Hochberg [51] and Tamhane [54].

2.3 Conditional confidence levels

Olshen [77] proved the following interesting result under mild conditions on n, p, and α:

$$P\left\{\frac{1}{p}\frac{(\hat{\boldsymbol{\beta}} - \boldsymbol{\beta})^T(\mathbf{X}^T\mathbf{X})(\hat{\boldsymbol{\beta}} - \boldsymbol{\beta})}{\hat{\sigma}^2} \leq F_{\alpha, p, n-p} \,\middle|\, \frac{1}{p}\frac{\hat{\boldsymbol{\beta}}^T(\mathbf{X}^T\mathbf{X})\hat{\boldsymbol{\beta}}}{\hat{\sigma}^2} > F_{\alpha, p, n-p}\right\} < 1 - \alpha \tag{2.5}$$

† Footnote to second edition: This table is reproduced in the Addendum.

for all $\boldsymbol{\beta}$ and σ^2, where $\mathbf{Y} \sim N(\mathbf{X}\boldsymbol{\beta}, \sigma^2\mathbf{I})$, $\mathbf{X}$ is an $n \times p$ matrix of full rank, $\hat{\boldsymbol{\beta}} = (\mathbf{X}^T\mathbf{X})^{-1}\mathbf{X}^T\mathbf{Y}$,

$$\hat{\sigma}^2 = \mathbf{Y}^T(\mathbf{I} - \mathbf{X}(\mathbf{X}^T\mathbf{X})^{-1}\mathbf{X}^T)\mathbf{Y}/(n-p),$$

and $F_{\alpha,\, p,\, n-p}$ is the $1 - \alpha$ percentile point of an F distribution with p and $n - p$ degrees of freedom. This means that, if Scheffé simultaneous confidence intervals are computed only when the F test is significant, then the conditional probability of coverage for the confidence intervals is always less than the nominally stated unconditional probability. Monte Carlo experiments indicate that the discrepancy can be substantial.

Similar effects on conditional error rates for other multiple comparisons procedures have been established by Bernhardson [76].

2.4 Empirical Bayes approach

In a series of papers [85, 86, 87, 88, 89] Duncan and his associates have developed the Bayesian approach to multiple comparisons in a balanced one-way classification.

Specifically, they assume an additive linear loss model. For a component problem involving a decision of whether $\mu_i > \mu_j$ or $\mu_i \leq \mu_j$, the linear loss structure is

$$\begin{aligned} L(d^+, \Delta) &= k_1|\Delta|, \quad \Delta \leq 0, & L(d^0, \Delta) &= 0, \quad \Delta \leq 0 \\ &= 0, \quad \Delta > 0, & &= k_2\Delta, \quad \Delta > 0, \end{aligned} \tag{2.6}$$

where $\Delta = \mu_i - \mu_j$, d^+ is the decision $\Delta > 0$, and d^0 is the decision $\Delta \leq 0$. The ratio $k = k_1/k_2$ measures the relative seriousness of type I errors with respect to type II errors. The overall loss is the sum of the component losses over the decisions $\{\mu_i > \mu_j\}$ vs. $\{\mu_i \leq \mu_j\}$ and $\{\mu_j > \mu_i\}$ vs. $\{\mu_j \leq \mu_i\}$ for all pairs i, j.

The usual normal theory model $Y_{ij} \sim N(\mu_i, \sigma_e^2)$, $i = 1, \ldots, r$, $j = 1, \ldots, n$, is adopted for the observations. The prior distribution assumes that any $r - 1$ orthonormal contrasts among the μ_i are independently distributed as $N(0, \sigma_\mu^2)$, and a truncated product of two independent conjugate χ^2 densities is assumed for σ_e^2 and $\sigma_a^2 = \sigma_e^2 + n\sigma_\mu^2$.

Because of the additivity of the loss structure, the overall risk is minimized by minimizing each component risk, and this leads to the k-ratio t test, which is to decide

$$\begin{aligned} \mu_i > \mu_j \quad &\text{if} \quad \bar{Y}_{i\cdot} - \bar{Y}_{j\cdot} > t(k, F, q, f)S_d, \\ \mu_i \approx \mu_j \quad &\text{if} \quad |\bar{Y}_{i\cdot} - \bar{Y}_{j\cdot}| < t(k, F, q, f)S_d, \end{aligned} \tag{2.7}$$

and

$$\mu_i < \mu_j \quad \text{if} \quad \bar{Y}_{i\cdot} - \bar{Y}_{j\cdot} < -t(k, F, q, f)S_d,$$

where $\mu_i \approx \mu_j$ indicates that μ_i is unordered relative to μ_j. The critical t value is a function of the loss ratio $k = k_1/k_2$, the Bayesian F ratio

$$F = \frac{f}{q} \cdot \frac{n \sum_1^r (\bar{Y}_{i\cdot} - \bar{Y}_{\cdot\cdot})^2 + q_p \bar{\sigma}_a^2}{\sum_1^r \sum_1^n (Y_{ij} - \bar{Y}_{i\cdot})^2 + f_p \bar{\sigma}_e^2}, \tag{2.8}$$

where $\bar{\sigma}_e^2$ and $\bar{\sigma}_a^2$ are the prior values for σ_e^2 and $\sigma_a^2 = \sigma_e^2 + n\sigma_\mu^2$, and the Bayesian degrees of freedom

$$f = r(n-1) + f_p \qquad q = (r-1) + q_p, \tag{2.9}$$

where f_p and q_p are the degrees of freedom for the conjugate χ^2 prior distributions. Tables of $t(k, F, q, f)$ are available in the Corrigenda of [87]. The Bayesian standard deviation estimate, S_d, used in (2.7), is the square root of

$$S_d^2 = \frac{2}{n} \cdot \frac{\sum_1^r \sum_1^n (Y_{ij} - \bar{Y}_{i\cdot})^2 + f_p \bar{\sigma}_e^2}{f}. \tag{2.10}$$

If, as can often happen in practice, neither of the prior values $\bar{\sigma}_e^2$ and $\bar{\sigma}_a^2$ are available, then the prior degrees of freedom are each zero, the conjugate χ^2 prior distributions are the Jeffrey's indifference priors, and the F and S_d^2 statistics in (2.8) and (2.10) have their usual non-Bayesian form.

The innovative feature of this approach is the adaptive nature of the decision rule. How large each difference $|\bar{Y}_{i\cdot} - \bar{Y}_{j\cdot}|$ must be in order for the population means to be declared significantly different depends upon the F ratio of between to within population variation. The critical value $t(k, F, q, f)$ is a decreasing function of F, so the larger the F ratio for the test of the equality of all the means, the easier it is to reject the equality of an individual pair of means. Thus this approach accomplishes in a smooth fashion what the Fisher protected least significant difference (LSD) test attempts to do in a two stage manner.

Confidence intervals associated with the previously mentioned k-ratio t test have been derived in [85]. In the large sample case ($q = f = \infty$), they take the form

$$\mu_i - \mu_j \in (1 - 1/F)(\bar{Y}_{i\cdot} - \bar{Y}_{j\cdot}) \pm (1 - 1/F)^{1/2} S_d\, t(k), \tag{2.11}$$

where $t(k)$ is tabled in [85]. Dixon and Duncan have established that these intervals are large sample approximations to the minimum Bayes risk intervals for additive squared-error loss.

The intervals (2.11) provide a link between the theories of simultaneous confidence intervals and simultaneous estimation. The form of the point estimator in (2.11) is nearly identical to the James-Stein estimator for a normal mean in a one-way classification

$$\hat{\mu}_i = \bar{Y}_{\cdot\cdot} + \left(1 - \frac{r-3}{r(n-1)+2} \frac{\sum_1^r \sum_1^n (Y_{ij} - \bar{Y}_{i\cdot})^2}{n \sum_1^r (\bar{Y}_{i\cdot} - \bar{Y}_{\cdot\cdot})^2}\right)(\bar{Y}_{i\cdot} - \bar{Y}_{\cdot\cdot}). \tag{2.12}$$

Efron and Morris [2] have given an empirical Bayes interpretation to the James-Stein estimator under additive squared-error loss which ties in with the approach of Dixon and Duncan.

For additional work on Bayes confidence sets with implications for multiple comparisons problems, the reader is referred to Faith [3].

2.5 Confidence bands in regression

For the simple linear regression model $Y_i = \alpha + \beta(x_i - \bar{x}) + e_i$, with $e_i \sim N(0, \sigma^2)$, the principal confidence band available in 1966 for bounding the regression line $\alpha + \beta(x - \bar{x})$ was the Working-Hotelling-Scheffé hyperbolic band (see [6, pp. 110–114]). Since then, a plethora of confidence bands have appeared, with many different shapes and many different purposes.

Bowden and Graybill [132] extended and simplified the earlier work of Gafarian [4] on confidence bands of uniform width over an interval $[a, b]$. The uniform width confidence band is

$$\alpha + \beta(x - \bar{x}) \in \hat{\alpha} + \hat{\beta}(x - \bar{x}) \pm C_\alpha^* \hat{\sigma}, \; x \in [a, b], \tag{2.13}$$

where $\hat{\alpha}$, $\hat{\beta}$, $\hat{\sigma}^2$ are the usual least-squares estimators. The critical constant C_α^* is computable from the bivariate t distribution, but depends on $a, b, n, \bar{x}$, $\sum_1^n (x_i - \bar{x})^2$ as well as the significance level α. Bowden and Graybill give tables of C_α^* which, with interpolation, will cover a wide range of cases. For applications outside the scope of the tables conservative bands can be achieved by constructing a confidence rectangle for

$$\alpha + \beta(a - \bar{x}) \quad \text{and} \quad \alpha + \beta(b - \bar{x}) \tag{2.14}$$

from the Bonferroni or Šidák inequalities applied to the univariate t distribution. Trapezoidal bands could be obtained instead by allowing the confidence intervals placed on the two quantities in (2.14) to vary in width.

Graybill and Bowden [135] derived a linear segment confidence band

$$\alpha + \beta(x - \bar{x}) \in \hat{\alpha} + \hat{\beta}(x - \bar{x}) \pm |m|_{\alpha, 2, n-2}\, \hat{\sigma} \left[\frac{1}{\sqrt{n}} + \frac{|x - \bar{x}|}{(\sum_1^n (x_i - \bar{x})^2)^{1/2}} \right] \tag{2.15}$$

for $-\infty < x < +\infty$. The band (2.15) is obtained from projections of a rectangular confidence region for α and β based on the independent normal estimates $\hat{\alpha}$ and $\hat{\beta}$. This is why the $1 - \alpha$ percentile point $|m|_{\alpha, 2, n-2}$ of a studentized maximum modulus distribution on 2 and $n - 2$ degrees of freedom enters as the critical constant. Again, the improved tables for this distribution by Hahn and Hendrickson [34] are useful.† The Graybill-

† Footnote to second edition: This table is reproduced in the Addendum.

Bowden paper improves on the work of Folks and Antle [134], who derived conservative linear segment bands.

Halperin and Guarian [136] studied the distribution theory when the Working-Hotelling-Scheffé hyperbolic band is constrained to a finite interval $[a, b]$. For an interval symmetric about $\bar{x}$, the special tables which are required for the determination of the appropriate critical constant appear in an earlier work [137]. Although the Halperin-Guarian band has smaller average width than the Bowden-Graybill uniform width band, the latter band and the Graybill-Bowden linear segment band have the advantage of tables which cover a broader range and are easier to use. Both linear bands also enjoy the convenience of being easier to graph.

Dunn [133] modified the Graybill-Bowden linear segment band by considering its restriction to a finite interval $[a, b]$ symmetric about $\bar{x}$. A factor involving the length of the interval $[a, b]$ enters into the expression (2.15), and the critical constant comes from a studentized maximum modulus distribution with three degrees of freedom in the numerator.

For the quadratic regression model $\beta_0 + \beta_1 x + \beta_2 x^2$, the only tool for constructing a confidence band until recently has been the F projection method of Scheffé (see [6, pp. 110–114]). This band is conservative because the vectors $(1, x, x^2)$ do not map out a full three-dimensional space as x varies from $-\infty$ to $+\infty$. Wynn and Bloomfield [141] studied how to sharpen the Scheffé-type band and provided some tables for doing so. Trout and Chow [139], on the other hand, considered a uniform width band restricted to $x \in [a, b]$. Unfortunately, tables for applying this procedure are rather limited.

Bohrer [125] studied sharpening of the Scheffé band in the multiple regression model

$$\mathbf{x\beta} = (x_1, \ldots, x_p)(\beta_1, \ldots, \beta_p)^T,$$

where $\mathbf{x}$ is restricted to the nonnegative orthant $R_+ = \{\mathbf{x} \mid x_i \geq 0, i = 1, \ldots, p\}$.

There may be instances in which only a one-sided confidence band is required for the regression surface. Presumably, then, one could improve over the two-sided band. Bohrer [126] proved that this cannot be accomplished for a Scheffé-type band in a multiple regression model $\mathbf{x\beta}$ with no intercept (i.e., no $x_i \equiv 1$). However, for $\mathbf{x}$ restricted to a subset of the nonnegative orthant R_+, improvement is possible and is sometimes substantial, as demonstrated by Bohrer and Francis [129]. Also, Bohrer and Francis [130] showed how to obtain a sharper one-sided band for simple linear regression $\alpha + \beta(x - \bar{x})$ when x is confined to an interval $[a, b]$. Finally, Hochberg and Quade [138] gave a sharpened one-sided band for multiple regression $\beta_0 + \mathbf{x\beta}$ with an intercept β_0.

3 REFERENCES

[1] Dunn, O. J. (1958). Estimation of the means of dependent variables, *Annals of Mathematical Statistics*, **29**, 1095–1111.

[2] Efron, B., and Morris, C. (1973). Stein's estimation rule and its competitors—an empirical Bayes approach, *Journal of the American Statistical Association*, **68**, 117–130.

[3] Faith, R. E. (1976). Minimax Bayes set and point estimators of a multivariate normal mean, Technical Report No. 66, Department of Statistics, University of Michigan.

[4] Gafarian, A. V. (1964). Confidence bands in straight line regression, *Journal of the American Statistical Association*, **59**, 182–213.

[5] Lieberman, G. J., and Miller, R. G., Jr. (1963). Simultaneous tolerance intervals in regression, *Biometrika*, **50**, 155–168.

[6] Miller, R. G., Jr. (1966). *Simultaneous Statistical Inference*, New York: McGraw-Hill Book Co.

[7] Slepian, D. (1962). The one-sided barrier problem for Gaussian noise, *Bell System Technical Journal*, **41**, 463–501.

4 BIBLIOGRAPHY 1966–1976

4.1 Survey articles

[8] Dunnett, C. W. (1970). Multiple comparisons, in *Statistics in Endocrinology*, eds. J. W. McArthur and T. Colton, Cambridge, Mass: MIT Press, 79–103.

[9] Games, P. A. (1971). Multiple comparisons of means, *American Educational Research Journal*, **8**, 531–565.

[10] Gill, J. L. (1973). Current status of multiple comparisons of means in designed experiments, *Journal of Dairy Science*, **56**, 973–977.

[11] O'Neill, R. T., and Wetherill, B. G. (1971). The present state of multiple comparisons methods, *Journal of the Royal Statistical Society*, Ser. B, **33**, 218–241.

[12] Spjøtvoll, E. (1974). Multiple testing in the analysis of variance, *Scandinavian Journal of Statistics*, **1**, 97–114.

[13] Thomas, D. A. H. (1973). Multiple comparisons among means, a review, *The Statistician*, **22**, 16–42.

4.2 Probability inequalities

[14] Dykstra, R. L., Hewett, J. E., and Thompson, W. A., Jr. (1973). Events which are almost independent, *Annals of Statistics*, **1**, 674–681.

[15] Esary, J. D., Proschan, F., and Walkup, D. W. (1967). Associated random variables with applications, *Annals of Mathematical Statistics*, **38**, 1466–1474.

[16] Halperin, M. (1967). An inequality on a bivariate Student's "t" distribution, *Journal of the American Statistical Association*, **62**, 603–606.

[17] Jensen, D. R. (1969). An inequality for a class of bivariate chi-square distributions, *Journal of the American Statistical Association*, **64**, 333–336.

[18] Jogdeo, K. (1970). A simple proof of an inequality for multivariate normal probabilities of rectangles, *Annals of Mathematical Statistics*, **41**, 1357–1359.
[19] Khatri, C. G. (1967). On certain inequalities for normal distributions and their applications to simultaneous confidence bounds, *Annals of Mathematical Statistics*, **38**, 1853–1867.
[20] ——— (1970). Further contributions to some inequalities for normal distributions and their applications to simultaneous confidence bounds, *Annals of the Institute of Statistical Mathematics*, **22**, 451–458.
[21] Olkin, I. (1972). Monotonicity properties of Dirichlet integrals with applications to the multinomial distribution and the analysis of variance, *Biometrika*, **59**, 303–307.
[22] Scott, A. (1967). A note on conservative confidence regions for the mean of a multivariate normal, *Annals of Mathematical Statistics*, **38**, 278–280. Correction: **39** (1968), 2161.
[23] Sen, P. K. (1971). A Hájek-Rényi type inequality for stochastic vectors with applications to simultaneous confidence regions, *Annals of Mathematical Statistics*, **42**, 1132–1134.
[24] Šidák, Z. (1967). Rectangular confidence regions for the means of multivariate normal distributions, *Journal of the American Statistical Association*, **62**, 626–633.
[25] ——— (1968). On multivariate normal probabilities of rectangles: their dependence on correlations, *Annals of Mathematical Statistics*, **39**, 1425–1434.
[26] ——— (1971). On probabilities of rectangles in multivariate Student distributions: their dependence on correlations, *Annals of Mathematical Statistics*, **42**, 169–175.
[27] Tong, Y. L. (1970). Some probability inequalities of multivariate normal and multivariate t, *Journal of the American Statistical Association*, **65**, 1243–1247.

4.3 Tables

[28] Beckman, R. J., and Tietjen, G. L. (1973). Upper 10% and 25% points of the maximum F ratio, *Biometrika*, **60**, 213–214.
[29] Chambers, C. (1967). Extension of tables of percentage points of the largest variance ratio, $s_{\max}^2/s_0^2$, *Biometrika*, **54**, 225–227.
[30] Dayton, C. M., and Schafer, W. D. (1973). Extended tables of t and chi-square for Bonferroni tests with unequal error allocation, *Journal of the American Statistical Association*, **68**, 78–83.
[31] Dudewicz, E. J., Ramberg, J. S., and Chen, H. J. (1975). New tables for multiple comparison with a control (unknown variances), *Biometrische Zeitschrift*, **17**, 13–26.
[32] Dunn, O. J., Kronmal, R. A., and Yee, W. J. (1968). Tables of the multivariate t-distribution, School of Public Health, University of California, Los Angeles.
[33] Dutt, J. E., Mattes, K. D., and Tao, L. C. (1975). Tables of the trivariate t for comparing three treatments to a control with unequal sample sizes, Technical Report No. 3, Math/Stat Services, G. D. Searle & Co., Chicago.
[34] Hahn, G. J., and Hendrickson, R. W. (1971). A table of percentage points of the distribution of the largest absolute value of k Student t variates and its applications, *Biometrika*, **58**, 323–332.
[35] Hanumara, R. C., and Thompson, W. A., Jr. (1968). Percentage points of the extreme roots of a Wishart matrix, *Biometrika*, **55**, 505–512.

[36] Harter, H. L. (1969). *Order Statistics and Their Use in Testing and Estimation, Vol. 1, Tests Based on Range and Studentized Range of Samples from a Normal Population*, Aerospace Research Laboratories. (Available from Superintendent of Documents, U.S. Government Printing Office, Washington, D.C. 20402.)
[37] Krishnaiah, P. R., and Armitage, J. V. (1966). Tables for multivariate t-distribution, *Sankhyā*, B, **28**, 31–56.
[38] ———, and Armitage, J. V. (1970). On a multivariate F distribution, in *Essays in Probability and Statistics*, eds. R. C. Bose et al., Chapel Hill, N.C.: University of North Carolina Press, 439–468.
[39] Lachenbruch, P. A. (1969). Tables of simultaneous confidence limits for the binomial and Poisson distributions, *Biometrika*, **56**, 452.
[40] Pillai, K. C. S. (1967). Upper percentage points of the largest root of a matrix in multivariate analysis, *Biometrika*, **54**, 189–194.
[41] Steffens, F. E. (1969). Critical values for bivariate Student t-tests, *Journal of the American Statistical Association*, **64**, 637–646.
[42] Tietjen, G. L., and Beckman, R. J. (1972). Tables for using the maximum F-ratio in multiple comparison procedures, *Journal of the American Statistical Association*, **67**, 581–583.
[43] Tobach, E., Smith, M., Rose, G., and Richter, D. (1967). A table for making rank sum multiple paired comparisons, *Technometrics*, **9**, 561–567.
[44] Trout, J. R., and Chow, B. (1972). Table of the percentage points of the trivariate t-distribution with an application to uniform confidence bands, *Technometrics*, **14**, 855–879.
[45] Vithayasai, C., and Robson, D. S. (1973). One- and two-tailed simultaneous orthogonal contrasts, *Communications in Statistics*, **1**, 459–470.

4.4 Normal multifactor methods

Unbalanced and Heteroscedastic ANOVA

[46] Banerjee, S. (1970). Simultaneous confidence intervals for two linear functions of population means when population variances are not equal, *Proceedings of the Cambridge Philosophical Society*, **67**, 365–370.
[47] Brown, M. B., and Forsythe, A. B. (1974). The ANOVA and multiple comparisons for data with heterogeneous variance, *Biometrics*, **30**, 719–724.
[48] Dunn, O. J. (1974). On multiple tests and confidence intervals, *Communications in Statistics*, **3**, 101–103.
[49] Hochberg, Y. (1974). Some conservative generalizations of the T-method in simultaneous inference, *Journal of Multivariate Analysis*, **4**, 224–234.
[50] ——— (1975). An extension of the T-method to general unbalanced models of fixed effects, *Journal of the Royal Statistical Society*, Ser. B, **37**, 426–433.
[51] ——— (1976). A modification of the T-method of multiple comparisons for a one-way layout with unequal variances, *Journal of the American Statistical Association*, **71**, 200–203.
[52] Spjøtvoll, E. (1972). Joint confidence intervals for all linear functions of means in ANOVA with unknown variances, *Biometrika*, **59**, 638–685.
[53] ———, and Stoline, M. R. (1973). An extension of the T-method of multiple comparisons to include the cases with unequal sample sizes, *Journal of the American Statistical Association*, **68**, 975–978.
[54] Tamhane, A. C. (1977). Multiple comparisons in model I one-way ANOVA with unequal variances, *Communications in Statistics*, **A6**, 15–32.

Comparison of Procedures

[55] Carmer, S. G., and Swanson, M. R. (1971). Detection of differences between means: a Monte Carlo study of five pairwise multiple comparisons procedures, *Agronomy Journal*, **63**, 940–945.

[56] ———, and Swanson, M. R. (1973). Evaluation of ten pairwise multiple comparisons procedures by Monte Carlo methods, *Journal of the American Statistical Association*, **68**, 66–74.

[57] Einot, I., and Gabriel, K. R. (1975). A study of the powers of several methods of multiple comparisons, *Journal of the American Statistical Association*, **70**, 574–583.

[58] Keselman, H. J., Toothaker, L. E., and Shooter, M. (1975). An evaluation of two unequal n_k forms of the Tukey multiple comparisons statistic, *Journal of the American Statistical Association*, **70**, 584–587.

[59] Perlmutter, J., and Myers, J. L. (1973). A comparison of two procedures for testing multiple contrasts, *Psychological Bulletin*, **79**, 181–184.

[60] Thomas, D. A. H. (1974). Error rates in multiple comparisons among means—a simulation, *Applied Statistics*, **23**, 284–294.

[61] Ury, H. K., and Wiggins, A. D. (1975). A comparison of three procedures for multiple comparisons among means, *British Journal of Mathematical and Statistical Psychology*, **28**, 88–102.

[62] ——— (1976). A comparison of four procedures for multiple comparisons among means (pairwise contrasts) for arbitrary sample sizes, *Technometrics*, **18**, 89–97.

Robustness

[63] Brown, R. A. (1974). Robustness of the studentized range statistic, *Biometrika*, **61**, 171–175.

[64] Games, P. A. (1971). The inverse relation between the risks of type I and type II errors and suggestions for the unequal *n* case in multiple comparisons, *Psychological Bulletin*, **75**, 97–102.

[65] ———, and Howell, J. F. (1976). Pairwise multiple comparison procedures with unequal *N*'s and/or variances: a Monte Carlo study, *Journal of Educational Statistics*, **1**, 113–125.

[66] Howell, J. F., and Games, P. A. (1973). The robustness of the analysis of variance and the Tukey WSD test under various patterns of heterogeneous variances, *Journal of Experimental Education*, **41**, 33–37.

[67] ———, and Games, P. A. (1974). The effects of variance heterogeneity on simultaneous multiple-comparison procedures with equal sample size, *British Journal of Mathematical and Statistical Psychology*, **27**, 72–81.

[68] Keselman, H. J. (1976). A power investigation of the Tukey multiple comparison statistic, *Educational and Psychological Measurement*, **36**, 97–104.

[69] ———, Murray, R., and Rogan, J. (1976). Effect of very unequal group sizes on Tukey's multiple comparison test, *Educational and Psychological Measurement*, **36**, 263–270.

[70] ———, and Toothaker, L. E. (1973). Error rates for multiple comparison methods: some evidence concerning the misleading conclusions of Petrinovich and Hardyck, *Psychological Bulletin*, **80**, 31–32.

[71] Petrinovich, L. F., and Hardyck, C. D. (1969). Error rates for multiple comparison methods: some evidence concerning the frequency of erroneous conclusions, *Psychological Bulletin*, **71**, 43–54.

[72] Ramseyer, G. C., and Tcheng, T. (1973). The robustness of the studentized range statistic to violations of the normality and homogeneity of variance assumptions, *American Educational Research Journal*, **10**, 235–240.
[73] Smith, R. A. (1971). The effect of unequal group size on Tukey's HSD procedure, *Psychometrika*, **36**, 31–34.

Power Functions

[74] David, H. A., Lachenbruch, P. A., and Brandis, H. P. (1972). The power function of range and studentized range tests in normal samples, *Biometrika*, **59**, 161–168.
[75] Steffens, F. E. (1970). Power of bivariate studentized maximum and minimum modulus tests, *Journal of the American Statistical Association*, **65**, 1639–1644.

Conditional Error Rates

[76] Bernhardson, C. S. (1975). Type I error rates when multiple comparison procedures follow a significant F test or ANOVA, *Biometrics*, **31**, 229–232.
[77] Olshen, R. A. (1973). The conditional level of the F-test, *Journal of the American Statistical Association*, **68**, 692–698.

Graphical Techniques

[78] Boardman, T. J., and Moffitt, D. R. (1971). Graphical Monte Carlo type I error rates for multiple comparison procedures, *Biometrics*, **27**, 738–744.
[79] Feder, P. I. (1974). Graphical techniques in statistical data analysis—tools for extracting information from data, *Technometrics*, **16**, 287–299.
[80] ——— (1975). Studentized range graph paper—a graphical tool for the comparison of treatment means, *Technometrics*, **17**, 181–188.
[81] Moses, L. E. (1976). Charts for finding upper percentage points of Student's t in the range .01 to .00001, Technical Report No. 24 (5 R01 GM21215-02), Stanford University. (Note to second edition: Published in *Communications in Statistics*, **B7** (1978), 479–490.)
[82] Thöni, H. (1968). A nomogram for testing multiple comparisons, *Biometrische Zeitschrift*, **10**, 219–221.
[83] Zahn, D. A. (1975). Modifications of and revised critical values for the half-normal plot, *Technometrics*, **17**, 189–200.
[84] ——— (1975). An empirical study of the half-normal plot, *Technometrics*, **17**, 201–211.

Bayes Techniques

[85] Dixon, D. O., and Duncan, D. B. (1975). Minimum Bayes risk t-intervals for multiple comparisons, *Journal of the American Statistical Association*, **70**, 822–831.
[86] Duncan, D. B. (1975). t tests and intervals for comparisons suggested by the data, *Biometrics*, **31**, 339–359.
[87] Waller, R. A., and Duncan, D. B. (1969). A Bayes rule for the symmetric multiple comparisons problem, *Journal of the American Statistical Association*, **64**, 1484–1503. Corrigenda: **67** (1972), 253–255.
[88] ———, and Duncan, D. B. (1974). A Bayes rule for the symmetric multiple comparisons problem II, *Annals of the Institute of Statistical Mathematics*, **26**, 247–264.

[89] ———, and Kemp, K. E. (1975). Computations of Bayesian t-values for multiple comparisons, *Journal of Statistical Computation and Simulation*, **4**, 169–171.

Several Treatments vs. Control

[90] Bechhofer, R. E. (1969). Optimal allocation of observations when comparing several treatments with a control, in *Multivariate Analysis II*, ed. P. R. Krishnaiah, New York: Academic Press, 463–473.
[91] ———, and Nocturne, D. J.-M. (1972). Optimal allocation of observations when comparing several treatments with a control, II: 2-sided comparisons, *Technometrics*, **14**, 423–436.
[92] Shaffer, J. P. (1977). Multiple comparisons emphasizing selected contrasts: an extension and generalization of Dunnett's procedure, *Biometrics*, **33**, 293–304.
[93] Steffens, F. E., and de Villiers, R. (1973). Sample sizes required for two-sided comparisons of two treatments with a control, *Technometrics*, **15**, 915–921.

Tests on Interactions

[94] Bradu, D., and Gabriel, K. R. (1974). Simultaneous statistical inference on interactions in two-way analysis of variance, *Journal of the American Statistical Association*, **69**, 428–436.
[95] Cicchetti, D. V. (1972). Extension of multiple-range tests to interaction tables in the analysis of variance: a rapid approximate solution, *Psychological Bulletin*, **77**, 405–408.
[96] Gabriel, K. R., Putter, J., and Wax, Y. (1973). Simultaneous confidence intervals for product-type interaction contrasts, *Journal of the Royal Statistical Society*, Ser. B, **35**, 234–244.
[97] Harter, H. L. (1970). Multiple comparison procedures for interactions, *American Statistician*, **24**, 30–32.
[98] Sen, P. K. (1969). A generalization of the T-method of multiple comparisons for interactions, *Journal of the American Statistical Association*, **64**, 290–295.

Tests on Variances

[99] Bechhofer, R. E. (1968). Multiple comparisons with a control for multiple-classified variances of normal populations, *Technometrics*, **10**, 715–718.
[100] Broemeling, L. D. (1969). Confidence regions for variance ratios of random models, *Journal of the American Statistical Association*, **64**, 660–664.
[101] ———, and Bee, D. E. (1976). Simultaneous confidence intervals for parameters of a balanced incomplete block, *Journal of the American Statistical Association*, **71**, 425–428.
[102] Levy, K. J. (1975). A multiple range procedure for correlated variances in a two-way classification, *Biometrics*, **31**, 243–245.
[103] ——— (1975). An empirical comparison of several multiple range tests for variances, *Journal of the American Statistical Association*, **70**, 180–183.
[104] Jensen, D. R., and Jones, M. O. (1969). Simultaneous confidence intervals for variances, *Journal of the American Statistical Association*, **64**, 324–332.
[105] Sahai, H. (1974). Simultaneous confidence intervals for variance components in some balanced random effects models, *Sankhyā*, B, **36**, 278–287.
[106] ———, and Anderson, R. L. (1973). Confidence regions for variance ratios of random models for balanced data, *Journal of the American Statistical Association*, **68**, 951–952.

General

[107] Anderson, D. A. (1972). Overall confidence levels of the least significant difference procedure, *American Statistician*, **26**, 30–32.
[108] Bhargava, R. P., and Srivastava, M. S. (1973). On Tukey's confidence intervals for the contrasts in the means of the intraclass correlation model, *Journal of the Royal Statistical Society*, Ser. B, **35**, 147–152.
[109] Bock, J. (1974). Planung des Stichproben um Fanges beim Newman-Keuls-Test, *Biometrische Zeitschrift*, **16**, 417–422.
[110] Brown, G. H., and Robson, D. S. (1974). Joint confidence intervals for linear combinations of, at most, q of p means of normal distributions, *Communications in Statistics*, **3**, 213–222.
[111] Bryant, P. (1972). A short proof of a known lemma, *American Statistician*, **26**, 35–36.
[112] Dunnett, C. W. (1970). Answer to query 272: multiple comparison tests, *Biometrics*, **26**, 139–140.
[113] Enderlein, G. (1972). Die Maximum-Modulus-Methode zum Multiplen Vergleich von Gruppenmitteln mit dem Gesamtmittel, *Biometrische Zeitschrift*, **14**, 85–94.
[114] Frank, O. (1966). Simultaneous confidence intervals, *Skandinavisk Aktuarietidskrift*, **49**, 78–84.
[115] Gabriel, K. R., and Putter, J. (1973). On the differences of equicorrelated variables with equal variances, *Metron*, **31**, 1–5.
[116] Gebhardt, F. (1966). Approximations to the critical values for Duncan's multiple range test, *Biometrics*, **22**, 179–182.
[117] Hochberg, Y. (1974). The distribution of the range in general balanced models, *American Statistician*, **28**, 137–138.
[118] ——— (1975). Simultaneous inference under Behrens-Fisher conditions, a two sample approach, *Communications in Statistics*, **4**, 1109–1119.
[119] Johnson, D. E. (1973). A derivation of Scheffé's S-method by maximizing a quadratic form, *American Statistician*, **27**, 27–30.
[120] Klotz, J. (1969). A simple proof of Scheffé's multiple comparison theorem for contrasts in the one-way layout, *American Statistician*, **23**, 44–45.
[121] Marcus, R., and Peritz, E. (1976). Some simultaneous confidence bounds in normal models with restricted alternatives, *Journal of the Royal Statistical Society*, Ser. B, **38**, 157–165.
[122] Mudholkar, G. S., and Subbaiah, P. (1976). Unequal precision multiple comparisons for randomized block designs under nonstandard conditions, *Journal of the American Statistical Association*, **71**, 429–434.
[123] Schafer, W. D., and Macready, G. B. (1975). A modification of the Bonferroni procedure on contrasts which are grouped into internally independent sets, *Biometrics*, **31**, 227–228.
[124] Sverdrup, E. (1976). Significance testing in multiple statistical inference, *Scandinavian Journal of Statistics*, **3**, 73–78.

4.5 Regression

Confidence Bands

[125] Bohrer, R. (1967). On sharpening Scheffé bounds, *Journal of the Royal Statistical Society*, Ser. B, **29**, 110–114.
[126] ——— (1969). On one-sided confidence bounds for response surfaces, *Bulletin of the International Statistical Institute*, **43** (Book 2), 255–257.

[127] ——— (1973). An optimality property of Scheffé bounds, *Annals of Statistics*, **1**, 766–772.

[128] ——— (1973). A multivariate t probability integral, *Biometrika*, **60**, 647–654.

[129] ———, and Francis, G. K. (1972). Sharp one-sided confidence bounds over positive regions, *Annals of Mathematical Statistics*, **43**, 1541–1548.

[130] ———, and Francis, G. K. (1972). Sharp one-sided confidence bounds for linear regression over intervals, *Biometrika*, **59**, 99–107.

[131] Bowden, D. C. (1970). Simultaneous confidence bands for linear regression models, *Journal of the American Statistical Association*, **65**, 413–421.

[132] ———, and Graybill, F. A. (1966). Confidence bands of uniform and proportional width for linear models, *Journal of the American Statistical Association*, **61**, 182–198.

[133] Dunn, O. J. (1968). A note on confidence bands for a regression line over a finite interval, *Journal of the American Statistical Association*, **63**, 1028–1033.

[134] Folks, J. L., and Antle, C. E. (1967). Straight line confidence regions for linear models, *Journal of the American Statistical Association*, **62**, 1365–1374.

[135] Graybill, F. A., and Bowden, D. C. (1967). Linear segment confidence bands for simple linear models, *Journal of the American Statistical Association*, **62**, 403–408.

[136] Halperin, M., and Gurian, J. (1968). Confidence bands in linear regression with constraints, *Journal of the American Statistical Association*, **63**, 1020–1027.

[137] ———, Rastogi, S. C., Ho, I., and Yang, Y. Y. (1967). Shorter confidence bands in linear regression, *Journal of the American Statistical Association*, **62**, 1050–1067.

[138] Hochberg, Y., and Quade, D. (1975). One-sided simultaneous confidence bounds on regression surfaces with intercepts, *Journal of the American Statistical Association*, **70**, 889–891.

[139] Trout, J. R., and Chow, B. (1973). Uniform confidence bands for a quadratic model, *Technometrics*, **15**, 611–624.

[140] Wynn, H. P. (1975). Integrals for one-sided confidence bounds: a general result, *Biometrika*, **62**, 393–396.

[141] ———, and Bloomfield, P. (1971). Simultaneous confidence bands for regression analysis, *Journal of the Royal Statistical Society*, Ser. B, **33**, 202–217.

Calibration and Prediction

[142] Bowden, D. C. (1968). Answer to query 26: tolerance intervals in regression, *Technometrics*, **10**, 207–209.

[143] Chew, V. (1968). Simultaneous prediction intervals, *Technometrics*, **10**, 323–331.

[144] Hahn, G. J. (1972). Simultaneous prediction intervals for a regression model, *Technometrics*, **14**, 203–214.

[145] Laubscher, N. F. (1968). Single and multiple discrimination regions in multiple linear regression, *South African Statistical Journal*, **2**, 67–75.

[146] Lieberman, G. J., Miller, R. G., Jr., and Hamilton, M. A. (1967). Unlimited simultaneous discrimination intervals in regression, *Biometrika*, **54**, 133–145. Correction: **58** (1971), 687.

[147] Odén, A. (1973). Simultaneous confidence intervals in inverse linear regression, *Biometrika*, **60**, 339–343.

[148] Scheffé, H. (1973). A statistical theory of calibration, *Annals of Statistics*, **1**, 1–37.

[149] Steinhorst, R. K., and Bowden, D. C. (1971). Discrimination and confidence bands on percentiles, *Journal of the American Statistical Association*, **66**, 851–854.
[150] Wilson, A. L. (1967). An approach to simultaneous tolerance intervals in regression, *Annals of Mathematical Statistics*, **38**, 1536–1540.

General

[151] Aitken, M. A. (1974). Simultaneous inference and the choice of variable subsets in multiple regression, *Technometrics*, **16**, 221–227.
[152] Christensen, L. R. (1973). Simultaneous statistical inference in the normal multiple linear regression model, *Journal of the American Statistical Association*, **68**, 457–461.
[153] Duncan, D. B. (1970). Answer to query 273: multiple comparison methods for comparing regression coefficients, *Biometrics*, **26**, 141–143.
[154] Spjøtvoll, E. (1972). Multiple comparison of regression functions, *Annals of Mathematical Statistics*, **43**, 1076–1088.
[155] Steffens, F. E. (1968). On comparing two simple linear regression lines, *South African Statistical Journal*, **2**, 33–53.
[156] Tarone, R. E. (1976). Simultaneous confidence ellipsoids in the general linear model, *Technometrics*, **18**, 85–88.
[157] Thigpen, C. C., and Paulson, A. S. (1974). A multiple range test for analysis of covariance, *Biometrika*, **61**, 479–484.

4.6 Categorical data

[158] Anderson, D. A., McDonald, L. L., and Weaver, K. D. (1974). Tests on categorical data from the union-intersection principle, *Annals of the Institute of Statistical Mathematics*, **26**, 203–213.
[159] Bhapkar, V. P., and Somes, G. W. (1976). Multiple comparisons of matched proportions, *Communications in Statistics*, **A5**, 17–25.
[160] Brand, R. J., Pinnock, D. E., and Jackson, K. L. (1973). Large sample confidence bands for the logistic response curve and its inverse, *American Statistician*, **27**, 157–160.
[161] Cohen, J. (1967). An alternative to Marascuilo's "Large-sample multiple comparisons" for proportions, *Psychological Bulletin*, **67**, 199–201.
[162] Gabriel, K. R. (1966). Simultaneous test procedures for multiple comparisons on categorical data, *Journal of the American Statistical Association*, **61**, 1081–1096.
[163] Goodman, L. A. (1971). A simple simultaneous test procedure for quasi-independence in contingency tables, *Applied Statistics*, **20**, 165–177.
[164] Green, S-O. (1974). On simultaneous interval estimation of multinomial proportions, *Statistisktidskrift*, **3**, 237–245.
[165] Knoke, J. (1976). Multiple comparisons with dichotomous data, *Journal of the American Statistical Association*, **71**, 849–853.
[166] Renner, M. S. (1969). A graphical method of making multiple comparisons of frequencies, *Technometrics*, **11**, 321–330.
[167] Rodger, R. S. (1969). Linear hypotheses in $2 \times a$ frequency tables, *British Journal of Mathematical and Statistical Psychology*, **22**, 29–48.
[168] Shaffer, J. P. (1971). An exact multiple comparisons test for a multinomial distribution, *British Journal of Mathematical and Statistical Psychology*, **24**, 267–272.

[169] ——— (1973). Defining and testing hypotheses in multidimensional contingency tables, *Psychological Bulletin*, **79**, 127–141.

4.7 Nonparametric techniques

[170] Crouse, C. F. (1969). A multiple comparison rank procedure for a one-way analysis of variance, *South African Statistical Journal*, **3**, 35–48.

[171] Dunn-Rankin, P., and Wilcoxon, F. (1966). The true distributions of the range of rank totals in the two-way classification, *Psychometrika*, **31**, 573–580.

[172] Gabriel, K. R., and Lachenbruch, P. A. (1969). Non-parametric ANOVA in small samples: a Monte Carlo study of the adequacy of the asymptotic approximation, *Biometrics*, **25**, 593–596.

[173] Hollander, M. (1966). An asymptotically distribution-free multiple comparison procedure—treatment vs. control, *Annals of Mathematical Statistics*, **37**, 735–738.

[174] Marascuilo, L. A., and McSweeney, M. (1967). Nonparametric post hoc comparisons for trend, *Psychological Bulletin*, **67**, 401–412.

[175] McDonald, B. J., and Thompson, W. A., Jr. (1967). Rank sum multiple comparisons in one- and two-way classifications, *Biometrika*, **54**, 487–497.

[176] Odeh, R. E. (1967). The distribution of the maximum sum of ranks, *Technometrics*, **9**, 271–278.

[177] Rhyne, A. L., and Steel, R. G. D. (1967). A multiple comparisons sign test: all pairs of treatments, *Biometrics*, **23**, 539–549.

[178] Sen, P. K. (1966). On nonparametric simultaneous confidence regions and tests for the one criterion analysis of variance problem, *Annals of the Institute of Statistical Mathematics*, **18**, 319–336.

[179] ——— (1969). On nonparametric T-method of multiple comparisons for randomized blocks, *Annals of the Institute of Statistical Mathematics*, **21**, 329–333.

[180] Slivka, J. (1970). A one-sided nonparametric multiple comparison control percentile test: treatments versus control, *Biometrika*, **57**, 431–438.

4.8 Multivariate methods

[181] Gabriel, K. R. (1968). Simultaneous test procedures in multivariate analysis of variance, *Biometrika*, **55**, 489–504.

[182] ——— (1969). A comparison of some methods of simultaneous inference in MANOVA, in *Multivariate Analysis II*, ed. P. R. Krishnaiah, New York: Academic Press, 67–86.

[183] ———, and Sen, P. K. (1968). Simultaneous test procedures for one-way ANOVA and MANOVA based on rank scores, *Sankhyā, A*, **30**, 303–312.

[184] Grizzle, J. E. (1970). An example of the analysis of a series of response curves and an application of multivariate multiple comparisons, in *Essays in Probability and Statistics*, eds. R. C. Bose et al., Chapel Hill, N.C.: University of North Carolina Press, 311–326.

[185] Hummel, T. J., and Sligo, J. R. (1971). Empirical comparison of univariate and multivariate analysis of variance procedures, *Psychological Bulletin*, **76**, 49–51.

[186] Jensen, D. R. (1970). The joint distribution of traces of Wishart matrices and some applications, *Annals of Mathematical Statistics*, **41**, 133–145.

[187] ——— (1972). Some simultaneous multivariate procedures using Hotelling's T^2 statistics, *Biometrics*, **28**, 39–53.
[188] ——— (1974). On the joint distribution of Friedman's χ_r^2 statistics, *Annals of Statistics*, **2**, 311–322.
[189] Krishnaiah, P. R. (1968). Simultaneous tests for the equality of covariance matrices against certain alternatives, *Annals of Mathematical Statistics*, **39**, 1303–1309.
[190] ——— (1969). Simultaneous test procedures under general MANOVA models, in *Multivariate Analysis II*, ed. P. R. Krishnaiah, New York: Academic Press 121–143.
[191] ——— (1969). Further results on "Simultaneous test procedures under general MANOVA models," *Bulletin of the International Statistical Institute*, **43** (Book 2), 288–289.
[192] ———, and Schuurman, F. J. (1974). On the evaluation of some distributions that arise in simultaneous tests for the equality of the latent roots of the covariance matrix, *Journal of Multivariate Analysis*, **4**, 265–282.
[193] ———, and Waikar, V. B. (1971). Simultaneous tests for equality of latent roots against certain alternatives—I, *Annals of the Institute of Statistical Mathematics*, **23**, 451–468.
[194] ———, and Waikar, V. B. (1972). Simultaneous tests for equality of latent roots against certain alternatives—II, *Annals of the Institute of Statistical Mathematics*, **24**, 81–85.
[195] McKay, R. J. (1976). Simultaneous procedures in discriminant analysis involving two groups, *Technometrics*, **18**, 47–53.
[196] Mudholkar, G. S. (1966). On confidence bounds associated with multivariate analysis of variance and non-independence between two sets of variates, *Annals of Mathematical Statistics*, **37**, 1736–1746.
[197] ———, Davidson, M. L., and Subbaiah, P. (1974). Extended linear hypotheses and simultaneous tests in multivariate analysis of variance, *Biometrika*, **61**, 467–477.
[198] Sen, P. K., and Krishnaiah, P. R. (1974). On a class of simultaneous rank order tests in MANOCOVA, *Annals of the Institute of Statistical Mathematics*, **26**, 135–145.
[199] Tamura, R. (1969). Some multivariate comparison procedures based on ranks, *Annals of Mathematical Statistics*, **40**, 1486–1491.

4.9 Miscellaneous

[200] Aitken, M. A. (1969). Multiple comparisons in psychological experiments, *British Journal of Mathematical and Statistical Psychology*, **22**, 193–198.
[201] Carmer, S. G. (1976). Optimal significance levels for application of the least significant difference in crop performance trials, *Crop Science*, **16**, 95–99.
[202] Erlander, S., Gustavsson, J., and Svensson, A. (1972). On asymptotic simultaneous confidence regions for regression planes in a Poisson model, *Review of the International Statistical Institute*, **40**, 111–122.
[203] Gabriel, K. R. (1969). Simultaneous test procedures—some theory of multiple comparisons, *Annals of Mathematical Statistics*, **40**, 224–250.
[204] ——— (1970). On the relation between union intersection and likelihood ratio tests, in *Essays in Probability and Statistics*, eds. R. C. Bose et al., Chapel Hill, N.C.: University of North Carolina Press, 251–266.

[205] Ghosh, M., and Sen, P. K. (1973). On some sequential simultaneous confidence intervals procedures, *Annals of the Institute of Statistical Mathematics*, **25**, 123–133.
[206] Hahn, G. J. (1972). Simultaneous prediction intervals to contain the standard deviations or ranges of future samples from a normal distribution, *Journal of the American Statistical Association*, **67**, 938–939.
[207] ——— (1975). A simultaneous prediction limit on the means of future samples from an exponential distribution, *Technometrics*, **17**, 341–345.
[208] Hanumara, R. C. (1975). Estimating imprecisions of measuring instruments, *Technometrics*, **17**, 299–302.
[209] Hopkins, K. D., and Chadbourn, R. A. (1967). A schema for proper utilization of multiple comparisons in research and a case study, *American Educational Research Journal*, **4**, 407–412.
[210] Hurlburt, R. T., and Spiegel, D. K. (1976). Dependence of *F* ratios sharing a common denominator, *American Statistician*, **30**, 74–78.
[211] Jensen, D. R., Mayer, L. S., and Myers, R. H. (1975). Optimal designs and large-sample tests for linear hypotheses, *Biometrika*, **62**, 71–78.
[212] Kanofsky, P. (1968). Derivation of simultaneous confidence intervals for parametric functions from a parametric confidence region, *Sankhyā, A*, **30**, 379–386.
[213] Krishnaiah, P. R., and Murthy, V. K. (1966). Simultaneous tests for trend and serial correlations for Gaussian Markov residuals, *Econometrika*, **34**, 472–480. Erratum: **34** (1966), 908–909.
[214] Levy, K. J. (1975). Large-sample pair-wise comparisons involving correlations, proportions, or variances, *Psychological Bulletin*, **82**, 174–176.
[215] ——— (1975). Large-sample many-one comparisons involving correlations, proportions, or variances, *Psychological Bulletin*, **82**, 177–179.
[216] Likeš, J. (1968). Note on Tukey's method of multiple comparisons, *Biometrische Zeitschrift*, **10**, 19–24.
[217] ———, and Nedělka, S. (1973). Note on studentized range in samples from an exponential distribution, *Biometrische Zeitschrift*, **15**, 545–555.
[218] Mantel, N. (1968). Simultaneous confidence intervals and experimental design with normal correlation, *Biometrics*, **24**, 434–437.
[219] Marascuilo, L. A. (1966). Large-sample multiple comparisons, *Psychological Bulletin*, **65**, 280–290.
[220] McCool, J. I. (1975). Multiple comparisons for Weibull parameters, *IEEE Transactions on Reliability*, **R-24**, 186–192.
[221] Reading, J. C. (1975). A multiple comparison procedure for classifying all pairs out of *k* means as close or distant, *Journal of the American Statistical Association*, **70**, 832–838.
[222] Rodger, R. S. (1967). Type I errors and their decision basis, *British Journal of Mathematical and Statistical Psychology*, **20**, 51–62.
[223] ——— (1973). Confidence intervals for multiple comparisons and the misuse of the Bonferroni inequality, *British Journal of Mathematical and Statistical Psychology*, **26**, 58–60.
[224] ——— (1974). Multiple contrasts, factors, error rate and power, *British Journal of Mathematical and Statistical Psychology*, **27**, 179–198.
[225] ——— (1975). The number of non-zero, post hoc contrasts from ANOVA and error-rate, I, *British Journal of Mathematical and Statistical Psychology*, **28**, 71–78.
[226] ——— (1975). Setting rejection rate for contrasts selected post hoc when

some nulls are false, *British Journal of Mathematical and Statistical Psychology*, **28**, 214–232.
[227] Scheffé, H. (1970). Multiple testing versus multiple estimation, improper confidence sets, estimation of directions and ratios, *Annals of Mathematical Statistics*, **41**, 1–29.
[228] Seeger, P. (1968). A note on a method for the analysis of significances en masse, *Technometrics*, **10**, 586–593.
[229] Spjøtvoll, E. (1972). On the optimality of some multiple comparison procedures, *Annals of Mathematical Statistics*, **43**, 398–411.
[230] Srinivasan, R., and Wharton, R. M. (1976). Further results on simultaneous confidence intervals for the normal distribution, *Annals of the Institute of Statistical Mathematics*, **28**, 25–33.
[231] Ury, H. K., and Wiggins, A. D. (1971). Large sample and other multiple comparisons among means, *British Journal of Mathematical and Statistical Psychology*, **24**, 174–194.
[232] ———, and Wiggins, A. D. (1974). Use of the Bonferroni inequality for multiple comparisons among means with post hoc contrasts, *British Journal of Mathematical and Statistical Psychology*, **27**, 176–178.

4.10 Pre-1966 articles missed in [6]

[233] Balaam, L. N. (1963). Multiple comparisons—a sampling experiment, *Australian Journal of Statistics*, **5**, 62–84.
[234] Erlander, S., and Gustavsson, J. (1965). Simultaneous confidence regions in normal regression analysis with an application to road accidents, *Review of the International Statistical Institute*, **33**, 364–377.
[235] Krishnaiah, P. R. (1965). On the simultaneous ANOVA and MANOVA tests, *Annals of the Institute of Statistical Mathematics*, **17**, 35–53.
[236] ——— (1965). On a multivariate generalization of the simultaneous analysis of variance test, *Annals of the Institute of Statistical Mathematics*, **17**, 167–173.
[237] ——— (1965). Simultaneous tests for the equality of variances against certain alternatives, *Australian Journal of Statistics*, **7**, 105–109. Correction: **10** (1968), 43.
[238] ——— (1965). Multiple comparison tests in multiresponse experiments, *Sankhyā, A*, **27**, 65–72.
[239] Pillai, K. C. S. (1965). On the distribution of the largest characteristic root of a matrix in multivariate analysis, *Biometrika*, **52**, 405–414.
[240] Ryan, T. A. (1959). Multiple comparisons in psychological research, *Psychological Bulletin*, **56**, 26–47.
[241] ——— (1960). Significance tests for multiple comparison of proportions, variances, and other statistics, *Psychological Bulletin*, **57**, 318–328.

4.11 Late additions

[242] Chew, V. (1976). Comparing treatment means: a compendium, *Hortscience*, **11**, 348–357.
[243] Cima, J. A., and Hochberg, Y. (1976). On optimality criteria in simultaneous interval estimation, *Communications in Statistics*, **A5**, 875–882.
[244] Dutt, J. E., Mattes, K. D., Soms, A. P., and Tao, L. C. (1976). An approximation to the maximum modulus of the trivariate t with a comparison to the exact values, *Biometrics*, **32**, 465–469.

[245] Farebrother, R. W. (1976). The minimum proportional-variance unbiased linear estimator of β and simultaneous confidence intervals, *Journal of the American Statistical Association*, **71**, 761–762.

[246] Hackl, P. (1975). Simultane Inferenz von Abweichungen zwischen beobachteten und erwarteten Häufigkeiten aus Multinomialverteilungen, *Biometrische Zeitschrift*, **17**, 437–445.

[247] Hettmansperger, T. P. (1975). Non-parametric inference for ordered alternatives in a randomized block design, *Psychometrika*, **40**, 53–62.

[248] Hochberg, Y., and Lachenbruch, P. A. (1976). Two stage multiple comparison procedures based on the studentized range, *Communications in Statistics*, **A5**, 1447–1453.

[249] Jensen, D. R. (1976). The comparison of several response functions with a standard, *Biometrics*, **32**, 51–59.

[250] Johnson, D. E. (1976). Some new multiple comparison procedures for the two-way AOV model with interaction, *Biometrics*, **32**, 929–934.

[251] Levy, K. J. (1976). The randomized response technique and large sample pairwise comparisons among the parameters of k independent binomial populations, *British Journal of Mathematical and Statistical Psychology*, **29**, 257–262.

[252] Reinach, S. G. (1976). Multiple testing procedures for the k-sample runs test, *South African Statistical Journal*, **10**, 117–133.

[253] Waldo, D. R. (1976). An evaluation of multiple comparison procedures, *Journal of Animal Science*, **42**, 539–544.

[254] Das Gupta, S., Eaton, M. L., Olkin, I., Perlman, M., Savage, L. J., and Sobel, M. (1972). Inequalities on the probability content of convex regions for elliptically contoured distributions, *Proceedings of the Sixth Berkeley Symposium on Mathematical Statistics and Probability*, II, 241–265.

[255] Bryant, J. L., and Paulson, A. S. (1976). An extension of Tukey's method of multiple comparisons to experimental designs with random concomitant variables, *Biometrika*, **63**, 631–638.

Added to second edition

[256] Hochberg, Y. (1976). On simultaneous confidence intervals for linear functions of means when the estimators have equal variances and equal covariances, *Metron* **34**, 123–128.

[257] Srinivasan, R., Kanofsky, P., and Wharton, R. M. (1975). Simultaneous confidence intervals for exponential distributions, *Sankhyā, Ser. B*, **37**, 271–292.

[258] Mudholkar, G. S., and Subbaiah, P. (1975). A note on MANOVA multiple comparisons based upon step-down procedure, *Sankhyā, Ser. B*, **37**, 300–307.

[259] Schuurmann, F. J., Krishnaiah, P. R., and Chattopadhyay, A. K. (1975). Tables for a multivariate F distribution, *Sankhyā, Ser. B*, **37**, 308–331.

[260] Hopkins, C. E., and Gross, A. J. (1970). Significance levels in multiple comparison tests, *Health Services Research* **5**, 132–140.

5 LIST OF JOURNALS SCANNED

The American Statistician
Annals of the Institute of Statistical Mathematics
The Annals of (Mathematical) Statistics

Applied Statistics (Journal of the Royal Statistical Society, Series C)
The Australian Journal of Statistics
Biometrics
Biometrika
Biometrische Zeitschrift
The British Journal of Mathematical and Statistical Psychology
Bulletin of the International Statistical Institute
The Canadian Journal of Statistics
Communications in Statistics
Journal of the American Statistical Association
Journal of Multivariate Analysis
Journal of the Royal Statistical Society, Series A
Journal of the Royal Statistical Society, Series B
Journal of Statistical Computation and Simulation
Metrika
Psychological Bulletin
Psychometrika
Review of the International Statistical Institute
Sankhyā
Scandinavian Actuarial Journal (Skandinavisk Aktuarietidskrift)
Scandinavian Journal of Statistics
South African Statistical Journal
Statistica Neerlandica
The Statistician (Journal of the Institute of Statisticians)
Technometrics

ADDENDUM

New Table of the Studentized Maximum Modulus

In the original edition the Pillai and Ramachandran (1954) table of critical points for the studentized maximum modulus was included in Appendix B as Table III. These critical points and a few additional points computed by Dunn and Massey (1965) were the only ones readily available at the time. The need for a more comprehensive table was filled by Hahn and Hendrickson in 1971. They tabulated critical points for the multivariate t distribution with common correlation $\rho = 0$, .2, .4, and .5. Their Table 1 for the studentized maximum modulus ($\rho = 0$) is reproduced from *Biometrika* in this second edition as Table IIIA with the permission of the authors and the editor.

REFERENCE

Hahn, G. J., and R. W. Hendrickson (1971): A table of percentage points of the distribution of the largest absolute value of k Student t variates and its applications, *Biometrika*, **58**, 323–332.

Table IIIA

PERCENTAGE POINTS OF THE STUDENTIZED MAXIMUM MODULUS $|m|^{\alpha}_{k,\nu}$

$\alpha = .10$

ν \ k	1	2	3	4	5	6	8	10	12	15	20
3	2.353	2.989	3.369	3.637	3.844	4.011	4.272	4.471	4.631	4.823	5.066
4	2.132	2.662	2.976	3.197	3.368	3.506	3.722	3.887	4.020	4.180	4.383
5	2.015	2.491	2.769	2.965	3.116	3.239	3.430	3.576	3.694	3.837	4.018
6	1.943	2.385	2.642	2.822	2.961	3.074	3.249	3.384	3.493	3.624	3.790
7	1.895	2.314	2.556	2.725	2.856	2.962	3.127	3.253	3.355	3.478	3.635
8	1.860	2.262	2.494	2.656	2.780	2.881	3.038	3.158	3.255	3.373	3.522
9	1.833	2.224	2.447	2.603	2.723	2.819	2.970	3.086	3.179	3.292	3.436
10	1.813	2.193	2.410	2.562	2.678	2.771	2.918	3.029	3.120	3.229	3.368
11	1.796	2.169	2.381	2.529	2.642	2.733	2.875	2.984	3.072	3.178	3.313
12	1.782	2.149	2.357	2.501	2.612	2.701	2.840	2.946	3.032	3.136	3.268
15	1.753	2.107	2.305	2.443	2.548	2.633	2.765	2.865	2.947	3.045	3.170
20	1.725	2.065	2.255	2.386	2.486	2.567	2.691	2.786	2.863	2.956	3.073
25	1.708	2.041	2.226	2.353	2.450	2.528	2.648	2.740	2.814	2.903	3.016
30	1.697	2.025	2.207	2.331	2.426	2.502	2.620	2.709	2.781	2.868	2.978
40	1.684	2.006	2.183	2.305	2.397	2.470	2.585	2.671	2.741	2.825	2.931
60	1.671	1.986	2.160	2.278	2.368	2.439	2.550	2.634	2.701	2.782	2.884

$\alpha = .05$

ν \ k	1	2	3	4	5	6	8	10	12	15	20
3	3.183	3.960	4.430	4.764	5.023	5.233	5.562	5.812	6.015	6.259	6.567
4	2.777	3.382	3.745	4.003	4.203	4.366	4.621	4.817	4.975	5.166	5.409
5	2.571	3.091	3.399	3.619	3.789	3.928	4.145	4.312	4.447	4.611	4.819
6	2.447	2.916	3.193	3.389	3.541	3.664	3.858	4.008	4.129	4.275	4.462
7	2.365	2.800	3.056	3.236	3.376	3.489	3.668	3.805	3.916	4.051	4.223
8	2.306	2.718	2.958	3.128	3.258	3.365	3.532	3.660	3.764	3.891	4.052
9	2.262	2.657	2.885	3.046	3.171	3.272	3.430	3.552	3.651	3.770	3.923
10	2.228	2.609	2.829	2.984	3.103	3.199	3.351	3.468	3.562	3.677	3.823
11	2.201	2.571	2.784	2.933	3.048	3.142	3.288	3.400	3.491	3.602	3.743
12	2.179	2.540	2.747	2.892	3.004	3.095	3.236	3.345	3.433	3.541	3.677
15	2.132	2.474	2.669	2.805	2.910	2.994	3.126	3.227	3.309	3.409	3.536
20	2.086	2.411	2.594	2.722	2.819	2.898	3.020	3.114	3.190	3.282	3.399
25	2.060	2.374	2.551	2.673	2.766	2.842	2.959	3.048	3.121	3.208	3.320
30	2.042	2.350	2.522	2.641	2.732	2.805	2.918	3.005	3.075	3.160	3.267
40	2.021	2.321	2.488	2.603	2.690	2.760	2.869	2.952	3.019	3.100	3.203
60	2.000	2.292	2.454	2.564	2.649	2.716	2.821	2.900	2.964	3.041	3.139

$\alpha = .01$

ν \ k	1	2	3	4	5	6	8	10	12	15	20
3	5.841	7.127	7.914	8.479	8.919	9.277	9.838	10.269	10.616	11.034	11.559
4	4.604	5.462	5.985	6.362	6.656	6.897	7.274	7.565	7.801	8.087	8.451
5	4.032	4.700	5.106	5.398	5.625	5.812	6.106	6.333	6.519	6.744	7.050
6	3.707	4.271	4.611	4.855	5.046	5.202	5.449	5.640	5.796	5.985	6.250
7	3.500	3.998	4.296	4.510	4.677	4.814	5.031	5.198	5.335	5.502	5.716
8	3.355	3.809	4.080	4.273	4.424	4.547	4.742	4.894	5.017	5.168	5.361
9	3.250	3.672	3.922	4.100	4.239	4.353	4.532	4.672	4.785	4.924	5.103
10	3.169	3.567	3.801	3.969	4.098	4.205	4.373	4.503	4.609	4.739	4.905
11	3.106	3.485	3.707	3.865	3.988	4.087	4.247	4.370	4.470	4.593	4.750
12	3.055	3.418	3.631	3.782	3.899	3.995	4.146	4.263	4.359	4.475	4.625
15	2.947	3.279	3.472	3.608	3.714	3.800	3.935	4.040	4.125	4.229	4.363
20	2.845	3.149	3.323	3.446	3.541	3.617	3.738	3.831	3.907	3.999	4.117
25	2.788	3.075	3.239	3.354	3.442	3.514	3.626	3.713	3.783	3.869	3.978
30	2.750	3.027	3.185	3.295	3.379	3.448	3.555	3.637	3.704	3.785	3.889
40	2.705	2.969	3.119	3.223	3.303	3.367	3.468	3.545	3.607	3.683	3.780
60	2.660	2.913	3.055	3.154	3.229	3.290	3.384	3.456	3.515	3.586	3.676

Bibliography

References preceded by an asterisk (*) have not been referred to directly in the text. They are included because they have bearing on simultaneous inference, and the reader may want to examine them.

*Acton, F. S. (1959): "Analysis of Straight-Line Data," John Wiley & Sons, Inc., New York.

*Aitchison, J. (1964): Confidence-region tests, *J. Roy. Statist. Soc., Ser. B,* **26**:462–476.

Anderson, R. L., and T. A. Bancroft (1952): "Statistical Theory in Research," McGraw-Hill Book Company, New York.

Anderson, T. W. (1958): "An Introduction to Multivariate Statistical Analysis," John Wiley & Sons, Inc., New York.

——— (1965): Some optimum confidence bounds for roots of determinantal equations, *Ann. Math. Statist.,* **36**:468–488.

Anscombe, F. J. (1960): Rejection of outliers, *Technometrics,* **2**:123–147.

*——— (1965): Comments on Kurtz-Link-Tukey-Wallace paper, *Technometrics,* **7**:167–168.

*Antle, C. E. (1965): *See* Folks, J. L.

*Armitage, J. V. (1964): *See* Krishnaiah, P. R.

*——— (1965): *See* Krishnaiah, P. R.

*———, and P. R. Krishnaiah (1964): Tables for the studentized largest chi-square distribution and their applications, *Aerospace Res. Lab. Tech. Rept.,* Wright-Patterson AFB, Ohio.

*Balaam, L. N., and W. T. Federer (1965): Query 11: Error rate bases, *Technometrics,* **7**:260–262.

Bancroft, T. A. (1952): *See* Anderson, R. L.

Bantegui, C. G. (1959): *See* Pillai, K. C. S.

Bartlett, M. S. (1937): Properties of sufficiency and statistical tests, *Proc. Roy. Soc. (London), Ser. A,* **160**:268–282.

Bechhofer, R. E. (1954): A single-sample multiple decision procedure for ranking means of normal populations with known variances, *Ann. Math. Statist.*, **25**:16–39.

——— (1958): A sequential multiple-decision procedure for selecting the best one of several normal populations with a common unknown variance, and its use with various experimental designs, *Biometrics*, **14**:408–429.

———, and S. Blumenthal (1962): A sequential multiple-decision procedure for selecting the best of several normal populations with a common unknown variance, II: Monte Carlo sampling results and new computing formulae, *Biometrics*, **18**:52–67.

———, C. W. Dunnett, and M. Sobel (1954): A two-sample multiple decision procedure for ranking means of normal populations with a common unknown variance, *Biometrika*, **41**:170–176.

———, S. Elmaghraby, and N. Morse (1959): A single-sample multiple-decision procedure for selecting the multinomial event which has the highest probability, *Ann. Math. Statist.*, **30**:102–119.

———, and M. Sobel (1954): A single sample decision procedure for ranking variances of normal populations, *Ann. Math. Statist.*, **25**:273–289.

Bennett, B. M. (1963): *See* Finney, D. J., et al.

Beyer, W. H. (1953): Certain percentage points of the distribution of the studentized range of large samples. Unpublished master's thesis, Virginia Polytechnic Institute, Blacksburg, Va.

Birnbaum, Z. W. (1952): Numerical tabulation of the distribution of Kolmogorov's statistic for finite sample size, *J. Am. Statist. Assoc.*, **47**:425–441.

———, and R. A. Hall (1960): Small sample distributions for multi-sample statistics of the Smirnov type, *Ann. Math. Statist.*, **31**:710–720.

———, and F. H. Tingey (1951): One-sided confidence contours for probability distribution functions, *Ann. Math. Statist.*, **22**:592–596.

Bleicher, E. (1954): The extension of the multiple comparisons test to lattice designs. Unpublished master's thesis, Virginia Polytechnic Institute, Blacksburg, Va.

Bliss, C. I., W. G. Cochran, and J. W. Tukey (1956): A rejection criterion based upon the range, *Biometrika*, **43**:418–422.

Blumenthal, S. (1962): *See* Bechhofer, R. E.

*Bonner, R. G. (1954): *See* Duncan, D. B.

Borenius, G. (1958): On the distribution of the extreme values in a sample from a normal distribution, *Skand. Aktuarietidskrift*, **41**:131–166.

Bose, R. C. (1953): *See* Roy, S. N.

Box, G. E. P. (1953): Non-normality and tests on variances, *Biometrika*, **40**:318–335.

Clemm, D. S. (1959): *See* Harter, H. L.; also Harter, H. L., et al.

Cochran, W. G. (1941): The distribution of the largest of a set of estimated variances as a fraction of their total, *Ann. Eugenics*, **11**:47–52.

——— (1956): *See* Bliss, C. I., et al.

*Conover, W. J. (1965): Several k-sample Kolmogorov-Smirnov tests, *Ann. Math. Statist.*, **36**:1019–1026.

Cornfield, J. (1955): *See* Halperin, M., et al.

*Cornfield, J., M. Halperin, and S. W. Greenhouse (1953): Simultaneous tests of significance and simultaneous confidence intervals for comparisons of many means. Unpublished manuscript.

*Cox, D. R. (1965): A remark on multiple comparison methods, *Technometrics*, **7**:223–224.

*Daniel, C. (1960): Locating outliers in factorial experiments, *Technometrics*, **2**:149–156.

Darling, D. A. (1957): The Kolmogorov-Smirnov, Cramér-von Mises tests, *Ann. Math. Statist.*, **28**:823–838.

David, H. A. (1952): Upper 5 and 1% points of the maximum F-ratio, *Biometrika*, **39**:422–424.

——— (1956): Revised upper percentage points of the extreme studentized deviate from the sample mean, *Biometrika*, **43**:449–451.

——— (1961): *See* Quesenberry, C. P.; also Thigpen, C. C.

———, H. O. Hartley, and E. S. Pearson (1954): The distribution of the ratio, in a single normal sample, of range to standard deviation, *Biometrika*, **41**:482–493.

David, H. T. (1958): A three-sample Kolmogorov-Smirnov test, *Ann. Math. Statist.*, **29**:842–851.

Dixon, W. J. (1950): Analysis of extreme values, *Ann. Math. Statist.*, **21**:488–506.

——— (1951): Ratios involving extreme values, *Ann. Math. Statist.*, **22**:68–78.

——— (1953): Processing data for outliers, *Biometrics*, **9**:74–89.

——— (1960): Simplified estimation from censored normal samples, *Ann. Math. Statist.*, **31**:385–391.

Duncan, D. B. (1947): Significance tests for differences between ranked variates drawn from normal populations. Unpublished doctoral thesis, Iowa State College, Ames, Iowa.

——— (1951): A significance test for differences between ranked treatments in an analysis of variance, *Virginia J. Sci.*, **2**:171–189.

——— (1952): On the properties of the multiple comparisons test, *Virginia J. Sci.*, **3**:49–67.

*——— (1953): Significance tests for differences between ranked treatments in an analysis of variance, *Virginia Poly. Inst. Tech. Rept.*, Blacksburg, Va.

——— (1955): Multiple range and multiple F tests, *Biometrics*, **11**:1–42.

——— (1957): Multiple range tests for correlated and heteroscedastic means, *Biometrics*, **13**:164–176.

——— (1961): Bayes rules for a common multiple comparisons problem and related Student-t problems, *Ann. Math. Statist.*, **32**:1013–1033.

——— (1965): A Bayesian approach to multiple comparisons, *Technometrics*, **7**:171–222.

*———, and R. G. Bonner (1954): Simultaneous confidence intervals derived from multiple range and multiple F tests, *Virginia Agri. Exp. Sta. Tech. Rept.*, Blacksburg, Va.

Dunn, O. J. (1958): Estimation of the means of dependent variables, *Ann. Math. Statist.*, **29**:1095–1111.

——— (1959): Confidence intervals for the means of dependent, normally distributed variables, *J. Am. Statist. Assoc.*, **54**:613–621.

Dunn, O. J. (1961): Multiple comparisons among means, *J. Am. Statist. Assoc.*, **56**:52–64.

——— (1964): Multiple comparisons using rank sums, *Technometrics*, **6**:241–252.

*——— (1965): A property of the multivariate *t*-distribution, *Ann. Math. Statist.*, **36**:712–714.

———, and F. J. Massey, Jr. (1965): Estimation of multiple contrasts using *t*-distributions, *J. Am. Statist. Assoc.*, **60**:573–583.

Dunnett, C. W. (1954): *See* Bechhofer, R. E., et al.

——— (1955): A multiple comparisons procedure for comparing several treatments with a control, *J. Am. Statist. Assoc.*, **50**:1096–1121.

*——— (1955): Multiple comparisons with a standard, *Am. Soc. Qual. Cont. 9th Ann. Conv. Trans.*, 485–492.

*——— (1961): Query 162: Multiple comparisons between treatments and a control, *Biometrics*, **17**:324–326.

——— (1964): New tables for multiple comparisons with a control, *Biometrics*, **20**:482–491.

———, and M. Sobel (1954): A bivariate generalization of Student's *t*-distribution, with tables for certain special cases, *Biometrika*, **41**:153–169.

———, and ——— (1955): Approximations to the probability integral and certain percentage points of a multivariate analogue of Student's *t*-distribution, *Biometrika*, **42**:258–260.

Durand, D. (1954): Joint confidence regions for multiple regression coefficients, *J. Am. Statist. Assoc.*, **49**:130–146.

Dwass, M. (1955): A note on simultaneous confidence intervals, *Ann. Math. Statist.*, **26**:146–147.

*——— (1959): Multiple confidence procedures, *Ann. Inst. Statist. Math.*, **10**:277–282.

——— (1960): Some *k*-sample rank-order tests, in I. Olkin, et al., (eds.), "Contributions to Probability and Statistics," Stanford University Press, Stanford, Calif.

Elmaghraby, S. (1959): *See* Bechhofer, R. E., et al.

Federer, W. T. (1955): "Experimental Design, Theory and Applications," The Macmillan Company, New York.

*——— (1965): *See* Balaam, L. N.

Feller, W. (1957): "An Introduction to Probability Theory and its Applications," 2d ed., vol. 1, John Wiley & Sons, Inc., New York.

Fieller, E. C. (1940): The biological standardization of insulin, *J. Roy. Statist. Soc., Suppl.*, **7**:1–54.

Finney, D. J., R. Latscha, B. M. Bennett, P. Hsu, and E. S. Pearson (1963): "Tables for Testing Significance in a 2 × 2 Contingency Table," Cambridge University Press, London.

Fisher, R. A. (1929): Tests of significance in harmonic analysis, *Proc. Roy. Soc. (London), Ser. A*, **125**:54–59.

——— (1934): "Statistical Methods for Research Workers," 5th ed., Oliver & Boyd, Ltd., Edinburgh and London.

——— (1935): "The Design of Experiments," Oliver & Boyd, Ltd., Edinburgh and London.

Fisher, R. A. (1940): On the similarity of the distributions found for the test of significance in harmonic analysis and in Steven's problem in geometrical probability, *Ann. Eugenics*, **10**:14–17.

———, and F. Yates (1963): "Statistical Tables for Biological, Agricultural, and Medical Research," 6th ed., Oliver & Boyd, Ltd., Edinburgh and London.

*Folks, J. L., and C. E. Antle (1965): Confidence regions for linear models. Unpublished manuscript.

Foster, F. G. (1957): Upper percentage points of the generalized beta distribution, II, *Biometrika*, **44**:441–453.

——— (1958): Upper percentage points of the generalized beta distribution, III, *Biometrika*, **45**:492–503.

———, and D. H. Rees (1957): Upper percentage points of the generalized beta distribution, I, *Biometrika*, **44**:237–247.

Fox, M. (1956): Charts of the power of the F-test, *Ann. Math. Statist.*, **27**:484–497.

Fraser, D. A. S. (1957): "Nonparametric Methods in Statistics," John Wiley & Sons, Inc., New York.

Friedman, M. (1937): The use of ranks to avoid the assumption of normality implicit in the analysis of variance, *J. Am. Statist. Assoc.*, **32**:675–701.

*Gabriel, K. R. (1963): A procedure for testing the homogeneity of all sets of means in analysis of variance, *Biometrics*, **20**:459–477.

*——— (1963): Simultaneous test procedures, with applications to the general linear hypothesis, univariate and multivariate. Unpublished manuscript.

*——— (1964): Simultaneous test procedures for multiple comparisons on categorical data. Unpublished manuscript.

Gafarian, A. V. (1964): Confidence bands in straight line regression, *J. Am. Statist. Assoc.*, **59**:182–213.

Galton, F. (1902): The most suitable proportion between the values of first and second prizes, *Biometrika*, **1**:385–390.

Gantmacher, F. R. (1959): "The Theory of Matrices," vol. 1, Chelsea Publishing Company, New York.

*Ghosh, M. N. (1955): Simultaneous tests of linear hypotheses, *Biometrika*, **42**:441–449.

*Gnanadesikan, R. (1957): *See* Roy, S. N.

*——— (1959): *See* Roy, S. N.

Gold, R. Z. (1963): Tests auxiliary to χ^2 tests in a Markov chain, *Ann. Math. Statist.*, **34**:56–74.

Goodman, L. A. (1964a): Simultaneous confidence limits for cross-product ratios in contingency tables, *J. Roy. Statist. Soc., Ser. B*, **26**:86–102.

——— (1964b): Simultaneous confidence intervals for contrasts among multinomial populations, *Ann. Math. Statist.*, **35**:716–725.

——— (1965): On simultaneous confidence intervals for multinomial proportions, *Technometrics*, **7**:247–254.

Grad, A., and H. Solomon (1955): Distribution of quadratic forms and some applications, *Ann. Math. Statist.*, **26**:464–477.

Graybill, F. A. (1961): "An Introduction to Linear Statistical Models," vol. 1, McGraw-Hill Book Company, New York.

——— (1963): *See* Mood, A. M.

Graybill, F. A., and G. Marsaglia (1957): Idempotent matrices and quadratic forms in the general linear hypothesis, *Ann. Math. Statist.*, **28**:678–686.

*Greenhouse, S. W. (1953): *See* Cornfield, J., et al.

——— (1955): *See* Halperin, M., et al.

——— (1958): *See* Halperin, M.

Grenander, U., and M. Rosenblatt (1957): "Statistical Analysis of Stationary Time Series," John Wiley & Sons, Inc., New York.

Grubbs, F. E. (1950): Sample criteria for testing outlying observations, *Ann. Math. Statist.*, **21**:27–58.

Gupta, S. S. (1956): On a decision rule for a problem in ranking means, *Univ. North Carolina Tech. Rept.*, Chapel Hill, N.C.

——— (1962): On a selection and ranking procedure for gamma populations, *Ann. Inst. Statist. Math.*, **14**:199–216.

——— (1963): Probability integrals of multivariate normal and multivariate t, *Ann. Math. Statist.*, **34**:792–828.

——— (1965): On some multiple decision (selection and ranking) rules, *Technometrics*, **7**:225–245.

———, and M. Sobel (1957): On a statistic which arises in selection and ranking problems, *Ann. Math. Statist.*, **28**:957–967.

———, and ——— (1958): On selecting a subset which contains all populations better than a standard, *Ann. Math. Statist.*, **29**:235–244.

———, and ——— (1960): Selecting a subset containing the best of several binomial populations, in I. Olkin, et al., (eds.), "Contributions to Probability and Statistics," Stanford University Press, Stanford, Calif.

———, and ——— (1962): On selecting a subset containing the population with the smallest variance, *Biometrika*, **49**:495–507.

*———, and ——— (1962): On the smallest of several correlated F statistics, *Biometrika*, **49**:509–523.

Guthrie, E. H. (1959): *See* Harter, H. L., et al.

Hall, R. A. (1960): *See* Birnbaum, Z. W.

*Halperin, M. (1953): *See* Cornfield, J., et al.

———, and S. W. Greenhouse (1958): A note on multiple comparisons for adjusted means in the analysis of covariance, *Biometrika*, **45**:256–259.

———, ———, J. Cornfield, and J. Zalokar (1955): Tables of percentage points for the studentized maximum absolute deviate in normal samples, *J. Am. Statist. Assoc.*, **50**:185–195.

Hannan, E. J. (1960): "Time Series Analysis," Methuen & Co., Ltd., London.

Harter, H. L. (1957): Error rates and sample sizes for range tests in multiple comparisons, *Biometrics*, **13**:511–536.

——— (1960a): Tables of range and studentized range, *Ann. Math. Statist.*, **31**:1122–1147.

——— (1960b): Critical values for Duncan's new multiple range test, *Biometrics*, **16**:671–685.

*——— (1961): Corrected error rates for Duncan's new multiple range test, *Biometrics*, **17**:321–324.

*——— (1961): Use of tables of percentage points of range and studentized range, *Technometrics*, **3**:407–411.

*Harter, H. L. (1962): Contributions to multiple comparisons tests, *Aerospace Res. Lab. Tech. Rept.*, Wright-Patterson AFB, Ohio.

———, and D. S. Clemm (1959): The probability integrals of the range and of the studentized range—probability integral, percentage points and moments of the range, *Wright Air Dev. Center Tech. Rept.*, Wright-Patterson AFB, Ohio.

———, ———, and E. H. Guthrie (1959): The probability integrals of the range and of the studentized range—probability integral and percentage points of the studentized range; critical values for Duncan's new multiple range test, *Wright Air Dev. Center Tech. Rept.*, Wright-Patterson AFB, Ohio.

Hartley, H. O. (1943): *See* Pearson, E. S.

——— (1950): The maximum F-ratio as a short-cut test for heterogeneity of variance, *Biometrika*, **37**:308–312.

——— (1951): *See* Pearson, E. S.

——— (1953): Corrigenda: Tables of percentage points of the "studentized" range, *Biometrika*, **40**:236.

——— (1954): *See* David, H. A., et al.

*——— (1955): Some recent developments in analysis of variance, *Comm. Pure Appl. Math.*, **8**:47–72.

——— (1962): *See* Pearson, E. S.

Healy, W. C., Jr. (1956): Two-sample procedures in simultaneous estimation, *Ann. Math. Statist.*, **27**:687–702.

Heck, D. L. (1960): Charts of some upper percentage points of the distribution of the largest characteristic root, *Ann. Math. Statist.*, **31**:625–642.

Hodges, J. L., Jr. and E. L. Lehmann (1956): The efficiency of some nonparametric competitors of the t-test, *Ann. Math. Statist.*, **27**:324–335.

Hoeffding, W. (1948): A class of statistics with asymptotically normal distribution, *Ann. Math. Statist.*, **19**:293–325.

——— (1952): The large sample power of tests based on permutations of observations, *Ann. Math. Statist.*, **23**:169–192.

Hollander, M. (1965): Rank tests for randomized blocks when the alternatives have an a priori ordering. Unpublished doctoral thesis, Stanford University, Stanford, Calif.

Hotelling, H. (1929): *See* Working, H.

Hsu, P. (1963): *See* Finney, D. J., et al.

Huber, P. J. (1964): Robust estimation of a location parameter, *Ann. Math. Statist.*, **35**:73–101.

Hurst, D. C. (1964): *See* Quesenberry, C. P.

Irwin, J. O. (1925): On a criterion for the rejection of outlying observations, *Biometrika*, **17**:238–250.

*Jackson, J. E. (1965): Comments on a paper by Kurtz, Link, Tukey, and Wallace, *Technometrics*, **7**:163–165.

Johnson, N. L., and B. L. Welch (1940): Applications of the non-central t-distribution, *Biometrika*, **31**:362–389.

Karlin, S., and D. Truax (1960): Slippage problems, *Ann. Math. Statist.*, **31**:296–324.

Kempthorne, O. (1952): "The Design and Analysis of Experiments," John Wiley & Sons, Inc., New York.

Keuls, M. (1952): The use of the "studentized range" in connection with an analysis of variance, *Euphytica*, **1**:112–122.

Kiefer, J. (1959): *K*-sample analogues of the Kolmogorov-Smirnov and Cramér-v. Mises tests, *Ann. Math. Statist.*, **30**:420–447.

Kimball, A. W. (1951): On dependent tests of significance in the analysis of variance, *Ann. Math. Statist.*, **22**:600–602.

Kramer, C. Y. (1956): Extension of multiple range tests to group means with unequal numbers of replications, *Biometrics*, **12**:307–310.

——— (1957): Extension of multiple range tests to group correlated adjusted means, *Biometrics*, **13**:13–18.

*Krishnaiah, P. R. (1963): Simultaneous tests and the efficiency of generalized balanced incomplete block designs, *Aerospace Res. Lab. Tech. Rept.*, Wright-Patterson AFB, Ohio.

*——— (1964): *See* Armitage, J. V.

*——— (1964): Multiple comparison tests in multivariate case, *Aerospace Res. Lab. Tech. Rept.*, Wright-Patterson AFB, Ohio.

*———, and J. V. Armitage (1964): Distribution of the studentized smallest chi-square, with tables and applications, *Aerospace Res. Lab. Tech. Rept.*, Wright-Patterson AFB, Ohio.

*———, and ——— (1965): Tables for the distribution of the maximum of correlated chi-square variates with one degree of freedom, *Aerospace Res. Lab. Tech. Rept.*, Wright-Patterson AFB, Ohio.

*———, and V. K. Murthy (1964): Simultaneous tests for trend and serial correlations for Gaussian Markov residuals, *Aerospace Res. Lab. Tech. Rept.*, Wright-Patterson AFB, Ohio.

Kruskal, W. H. (1952): A nonparametric test for the several sample problem, *Ann. Math. Statist.*, **23**:525–540.

———, and W. A. Wallis (1952): Use of ranks in one-criterion variance analysis, *J. Am. Statist. Assoc.*, **47**:583–621.

———, and ——— (1953): Errata to "Use of ranks in one-criterion variance analysis," *J. Am. Statist. Assoc.*, **48**:907–911.

*Kurtz, T. E. (1956): An extension of a multiple comparison procedure. Unpublished doctoral thesis, Princeton University, Princeton, N.J.

———, R. F. Link, J. W. Tukey, and D. L. Wallace (1965): Short-cut multiple comparisons for balanced single and double classifications: Part 1, Results, *Technometrics*, **7**:95–161.

*———, ———, ———, and ——— (1965): Authors' reply to Anscombe's comments, *Technometrics*, **7**:169.

*———, ———, ———, and ——— (1965): Short-cut multiple comparisons for balanced single and double classifications: Part 2, Derivations and approximations, *Biometrika*, **52**:485–498.

Latscha, R. (1963): *See* Finney, D. J., et al.

Lehmann, E. L. (1951): Consistency and unbiasedness of certain nonparametric tests, *Ann. Math. Statist.*, **22**:165–179.

——— (1953): The power of rank tests, *Ann. Math. Statist.*, **24**:23–43.

——— (1956): *See* Hodges, J. L., Jr.

Lehmann, E. L. (1957): A theory of some multiple decision problems, I, *Ann. Math. Statist.*, **28**:1–25.
——— (1958): Significance level and power, *Ann. Math. Statist.*, **29**:1167–1176.
——— (1961): Some Model I problems of selection, *Ann. Math. Statist.*, **32**:990–1012.
——— (1963a): Robust estimation in analysis of variance, *Ann. Math. Statist.*, **34**:957–966.
——— (1963b): Nonparametric confidence intervals for a shift parameter, *Ann. Math. Statist.*, **34**:1507–1512.
——— (1964): Asymptotically nonparametric inference in some linear models with one observation per cell, *Ann. Math. Statist.*, **35**:726–734.
Lehmer, E. (1944): Inverse tables of probabilities of errors of the second kind, *Ann. Math. Statist.*, **15**:388–398.
Lev, J. (1953): *See* Walker, H. M.
Lieberman, G. J. (1957): *See* Resnikoff, G. J.
——— (1961): Prediction regions for several predictions from a single regression line, *Technometrics*, **3**:21–27.
———, and R. G. Miller, Jr. (1963): Simultaneous tolerance intervals in regression, *Biometrika*, **50**:155–168.
Link, R. F. (1965): *See* Kurtz, T. E., et al; also *Kurtz, T. E., et al.
———, and D. L. Wallace (1952): Some short cuts to allowances. Unpublished manuscript.
Loève, M. (1963): "Probability Theory," 3d ed., D. Van Nostrand Company, Inc., Princeton, N.J.
McKay, A. T. (1935): The distribution of the difference between the extreme observation and the sample mean in samples of n from a normal universe, *Biometrika*, **27**:466–471.
McLaughlin, D. H. (1963): *See* Tukey, J. W.
Mandel, J. (1958): A note on confidence intervals in regression problems, *Ann. Math. Statist.*, **29**:903–907.
Mann, H. B., and D. R. Whitney (1947): On a test of whether one of two random variables is stochastically larger than the other, *Ann. Math. Statist.*, **18**:50–60.
Marsaglia, G. (1957): *See* Graybill, F. A.
Massey, F. J., Jr. (1951): The Kolmogorov-Smirnov test for goodness of fit, *J. Am. Statist. Assoc.*, **46**:68–78.
——— (1965): *See* Dunn, O. J.
May, J. M. (1952): Extended and corrected tables of the upper percentage points of the "studentized" range, *Biometrika*, **39**:192–193.
Miller, L. H. (1956): Table of percentage points of Kolmogorov statistics, *J. Am. Statist. Assoc.*, **51**:111–121.
Miller, R. G., Jr. (1963): *See* Lieberman, G. J.
Mood, A. M. (1950): "Introduction to the Theory of Statistics," McGraw-Hill Book Company, New York.
———, and F. A. Graybill (1963): "Introduction to the Theory of Statistics," 2d ed., McGraw-Hill Book Company, New York.
Morse, N. (1959): *See* Bechhofer, R. E., et al.

Moshman, J. (1952): Testing a straggler mean in a two-way classification using the range, *Ann. Math. Statist.*, **23**:126–132.

Mosteller, F. (1948): A k-sample slippage test for an extreme population, *Ann. Math. Statist.*, **19**:58–65.

———, and J. W. Tukey (1950): Significance levels for a k-sample slippage test, *Ann. Math. Statist.*, **21**:120–123.

*Murthy, V. K. (1964): *See* Krishnaiah, P. R.

Nair, K. R. (1948): Distribution of the extreme deviate from the sample mean, *Biometrika*, **35**:118–144.

*——— (1948): The studentized form of the extreme mean square test in the analysis of variance, *Biometrika*, **35**:16–31.

——— (1952): Tables of percentage points of the "studentized" extreme deviate from the sample mean, *Biometrika*, **39**:189–191.

Nemenyi, P. (1953): Percentage points of the studentized maximum modulus, *Statist. Res. Group Memo. Rept.*, Princeton University, Princeton, N.J.

——— (1961): Some distribution-free multiple comparison procedures in the asymptotic case (abstract), *Ann. Math. Statist.*, **32**:921–922.

——— (1963): Distribution-free multiple comparisons. Unpublished doctoral thesis, Princeton University, Princeton, N.J.

Newman, D. (1939): The distribution of the range in samples from a normal population, expressed in terms of an independent estimate of standard deviation, *Biometrika*, **31**:20–30.

Neyman, J., and B. Tokarska (1936): Errors of the second kind in testing "Student's" hypothesis, *J. Am. Statist. Assoc.*, **31**:318–326.

Noether, G. E. (1949): On a theorem by Wald and Wolfowitz, *Ann. Math. Statist.*, **20**:455–458.

Owen, D. B. (1962): "Handbook of Statistical Tables," Addison-Wesley Publishing Company, Inc., Reading, Mass.

Pachares, J. (1959): Table of the upper 10% points of the studentized range, *Biometrika*, **46**:461–466.

Paulson, E. (1942): A note on the estimation of some mean values for a bivariate distribution, *Ann. Math. Statist.*, **13**:440–445.

——— (1949): A multiple decision procedure for certain problems in the analysis of variance, *Ann. Math. Statist.*, **20**:95–98.

——— (1952a): On the comparison of several experimental categories with a control, *Ann. Math. Statist.*, **23**:239–246.

——— (1952b): An optimum solution to the k-sample slippage problem for the normal distribution, *Ann. Math. Statist.*, **23**:610–616.

——— (1962): A sequential procedure for comparing several experimental categories with a standard or control, *Ann. Math. Statist.*, **33**:438–443.

——— (1964): A sequential procedure for selecting the population with the largest mean from k normal populations, *Ann. Math. Statist.*, **35**:174–180.

Pearson, E. S. (1926): Further note on the distribution of range in samples taken from a normal population, *Biometrika*, **18**:173–194.

——— (1929): The distribution of frequency constants in small samples from non-normal symmetrical and skew populations, *Biometrika*, **21**:259–286.

Pearson, E. S. (1931): The analysis of variance in cases of non-normal variation, *Biometrika,* **23**:114–133.
——— (1954): *See* David, H. A., et al.
——— (1963): *See* Finney, D. J., et al.
———, and H. O. Hartley (1943): Tables of the probability integral of the studentized range, *Biometrika,* **33**:89–99.
———, and ——— (1951): Charts of the power function of the analysis of variance tests, derived from the non-central F-distribution, *Biometrika,* **38**:112–130.
———, and ——— (1962): "Biometrika Tables for Statisticians," 2d ed., vol. 1, Cambridge University Press, London.
———, and C. Chandra Sekar (1936): The efficiency of statistical tools and a criterion for the rejection of outlying observations, *Biometrika,* **28**:308–320.
———, and M. A. Stephens (1964): The ratio of range to standard deviation in the same normal sample, *Biometrika,* **51**:484–487.
Pearson, K. (1902): Note on Francis Galton's problem, *Biometrika,* **1**:390–399.
——— (1956): "Tables of the Incomplete Beta-Function," Cambridge University Press, London. First published in 1934.
Peiser, A. M. (1943): Asymptotic formulas for significance levels of certain distributions, *Ann. Math. Statist.,* **14**:56–62. [Correction, *Ann. Math. Statist.,* **20**:128–129 (1949).]
Pillai, K. C. S. (1952): On the distribution of "studentized" range, *Biometrika,* **39**:194–195.
——— (1956): On the distribution of the largest or the smallest root of a matrix in multivariate analysis, *Biometrika,* **43**:122–127.
——— (1957): "Concise Tables for Statisticians," Statistical Center, University of the Philippines, Manila, P.I.
——— (1959): Upper percentage points of the extreme studentized deviate from the sample mean, *Biometrika,* **46**:473–474.
——— (1964): On the distribution of the largest of seven roots of a matrix in multivariate analysis, *Biometrika,* **51**:270–275.
———, and C. G. Bantegui (1959): On the distribution of the largest of six roots of a matrix in multivariate analysis, *Biometrika,* **46**:237–240.
———, and K. V. Ramachandran (1954): On the distribution of the ratio of the ith observation in an ordered sample from a normal population to an independent estimate of the standard deviation, *Ann. Math. Statist.,* **25**:565–572.
———, and B. P. Tienzo (1959): On the distribution of the extreme studentized deviate from the sample mean, *Biometrika,* **46**:467–472.
Pitman, E. J. G. (1937a): Significance tests which may be applied to samples from any population, *J. Roy. Statist. Soc., Suppl.,* **4**:119–130.
——— (1937b): Significance tests which may be applied to samples from any population. III. The analysis of variance test, *Biometrika,* **29**:322–335.
Quesenberry, C. P., and H. A. David (1961): Some tests for outliers, *Biometrika,* **48**:379–390.
———, and D. C. Hurst (1964): Large sample simultaneous confidence intervals for multinomial proportions, *Technometrics,* **6**:191–195.

Ramachandran, K. V. (1954): *See* Pillai, K. C. S.

*——— (1956): Contributions to simultaneous confidence interval estimation, *Biometrics*, **12**:51–56.

*——— (1956): On the simultaneous analysis of variance test, *Ann. Math. Statist.*, **27**:521–528.

*——— (1956): On the Tukey test for the equality of means and the Hartley test for the equality of variances, *Ann. Math. Statist.*, **27**:825–831.

*——— (1958): On the studentized smallest chi-square, *J. Am. Statist. Assoc.*, **53**:868–872.

Rees, D. H. (1957): *See* Foster, F. G.

Reiersol, O. (1961): Linear and nonlinear multiple comparisons in logit analysis, *Biometrika*, **48**:359–365.

Resnikoff, G. J., and G. J. Lieberman (1957): "Tables of the Non-central *t*-Distribution," Stanford University Press, Stanford, Calif.

*Rhyne, A. L., and R. G. D. Steel (1965): Tables for a treatments versus control multiple comparisons sign test, *Technometrics*, **7**:293–306.

*Robson, D. S. (1961): Multiple comparisons with a control in balanced incomplete block designs, *Technometrics*, **3**:103–105.

Rosenblatt, M. (1957): *See* Grenander, U.

Roy, S. N. (1954): Some further results in simultaneous confidence interval estimation, *Ann. Math. Statist.*, **25**:752–761.

——— (1956): A note on "Some further results in simultaneous confidence interval estimation," *Ann. Math. Statist.*, **27**:856–858.

——— (1957): "Some Aspects of Multivariate Analysis," John Wiley & Sons, Inc., New York.

*——— (1962): A survey of some recent results in normal multivariate confidence bounds, *Bull. Inst. Int. Statist.*, **39,** Part 2: 405–422.

———, and R. C. Bose (1953): Simultaneous confidence interval estimation, *Ann. Math. Statist.*, **24**:513–536.

*———, and R. Gnanadesikan (1957): Further contributions to multivariate confidence bounds, *Biometrika*, **44**:399–410. [Corrigenda, *Biometrika*, **48**:474 (1961).]

*———, and ——— (1958): A note on "Further contributions to multivariate confidence bounds," *Biometrika*, **45**:581.

*———, and ——— (1959): Some contributions to anova in one or more dimensions: I, *Ann. Math. Statist.*, **30**:304–317.

*———, and ——— (1959): Some contributions to anova in one or more dimensions: II, *Ann. Math. Statist.*, **30**:318–340.

Sanders, P. G. (1953): The extension of the multiple comparisons test to rectangular lattices. Unpublished master's thesis, Virginia Polytechnic Institute, Blacksburg, Va.

Scheffé, H. (1953): A method for judging all contrasts in the analysis of variance, *Biometrika*, **40**:87–104.

——— (1956): A "mixed model" for the analysis of variance, *Ann. Math. Statist.*, **27**:23–36.

——— (1958): Experiments with mixtures, *J. Roy. Statist. Soc., Ser. B*, **20**:344–360.

Scheffé, H. (1959): "The Analysis of Variance," John Wiley & Sons, Inc., N. Y.
*——— (1961): Simultaneous interval estimates of linear functions of parameters, *Bull. Inst. Int. Statist.*, **38,** Part 4: 245–253.
——— (1963): The simplex-centroid design for experiments with mixtures, *J. Roy. Statist. Soc., Ser. B,* **25**:235–251.
Schuster, A. (1898): On the investigation of hidden periodicities with application to a supposed 26-day period of meteorological phenomena, *Terrestrial Magnetism,* **3**:13–41.
Sekar, C. Chandra (1936): *See* Pearson, E. S.
Sherman, E. (1965): A note on multiple comparisons using rank sums, *Technometrics,* **7**:255–256.
Siegel, S. (1956): "Nonparametric Statistics for the Behavioral Sciences," McGraw-Hill Book Company, New York.
Smirnov, N. V. (1948): Table for estimating the goodness of fit of empirical distributions, *Ann. Math. Statist.,* **19**:279–281.
Sobel, M. (1954): *See* Bechhofer, R. E., et al; also Dunnett, C. W.
——— (1955): *See* Dunnett, C. W.
——— (1957): *See* Gupta, S. S.
——— (1958): *See* Gupta, S. S.
——— (1960): *See* Gupta, S. S.
——— (1962): *See* Gupta, S. S.; also *Gupta, S. S.
Solomon, H. (1955): *See* Grad, A.
——— (1960): Distribution of quadratic forms, tables and applications, *Stanford Univ. Tech. Rept.,* Stanford, Calif.
Steel, R. G. D. (1959a): A multiple comparison sign test: treatments versus control, *J. Am. Statist. Assoc.,* **54**:767–775.
——— (1959b): A multiple comparison rank sum test: treatments versus control, *Biometrics,* **15**:560–572.
——— (1960): A rank sum test for comparing all pairs of treatments, *Technometrics,* **2**:197–207.
——— (1961): Some rank sum multiple comparisons tests, *Biometrics,* **17**:539–552.
*——— (1961): Query 163: Error rates in multiple comparisons, *Biometrics,* **17**:326–328.
*——— (1965): *See* Rhyne, A. L.
Stein, C. (1945): A two-sample test for a linear hypothesis whose power is independent of the variance, *Ann. Math. Statist.,* **16**:243–258.
Stephens, M. A. (1964): *See* Pearson, E. S.
Student (1927): Errors of routine analysis, *Biometrika,* **19**:151–164.
"Tables of the Cumulative Binomial Distribution" (1955), Harvard University Press, Cambridge, Mass.
Tang, P. C. (1938): The power function of the analysis of variance tests with tables and illustrations of their use, *Statist. Res. Mem.,* **2**:126–149.
Thigpen, C. C., and H. A. David (1961): Distribution of the largest observation in normal samples under non-standard conditions, *Virginia Poly. Inst. Tech. Rept.,* Blacksburg, Va.
Thompson, W. R. (1935): On a criterion for the rejection of observations and the

distribution of the ratio of deviation to sample standard deviation, *Ann. Math. Statist.*, **6**:214–219.

Thomson, G. W. (1955): Bounds for the ratio of range to standard deviation, *Biometrika*, **42**:268–269.

Tienzo, B. P. (1959): *See* Pillai, K. C. S.

Tingey, F. H. (1951): *See* Birnbaum, Z. W.

Tippett, L. H. C. (1925): On the extreme individuals and the range of samples taken from a normal population, *Biometrika*, **17**:364–387.

Tokarska, B. (1936): *See* Neyman, J.

Truax, D. (1953): An optimum slippage test for the variances of k-normal populations, *Ann. Math. Statist.*, **24**:669–674.

——— (1960): *See* Karlin, S.

Tukey, J. W. (1949): Comparing individual means in the analysis of variance, *Biometrics*, **5**:99–114.

——— (1950): *See* Mosteller, F.

——— (1951): Quick-and-dirty methods in statistics, Part II. Simple analyses for standard designs, *Am. Soc. Qual. Cont. 5th Ann. Conv. Trans.*, 189–197.

———(1952a): Allowances for various types of error rates. Unpublished IMS address, Virginia Polytechnic Institute, Blacksburg, Va.

——— (1952b): Various methods from a unified point of view. Unpublished IMS address, Chicago, Illinois.

——— (1953a): Some selected quick and easy methods of statistical analysis, *Trans. N.Y. Acad. Sci., Ser. II*, **16**:88–97.

——— (1953b): The problem of multiple comparisons. Unpublished manuscript.

——— (1954): Examples of quick and easy comparisons. Unpublished manuscript.

——— (1956): *See* Bliss, C. I., et al.

——— (1962): The future of data analysis, *Ann. Math. Statist.*, **33**:1–67.

——— (1965): *See* Kurtz, T. E., et al; also *Kurtz, T. E., et al.

———, and D. H. McLaughlin (1963): Less vulnerable confidence and significance procedures for location based on a single sample: Trimming/Winsorization 1, *Sankhyā, Ser. A*, **25**:331–352.

Wald, A., and J. Wolfowitz (1944): Statistical tests based on permutations of the observations, *Ann. Math. Statist.*, **15**:358–372.

———, and ——— (1946): Tolerance limits for a normal distribution, *Ann. Math. Statist.*, **17**:208–215.

Walker, H. M., and J. Lev (1953): "Statistical Inference," Holt, Rinehart and Winston, Inc., New York.

Wallace, D. L. (1952): *See* Link, R. F.

——— (1959): Simplified Beta-approximations to the Kruskal-Wallis H-test, *J. Am. Statist. Assoc.*, **54**:225–230.

——— (1965): *See* Kurtz, T. E., et al; also *Kurtz, T. E., et al.

Wallis, W. A. (1951): Tolerance intervals for linear regression, *Proc. 2nd Berkeley Symp. Math. Stat. Prob.*, University of California Press, Berkeley, Calif.

——— (1952): *See* Kruskal, W. H.

——— (1953): *See* Kruskal, W. H.

Welch, B. L. (1937): On the *Z*-test in randomized blocks and Latin squares, *Biometrika*, **29**:21–52.

——— (1940): *See* Johnson, N. L.

Whitney, D. R. (1947): *See* Mann, H. B.

Whittle, P. (1951): Hypothesis testing in time series analysis. Doctoral thesis, University of Uppsala, Almqvist & Wiksells Boktr.

Wilks, S. S. (1962): "Mathematical Statistics," John Wiley & Sons, Inc., New York.

Wolfowitz, J. (1944): *See* Wald, A.

——— (1946): *See* Wald, A.

Woolf, B. (1955): On estimating the relation between blood group and disease, *Ann. Human Genetics*, **19**:251–253.

Working, H. and H. Hotelling (1929): Application of the theory of error to the interpretation of trends, *J. Am. Statist. Assoc., Suppl.* (*Proc.*), **24**:73–85.

Wormleighton, R. (1959): Some tests of permutation symmetry, *Ann. Math. Statist.*, **30**:1005–1017.

Yates, F. (1963): *See* Fisher, R. A.

Zalokar, J. (1955): *See* Halperin, M., et al.

Author Index

Subject Index

Lecture Notes in Statistics

Vol. 1: R. A. Fisher: An Appreciation. Edited by S. E. Fienberg and D. V. Hinkley. xi, 208 pages, 1980.

Vol. 2: Mathematical Statistics and Probability Theory. Proceedings 1978. Edited by W. Klonecki, A. Kozek, and J. Rosiński. xxiv, 373 pages, 1980.

Vol. 3: B. D. Spencer, Benefit-Cost Analysis of Data Used to Allocate Funds. viii, 296 pages, 1980.

Vol. 4: E. A. van Doorn, Stochastic Monotonicity and Queueing Applications of Birth-Death Processes. vi, 118 pages, 1981.

Vol. 5: T. Rolski, Stationary Random Processes Associated with Point Processes. vi, 139 pages, 1981.

Vol. 6: S. S. Gupta and D.-Y. Huang, Multiple Statistical Decision Theory: Recent Developments. viii, 104 pages, 1981.

Vol. 7: M. Akahira and K. Takeuchi, Asymptotic Efficiency of Statistical Estimators. viii, 242 pages, 1981.

Vol. 8: The First Pannonian Symposium on Mathematical Statistics. Edited by P. Révész, L. Schmetterer, and V. M. Zolotarev. vi, 308 pages, 1981.

Vol. 9: B. Jørgensen, Statistical Properties of the Generalized Inverse Gaussian Distribution. vi, 188 pages, 1981.

Vol. 10: A. A. McIntosh, Fitting Linear Models: An Application of Conjugate Gradient Algorithms. vi, 200 pages, 1982.

Vol. 11: D. F. Nicholls and B. G. Quinn, Random Coefficient Autoregressive Models: An Introduction. v, 154 pages, 1982.

Vol. 12: M. Jacobsen, Statistical Analysis of Counting Processes. vii, 226 pages, 1982.

Vol. 13: J. Pfanzagl (with the assistance of W. Wefelmeyer), Contributions to a General Asymptotic Statistical Theory. vii, 315 pages, 1982.

Vol. 14: GLIM 82: Proceedings of the International Conference on Generalised Linear Models. Edited by R. Gilchrist. v, 188 pages, 1982.

Vol. 15: K. R. W. Brewer and M. Hanif, Sampling with Unequal Probabilities. ix, 164 pages, 1983.

Vol. 16: Specifying Statistical Models: From Parametric to Non-Parametric, Using Bayesian or Non-Bayesian Approaches. Edited by J. P. Florens, M. Mouchart, J. P. Raoult, L. Simar, and A. F. M. Smith. xi, 204 pages, 1983.

Vol. 17: I. V. Basawa and D. J. Scott, Asymptotic Optimal Inference for Non-Ergodic Models. ix, 170 pages, 1983.

Lecture Notes in Statistics

Vol. 18: W. Britton, Conjugate Duality and the Exponential Fourier Spectrum. v, 226 pages, 1983.

Vol. 19: L. Fernholz, von Mises Calculus For Statistical Functionals. viii, 124 pages, 1983.

Vol. 20: Mathematical Learning Models—Theory and Algorithms: Proceedings of a Conference. Edited by U. Herkenrath, D. Kalin, W. Vogel. xiv, 226 pages, 1983.

Vol. 21: H. Tong, Threshold Models in Non-linear Time Series Analysis. x, 323 pages, 1983.

Vol. 22: S. Johansen, Functional Relations, Random Coefficients and Nonlinear Regression with Application to Kinetic Data. viii, 126 pages. 1984

Vol. 23: D. G. Saphire, Estimation of Victimization Prevalence Using Data from the National Crime Survey. v, 165 pages. 1984.

Vol. 24: S. Rao, M.M. Gabr, An Introduction to Bispectral Analysis and Bilinear Time Series Models. ix, 280 pages, 1984

Vol. 25: Times Series Analysis of Irregularly Observed Data. Proceedings, 1983. Edited by Parzen. vii, 363 pages, 1984.

Vol. 26: Robust and Nonlinear Time Series Analysis. ix, 296 pages, 1984.

Vol. 27: Infinitely Divisible Statistical Experiments. vi, 166 pages, 1984.

Vol. 28: Differential-Geometrical Methods in Statistics. v, 296 pages, 1984.